Deep Learning for Information Fusion and Pattern Recognition

Deep Learning for Information Fusion and Pattern Recognition

Guest Editors

Yufeng Zheng
Erik Blasch

Basel • Beijing • Wuhan • Barcelona • Belgrade • Novi Sad • Cluj • Manchester

Guest Editors

Yufeng Zheng
Department of Data Science
University of Mississippi
Medical Center
Jackson
United States

Erik Blasch
Department of Data Processing
Air Force Research
Laboratory
Arlington
United States

Editorial Office
MDPI AG
Grosspeteranlage 5
4052 Basel, Switzerland

This is a reprint of the Special Issue, published open access by the journal *Sensors* (ISSN 1424-8220), freely accessible at: www.mdpi.com/journal/sensors/special_issues/dlifpr_sensors.

For citation purposes, cite each article independently as indicated on the article page online and using the guide below:

Lastname, A.A.; Lastname, B.B. Article Title. *Journal Name* **Year**, *Volume Number*, Page Range.

ISBN 978-3-7258-3000-8 (Hbk)
ISBN 978-3-7258-2999-6 (PDF)
https://doi.org/10.3390/books978-3-7258-2999-6

© 2025 by the authors. Articles in this book are Open Access and distributed under the Creative Commons Attribution (CC BY) license. The book as a whole is distributed by MDPI under the terms and conditions of the Creative Commons Attribution-NonCommercial-NoDerivs (CC BY-NC-ND) license (https://creativecommons.org/licenses/by-nc-nd/4.0/).

Contents

About the Editors vii

Preface ix

Hamed Elwarfalli, Dylan Flaute and Russell C. Hardie
Exponential Fusion of Interpolated Frames Network (EFIF-Net): Advancing Multi-Frame Image Super-Resolution with Convolutional Neural Networks
Reprinted from: *Sensors* **2024**, *24*, 296, https://doi.org/10.3390/s24010296 1

Yanbin Peng, Zhinian Zhai and Mingkun Feng
SLMSF-Net: A Semantic Localization and Multi-Scale Fusion Network for RGB-D Salient Object Detection
Reprinted from: *Sensors* **2024**, *24*, 1117, https://doi.org/10.3390/s24041117 18

Zhengyu Xia and Joohee Kim
Enhancing Mask Transformer with Auxiliary Convolution Layers for Semantic Segmentation
Reprinted from: *Sensors* **2023**, *23*, 581, https://doi.org/10.3390/s23020581 38

Xuefei Huang, Ka-Hou Chan, Weifan Wu, Hao Sheng and Wei Ke
Fusion of Multi-Modal Features to Enhance Dense Video Caption
Reprinted from: *Sensors* **2023**, *23*, 5565, https://doi.org/10.3390/s23125565 52

Jenniffer Carolina Triana-Martinez, Julian Gil-González, Jose A. Fernandez-Gallego, Andrés Marino Álvarez-Meza and Cesar German Castellanos-Dominguez
Chained Deep Learning Using Generalized Cross-Entropy for Multiple Annotators Classification
Reprinted from: *Sensors* **2023**, *23*, 3518, https://doi.org/10.3390/s23073518 68

Rafał Kozik, Wojciech Mazurczyk, Krzysztof Cabaj, Aleksandra Pawlicka, Marek Pawlicki and Michał Choraś
Deep Learning for Combating Misinformation in Multicategorical Text Contents
Reprinted from: *Sensors* **2023**, *23*, 9666, https://doi.org/10.3390/s23249666 87

Yang Wang, Dekai Shi and Weibin Zhou
Convolutional Neural Network Approach Based on Multimodal Biometric System with Fusion of Face and Finger Vein Features
Reprinted from: *Sensors* **2022**, *22*, 6039, https://doi.org/10.3390/s22166039 100

Yufeng Zheng and Erik Blasch
Facial Micro-Expression Recognition Enhanced by Score Fusion and a Hybrid Model from Convolutional LSTM and Vision Transformer
Reprinted from: *Sensors* **2023**, *23*, 5650, https://doi.org/10.3390/s23125650 115

Lingjian Kong, Kai Xie, Kaixuan Niu, Jianbiao He and Wei Zhang
Remote Photoplethysmography and Motion Tracking Convolutional Neural Network with Bidirectional Long Short-Term Memory: Non-Invasive Fatigue Detection Method Based on Multi-Modal Fusion
Reprinted from: *Sensors* **2024**, *24*, 455, https://doi.org/10.3390/s24020455 132

Maram Saleh Alwagdani and Emad Sami Jaha
Deep Learning-Based Child Handwritten Arabic Character Recognition and Handwriting Discrimination
Reprinted from: *Sensors* **2023**, *23*, 6774, https://doi.org/10.3390/s23156774 157

John Anthony C. Jose, Christopher John B. Bertumen, Marianne Therese C. Roque, Allan Emmanuel B. Umali, Jillian Clara T. Villanueva and Richard Josiah TanAi et al.
Smart Shelf System for Customer Behavior Tracking in Supermarkets
Reprinted from: *Sensors* **2024**, *24*, 367, https://doi.org/10.3390/s24020367 . . . . **178**

Xing Xu, Zhenpeng Xue and Yun Zhao
Research on an Algorithm of Express Parcel Sorting Based on Deeper Learning and Multi-Information Recognition
Reprinted from: *Sensors* **2022**, *22*, 6705, https://doi.org/10.3390/s22176705 . . . . **199**

Yuya Moroto, Keisuke Maeda, Ren Togo, Takahiro Ogawa and Miki Haseyama
Multimodal Transformer Model Using Time-Series Data to Classify Winter Road Surface Conditions
Reprinted from: *Sensors* **2024**, *24*, 3440, https://doi.org/10.3390/s24113440 . . . . **215**

Zhenxing Cai, Jianhong Yang, Huaiying Fang, Tianchen Ji, Yangyang Hu and Xin Wang
Research on Waste Plastics Classification Method Based on Multi-Scale Feature Fusion
Reprinted from: *Sensors* **2022**, *22*, 7974, https://doi.org/10.3390/s22207974 . . . . **234**

About the Editors

Yufeng Zheng

Yufeng Zheng is an associate professor of data science at the University of Mississippi Medical Center. He received the Ph.D. in optical engineering/image processing in 1997 from Tianjin University, China. He was a postdoctoral research associate at the University of Louisville, Kentucky, from 2001-2005. Dr. Zheng holds a utility patent in face recognition. He is the author or coauthor of 3 books, 6 book chapters, 24 articles in peer-reviewed journals, and 54 papers in conference proceedings. He is the principal investigator of many funded projects such as cybersecurity enhancement with keyboard dynamics, canopy coverage estimation with neural networks, multisensory image fusion and colorization; thermal face recognition; and multispectral face recognition. Dr. Zheng is a Cisco Certified Network Professional (CCNP), a senior member of IEEE and Signal Processing Society, and a senior member of SPIE. His research interests include image processing and pattern recognition; neural networks and artificial intelligence; information fusion; biometrics (facial recognition); machine learning; computer vision; and computer-aided diagnosis.

Erik Blasch

Erik Blasch received his B.S. in mechanical engineering from the Massachusetts Institute of Technology in 1992 and M.S. degrees in mechanical engineering, health science, and industrial engineering (human factors) from Georgia Tech. He completed an M.B.A., M.S.E.E., M.S. econ, M.S./Ph.D. psychology (ABD), and a Ph.D. in electrical engineering from Wright State University and is a graduate of Air War College. From 2000 to 2010, Dr. Blasch was the information fusion evaluation tech lead for the Air Force Research Laboratory (AFRL) Sensors Directorate—COMprehensive Performance Assessment of Sensor Exploitation (COMPASE) Center and an adjunct professor with Wright State University. From 2010 to 2012, Dr. Blasch was an exchange scientist at Defence R&D Canada at Valcartier, Quebec, in the Future Command and Control (C2) Concepts group. He is currently with the AFRL supporting information fusion developments. He received the 2009 IEEE Russ Bioengineering, 2012 IEEE AESS Magazine Mimno, and 2014 Military Sensing Symposium Mignogna Data Fusion awards. He is a past president of the International Society of Information Fusion (ISIF), and recognized as as Fellow of AIAA (astronautics), IEEE (electrical), MSS (sensing), RAeS (aerospace), and SPIE (optical) societies. His research interests include target tracking, information/sensor/image fusion, pattern recognition, and biologically inspired applications.

Preface

There is a large amount of data from different types of sensors, for instance, multispectral electro-optical/infrared (EO/IR) and computed tomography/magnetic resonance (CT/MR) images, among others. How to take advantage of multimodal data for object detection and pattern recognition is an active field of research. Information fusion (IF) is used for enhancing the performance of pattern classification, while deep learning (DL) technologies, including convolutional neural networks (CNNs), are powerful tools for improving object detection, segmentation, and recognition. It is viable to combine DL and IF to boost the overall performance of pattern classification and target recognition. Such combinations of powerful techniques may exploit the deeply hidden features of the multimodal, spatial, or temporal data.

This Reprint presents cutting-edge research utilizing deep learning (DL) and information fusion (IF) techniques. Key research areas include image and video analysis, covering topics such as super-resolution, object detection, semantic segmentation, video captioning, and text processing, including labeling enhancement and screening misinformation (e.g., from social media). Biometric applications explore innovations in human identification using facial and finger vein recognition, facial micro-expression analysis, and fatigue detection. Advanced applications extend to handwritten recognition, tracking supermarket customer behavior, parcel sorting, predicting road surface conditions, and plastic waste classification.

Yufeng Zheng and Erik Blasch
Guest Editors

Article

Exponential Fusion of Interpolated Frames Network (EFIF-Net): Advancing Multi-Frame Image Super-Resolution with Convolutional Neural Networks

Hamed Elwarfalli [1], Dylan Flaute [1,2] and Russell C. Hardie [1,*]

1. Department of Electrical and Computer Engineering, University of Dayton, 300 College Park, Dayton, OH 45469, USA; elwarfallih1@udayton.edu (H.E.); flauted1@udayton.edu (D.F.)
2. Applied Sensing Division, University of Dayton Research Institute, 300 College Park, Dayton, OH 45469, USA
* Correspondence: rhardie1@udayton.edu

Abstract: Convolutional neural networks (CNNs) have become instrumental in advancing multi-frame image super-resolution (SR), a technique that merges multiple low-resolution images of the same scene into a high-resolution image. In this paper, a novel deep learning multi-frame SR algorithm is introduced. The proposed CNN model, named Exponential Fusion of Interpolated Frames Network (EFIF-Net), seamlessly integrates fusion and restoration within an end-to-end network. Key features of the new EFIF-Net include a custom exponentially weighted fusion (EWF) layer for image fusion and a modification of the Residual Channel Attention Network for restoration to deblur the fused image. Input frames are registered with subpixel accuracy using an affine motion model to capture the camera platform motion. The frames are externally upsampled using single-image interpolation. The interpolated frames are then fused with the custom EWF layer, employing subpixel registration information to give more weight to pixels with less interpolation error. Realistic image acquisition conditions are simulated to generate training and testing datasets with corresponding ground truths. The observation model captures optical degradation from diffraction and detector integration from the sensor. The experimental results demonstrate the efficacy of EFIF-Net using both simulated and real camera data. The real camera results use authentic, unaltered camera data without artificial downsampling or degradation.

Keywords: multiframe super-resolution; convolutional neural network; fusion of interpolated frames; image restoration; subpixel registration

Citation: Elwarfalli, H.; Flaute, D.; Hardie, R.C. Exponential Fusion of Interpolated Frames Network (EFIF-Net): Advancing Multi-Frame Image Super-Resolution with Convolutional Neural Networks. *Sensors* **2024**, *24*, 296. https://doi.org/10.3390/s24010296

Academic Editors: Erik Blasch and Yufeng Zheng

Received: 28 November 2023
Revised: 29 December 2023
Accepted: 2 January 2024
Published: 4 January 2024

Copyright: © 2024 by the authors. Licensee MDPI, Basel, Switzerland. This article is an open access article distributed under the terms and conditions of the Creative Commons Attribution (CC BY) license (https://creativecommons.org/licenses/by/4.0/).

1. Introduction

In most image acquisition scenarios, there is a desire for the highest spatial resolution possible. However, ultimately, image resolution is limited by the sensor technology and characteristics. Using image post-processing techniques, it is possible to mitigate some of the limitations of a given sensor that may include noise, blur, and aliasing from undersampling. Processing that seeks to improve the resolution of an image beyond that of the native sensor is referred to as super-resolution (SR) [1]. Methods that use one input low-resolution (LR) image and produce one output high-resolution (HR) image are referred to as single-image super-resolution (SISR), while methods where multiple frames are fused to produce one HR image are referred to as multi-frame SR (MFSR). Several significant works documented in the literature [2–8] address SR for high-dimensional inputs, such as videos and 3D scans. The generation of HR images offers enriched details of locations and inherent objects, proving crucial in various applications, including high-definition TV sets, larger computer screens, and portable devices, like cameras, laptops, and mobile phones.

Convolutional neural networks (CNNs) have successfully been applied to many image processing and analysis tasks in security, surveillance, satellite, and medical imaging [9–17]. Early CNN-based SISR methods were introduced by Dong et al. [18,19]. Kim et al. presented the VDSR and DRCN networks [20,21]. Tai et al. pioneered DRRN and introduced memory blocks in MemNet [22,23]. Furthermore, SRGAN, a GAN-based approach for photo-realistic SR, was proposed by Ledig et al. [11]. ResNet was introduced by K. He et al., which was later extended to SRResNet [24]. EnhanceNet utilized a GAN-based model to merge perceptual loss with automated texture synthesis [14]. Recent years have seen significant improvements in deep SISR algorithms, such as EDSR [13], wide feature models by Yu et al. [25], and the residual channel attention network (RCAN) by Zhang et al. [26].

In contrast to SISR, MFSR focuses on extracting information from multiple LR images of the same scene, proving effective in overcoming undersampling issues in imaging systems [1]. In cases with relative motion between the scene and camera during video acquisition, subpixel displacements are common between frames. The diversity in sampling due to this motion can be leveraged to increase the sampling rate, reducing aliasing and improving resolution [27]. With MFSR, there is a trade-off between the temporal and spatial resolution [28]. Deep learning-based approaches have recently been applied to MFSR problems [29–34]. Bhat et al. proposed a multi-frame burst SR method that uses PWCNet for feature alignment and an attention-based fusion mechanism [35]. Notably, deformable convolution has proven effective in addressing interframe alignment issues [36–39]. Several recent architectural advancements have been proposed for specific purposes. For example, the work by Cao et al. [40] focuses on enhancing the resolution of human facial images. This approach involves extracting distinctive features from each LR image and utilizing these features, along with the relative shifts among LR images, to reconstruct the final SR image. In another study by An et al. [41], deep learning methodologies are applied within satellite and remote sensing domains. Additionally, there have been endeavors to integrate SISR and MFSR approaches. For instance, Gonbadani et al. [42] proposed an optimization-based method leveraging LR images and their associated semi-HR images generated via SISR. This optimization yields a closed-form solution, constituting a weighted combination of LR and semi-HR images. A multiframe network has also been introduced by the current authors that uses a single CNN architecture to fuse and restore multiple interpolated input frames [43].

This paper introduces a novel MFSR CNN model, named the Exponential Fusion of Interpolated Frames Network (EFIF-Net). The EFIF-Net extends the authors' previous algorithm, the Fusion of Interpolated Frames Network (FIFNET) [43]. This new model seamlessly combines multi-frame image fusion and restoration within a single end-to-end network. Key enhancements in EFIF-Net include the introduction of a custom exponentially weighted fusion (EWF) layer for image fusion and a modification of the RCAN for restoration to reduce the blurring in the fused image. We employ subpixel affine motion registration that accounts for camera platform motion. The input frames are externally upsampled and aligned using single-image interpolation. The interpolated frames are then fused using the custom EWF layer, taking into account subpixel registration information to assign more weight to pixels with lower interpolation errors.

For the generation of training and testing datasets, we simulate realistic image acquisition conditions, incorporating ground truth data. Our observation model captures optical degradation from diffraction and detector integration in the sensor. The experimental results highlight the effectiveness of EFIF-Net, using both simulated and real camera data. Notably, the real camera results are based on unaltered, authentic camera data without artificial downsampling or degradation. Owing in large part to the newly introduced EWF layers, the EFIF-Net method surpasses the performance of the previous FIFNET, particularly when dealing with a higher number of input frames.

The remainder of this paper is organized into several sections as follows. In Section 2, we define our observation model and introduce the proposed EFIF-Net MFSR method. We also provide details about network training and the performance analysis conducted. Experimental results are presented in Section 3. Finally, we conclude the paper with a discussion in Section 4.

2. Materials and Methods

2.1. Observation Model

The observation model is a mathematical model that captures the nature of the degradation processes that occur during image acquisition. The observation model can be used to generate synthetic datasets with degraded images and corresponding ground truth images for deep learning. In order to ensure that a CNN can provide a useful SR solution, the degradation process applied to generate training data must imitate realistic degradation effects. In this work, the observation model relates a static 2D ideal scene image with a group of LR observed frames with affine motion, blur, undersampling, and noise.

A block diagram of our observation model is shown in Figure 1. This is similar to the one used in the author's prior work [43], except here we incorporate an affine motion model rather than simple translation as before. The affine model makes the current approach more widely applicable to realistic camera platform motion [44]. The observation model replicates the physical image acquisition process by starting with a Nyquist sampled desired 2D fixed scene $d(n_1, n_2)$. Next, the model uses affine warping with bicubic subpixel interpolation to model the relative motion between the camera and scene (usually the result of camera motion and static scene). The affine coordinate warping may be expressed as

$$\tilde{\mathbf{x}}(k) = \mathbf{A}_k \mathbf{x} + \mathbf{t}_k, \qquad (1)$$

where $\mathbf{x} = [x, y]^T$ are the reference spatial coordinates and $\tilde{\mathbf{x}}(k) = [\tilde{x}(k), \tilde{y}(k)]^T$ are the warped coordinates for frame k. Note that the rotation, zoom and shear for each frame relative to a reference are captured in the matrices

$$\mathbf{A}_k = \begin{bmatrix} A_{1,1}(k) & A_{1,2}(k) \\ A_{2,1}(k) & A_{2,2}(k) \end{bmatrix}, \qquad (2)$$

and translation is captured in $\mathbf{t}_k = [t_x(k), t_y(k)]^T$, for $k = 1, 2, ..., K$. The output of the warping process is represented as $d_k(n_1, n_2)$ as shown in Figure 1.

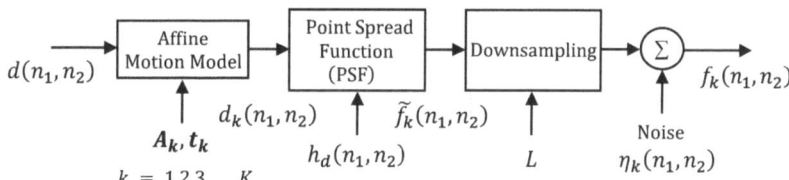

Figure 1. Observation model used to generate training data with known ground truth, as well as testing data for quantitative performance analysis. The input of the observation model is a single HR image $d(n_1, n_2)$, and its output is a sequence of K LR images $f_k(n_1, n_2)$.

The next step involves applying a realistic imaging-system point spread function (PSF), $h_d(n_1, n_2)$, using 2D convolution to create the blurred frames $\tilde{f}_k(n_1, n_2)$. The PSF modeling follows that reported previously by the authors [43] using the optical parameters in Table 1. The blurred image is then downsampled keeping every L'th pixel in the horizontal and vertical dimensions to produce a below Nyquist rate sampled image. This downsampled image is then corrupted with additive Gaussian noise, resulting in a set of K LR frames denoted by $f_k(n_1, n_2)$, where $k = 1, 2, \ldots, K$. While this one set of camera parameters is employed here, our observation model can be readily adapted to any camera, given the basic optical parameters of that system.

Table 1. Optical parameters used for training and testing for real and simulated images. These parameters are based on an Imaging Source DMK21BU04 visible USB camera with a Sony ICX098BL CCD sensor. The camera is equipped with a 5 mm focal length lens set to an f-number of $F = 5.6$. The camera/sensor and optics were obtained from The Imaging Source, LLC, Charlotte, NC, USA.

Parameter	Value
Aperture	$D = 0.893$ mm
Focal length	$l = 5.00$ mm
F-number	$F = 5.60$
Wavelength	$\lambda = 0.550$ µm
Optical cut-off frequency	$\rho_c = 324.68$ cyc/mm
Detector Pitch	$p = 5.6$ µm
Sampling frequency	$1/p = 178.57$ cyc/mm
Undersampling	$M = 3.636$
Upsampling factor	$L = 3.000$
Noise model	Additive Gaussian ($\sigma_\eta = 0.001$)
Dynamic range of dataset	0–1
Camera motion model	Affine

2.2. EFIF-Net Multiframe Super-Resolution

In this section, we present the EFIF-Net MFSR architecture and describe the preprocessing steps. Furthermore, we introduce and explore the significance of our unique innovation, the custom EWF fusion layer. This layer plays a pivotal role in seamlessly fusing interpolated frames, significantly enhancing the overall efficiency and performance of our model. Finally, we detail the network training process.

2.2.1. EFIF-Net Architecture

The EFIF-Net model is an end-to-end MFSR method as shown in Figure 2. The input is composed of the upsampled and aligned interpolated frames $g_k(n_1, n_2)$ and the corresponding subpixel registration arrays $R_k(n_1, n_2)$ for $k = 1, 2, \ldots K$. More will be said about the preprocessing required to produce this input in Section 2.2.2. The EFIF-Net output is one SR image estimate. We design the EFIF-Net model to accomplish two main aims: fusion and deconvolution. The fusion stage is completed by our custom EWF layer that is described in detail in Section 2.2.3. For deconvolution, we employ a modified version of the RCAN model. The RCAN model is a state-of-the-art CNN architecture originally designed for SISR [26]. In particular, we use only the residual in residual (RIR) structure of RCAN to form a very deep network to deconvolute/restore the imagery to produce the final SR image estimate. We remove the upscaling module from the RCAN network because the input size of the network matches the size of its output.

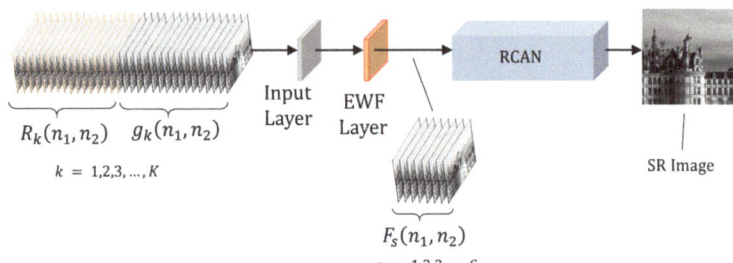

Figure 2. EFIF-Net architecture. Interpolated and aligned observed frames are combined with the subpixel registration information to form the input channels. The red layer represents our custom fusion layer. The fused feature images are then processed with a non-upsampling RCAN network to perform restoration. The output of the EFIF-Net is a single SR image.

2.2.2. Preprocessing

We preprocess the frames generated by the observation model depicted in Figure 1 using MATLAB Version R2022b to prepare it as the input for EFIF-Net. The LR frames, denoted as $f_k(n_1, n_2)$ and produced by the observation model, are individually subjected to upsampling and alignment using MATLAB's "interp2" function with bicubic interpolation. This step brings them onto a common $L\times$ upsampled HR grid. To automatically determine the necessary affine warping for alignment, we employ subpixel affine image registration using the method in [45], with the initial frame serving as the reference. These individually interpolated frames are represented as $g_k(n_1, n_2)$, where k ranges from 1 to K. Bicubic interpolation is chosen due to its ability to strike a balance between computational efficiency and performance as outlined in previous research by Hardie et al. [46].

Furthermore, we compute subpixel interframe registration information denoted as $R_k(n_1, n_2)$ for each interpolated pixel and introduce these data as additional input channels to the neural network. The distance computation is given by

$$R_k(n_1, n_2) = \sqrt{d_x^2(k, n_1, n_2) + d_y^2(k, n_1, n_2)}, \quad (3)$$

where $d_x(k, n_1, n_2)$, and $d_y(k, n_1, n_2)$ are the horizontal and vertical distances, respectively, of interpolated pixel $g_k(n_1, n_2)$ to the nearest LR pixel in the k'th input frame. A visual representation of Equation (3) is provided in Figure 3.

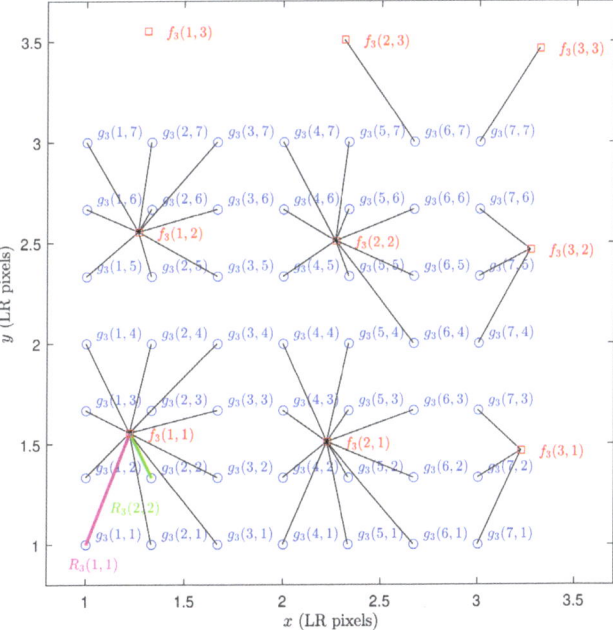

Figure 3. Spatial sampling grid shown in LR pixel spacings. The pixel positions of interpolated frame $g_3(n_1, n_2)$ are shown as blue circles for $L = 3$. The corresponding LR frame samples $f_3(n_1, n_2)$ are shown as red squares for a shift of $\mathbf{s}_3 = [0.18, 0.6]^T$ LR pixels and rotation of $\theta = -2.527$ degrees. The subpixel distances $R_3(n_1, n_2)$ are shown as black lines. An example of a large distance value is shown in magenta, and a small one is shown in green.

Note that the interframe registration information in Equation (3) has the same dimensions as the interpolated frames and may be viewed as images. This is illustrated in Figure 4 for $K = 3$ frames with shift and rotation frame motion and $L = 3$. In this figure, the darker pixels correspond to pixels with smaller interpolation distances, and therefore (presumably) lower interpolation error [44]. We designate the x- and y-shifts in LR pixels for frame k as $\mathbf{s}_k$ and the rotation in degrees as θ_k. Figure 4a corresponds to the interpolated reference frame with no motion (i.e., $\mathbf{s}_1 = [0.00, 0.00]^T$ and $\theta_0 = 0$). Here, every L'th pixel starting from pixel $n_1 = n_2 = 1$ lines up exactly with an input pixel and has an interpolation distance of 0. We present three example patches of subpixel registration arrays with motion in Figure 4b–d. They correspond to frame motions of $\mathbf{s}_2 = [0.52, -0.40]^T$, $\theta_2 = -0.37$, $\mathbf{s}_3 = [0.93, -1.08]^T$, $\theta_3 = -0.48$, and $\mathbf{s}_4 = [0.83, 1.28]^T$, $\theta_4 = -4.46$, respectively.

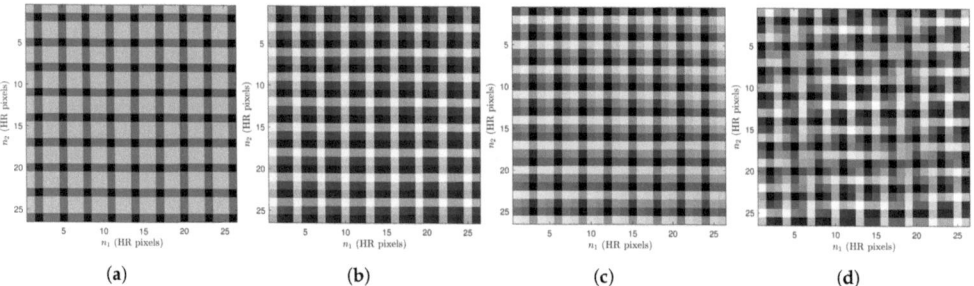

Figure 4. Visualization of $R_k(n_1, n_2)$ for a 25×25 patch size with different shifts and rotations for four frames and $L = 3$. The pixel brightness is proportional to the interpolation distance and presumed interpolation error. Pixel shifts and rotation are (**a**) $\mathbf{s}_1 = [0.00, 0.00]^T$, $\theta_1 = 0$, (**b**) $\mathbf{s}_2 = [0.52, -0.40]^T$, $\theta_2 = -0.37$, (**c**) $\mathbf{s}_3 = [0.93, -1.08]^T$, $\theta_3 = -0.48$, and (**d**) $\mathbf{s}_4 = [0.83, 1.28]^T$, $\theta_4 = -4.46$.

2.2.3. Exponential Weighted Fusion Layer

In our proposed EFIF-Net architecture, the custom EWF layer is located immediately after the input layer as shown in Figure 2. The EWF layer produces linear combinations of the input frames (fusion), pixel-by-pixel, with weights determined by a decreasing (negative-exponential) function of an interpolated frame's distance to the nearest measured sample point $R_k(n_1, n_2)$. That is, the EWF layer takes in the K interpolated and aligned input frames $g_k(n_1, n_2)$ along with the per-frame subpixel distance arrays $R_k(n_1, n_2)$ and outputs S different fused frames $F_s(n_1, n_2)$ as shown in Figure 5. Specifically, the fused images are given by

$$F_s(n_1, n_2) = \frac{\sum_{k=1}^{K} w_{k,s}(n_1, n_2) g_k(n_1, n_2)}{\sum_{k=1}^{k} w_{k,s}(n_1, n_2)}, \quad (4)$$

for $s = 1, 2, \ldots, S$, where the fusion weights are defined as

$$w_{k,s}(n_1, n_2) = e^{-R_k(n_1, n_2)^2 / \beta_s^2}. \quad (5)$$

The basic idea behind the EWF layer is that an interpolated pixel that is near an original LR pixel will have less interpolation error and should be given a higher weight in estimating the true intensity of the pixel based on the noisy interpolated intensities. This takes advantage of the multiframe setting since, if diverse camera motion is present, each HR interpolated pixel is likely to have an LR pixel nearby in at least one of the input frames. The rate of decay of the exponential with respect to distance $R_k(n_1, n_2)$ is a hyperparameter β_s that we control independently for each output fusion frame $F_s(n_1, n_2)$. Note that the fusion process produces a stack of fused frames, each with a different β_s, and these are used as input channels for RCAN restoration. Let us define the vector of all of these EWF hyperparameters as $\boldsymbol{\beta} = [\beta_1, \beta_2, \ldots, \beta_S]^T$. The optimal choice of $\boldsymbol{\beta}$ is unknown. But since the EWF layer is differentiable with respect to $\boldsymbol{\beta}$, we can optimize $\boldsymbol{\beta}$ via gradient descent as a learnable parameter of the EFIF-Net model during training. We randomly initialize these parameters and update them during the training process.

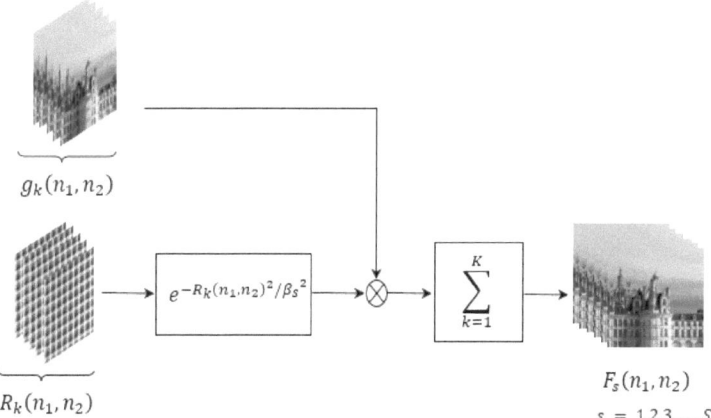

Figure 5. The EWF layer combines interpolated frames $g_k(n_1, n_2)$ with subpixel registration information $R_k(n_1, n_2)$ across various values of the parameter vector β to yield fused frames denoted as $F_S(n_1, n_2)$ for $s = 1, 2, \ldots, S$ as given by Equations (4) and (5).

2.2.4. Network Training

For training the EFIF-Net model, we use Python Version 3.8 with the PyTorch machine learning framework. The RCAN network model is modified from the publicly available version found in [47]. All pre- and post-processing is performed using MATLAB.

We chose images in the publicly available DIV2K Training dataset, which consists of 800 RGB images of 2K resolution [48]. These images are used as ground truth SR imagery. We converted all the images to grayscale with a dynamic range of 0–1. To simulate degradation during training and testing, we used the observation model explained in Section 2.1 and used SR with $L = 3$. Additionally, we added Gaussian noise with $\sigma_\eta = 0.001$ to prepare the network for simulated testing data. We used affine motion in the observation model to simulate the effects of camera motion in real-world image acquisition systems. In this study, we applied random affine motion to each image in the training and testing datasets using a transformation model that captures translation and rotation. Transformation parameters were randomly generated and applied to each image to simulate camera motion [45]. We then introduced PSF blur and additive Gaussian noise to the transformed image. By adding camera rotation to the observation model, we are able to address more realistic scenarios than translation alone as was done previously [43].

As the training patches are usually small compared to the original image size, we cropped the 2K resolution images to 480 × 480 sub-images. The size of the sub-images is different from the training patch size, which is 48 × 48, but this allows us to avoid reading the entire image when only a small part of it is needed. Each of the 800 images in the DIV2K dataset is cropped to create around 8 non-intersecting sub-images, resulting in a total of 6920 training sub-images of size 480 × 480. To determine the best length of the β vector S in the custom EWF layer, we tested various lengths and found that $S = 7$ is the most effective. The β vector is initialized randomly with values between 0 and 1.

We trained the EFIF-Net model using the default settings of the original RCAN network with 10 residual groups and 20 residual blocks as described in Section 4.1 from the original RCAN paper [26]. The batch size was set to 16 patches, and we used the Adam optimizer [49] with a learning rate of 10^{-4}. The model was trained for 104k updates using the $L1$ loss metric. To train the network, we used a Windows workstation with an AMD Ryzen 3970X 32-core processor running at 3.7 GHz and equipped with two NVIDIA GeForce RTX 3090 Graphics Processing Units. The workstation was sourced from Exxact Corp., Fremont, CA, USA. The training time varied between 8 h and 10 days, depending on the number of frames (K).

2.3. Performance Analysis

We evaluate the performance of the proposed EFIF-Net in comparison to benchmark methods. A quantitative performance analysis is conducted using images subjected to simulated degradation. Subsequently, a subjective assessment using authentic, unaltered camera data is conducted to evaluate the model's real-world performance.

2.3.1. Simulated Data

The imagery used for quantitative testing comes from three publicly available databases. These are the DIV2K Validation dataset [48] (100 images), the Set14 dataset [48] (14 images), and the BSDS100 dataset [50] (100 images). None of the images contained in these databases were used in training. The testing data underwent the same observation motion as the training data described in Section 2.2.4 so that we have objective ground truth for quantitative performance assessment. As benchmarks for the proposed EFIF-Net SR method, we also used single-frame bicubic interpolation, SISR with RCAN [26], and MFSR using FIFNET [43]. We employed two performance metrics, Peak Signal-to-Noise Ratio (PSNR) and the Structural Similarity Index (SSIM) [51]. We also present a number of processed images for subjective evaluation.

2.3.2. Real Camera Data

In addition to the quantitative performance analysis with simulated image degradation, we also applied the EFIF-Net and benchmark methods to real camera data acquired by the authors. The camera is an Imaging Source DMK21BU04 visible USB camera with a Sony ICX098BL CCD sensor with the optical parameters listed in Table 1. The camera is equipped with a 5 mm focal length lens set to an f-number of $F = 5.6$. The interframe motion was generated by manually panning and tilting the camera on a tripod during image acquisition, which was carried out at a rate of 30 frames per second. Using real-world images for testing SR models provides a more realistic evaluation of their performance.

Because the data were not artificially degraded, there are no corresponding ground truth images available for comparison. Thus, the results with real camera data are presented solely for subjective evaluation purpose. To facilitate this process, we selected familiar scene content and a well-defined test pattern. Since the observation model presented in Section 2.1 is based on a realistic camera PSF, the trained network can be applied directly to the data from that camera. However, to better match the observed signal-to-noise ratio for these real camera data, the networks were trained with appropriate values of σ_η defined in Section 3.2.

3. Results

The experimental results are presented in this section for the simulated data in Section 3.1 and real camera data in Section 3.2. The details of the experimental procedure regarding pre-processing are given in Section 2.2.2. Network training details are given in Section 2.2.4, and the performance analysis details are provided in Section 2.3.

3.1. Quantitative Results with Simulated Data

The graph in Figure 6 shows the average PSNR results for the DIV2K Validation dataset, based on the number of input frames used. The upsampling factor L is set to three, and the noise standard deviation is $\sigma_\eta = 0.001$. The graph shows results for EFIF-Net and benchmark methods. The results in this experiment show that EFIF-Net outperforms all of the benchmark methods for $K > 1$, with similar performance to the single-frame RCAN method for $K = 1$. Furthermore, increasing the number of input frames improves the performance of EFIF-Net. This is demonstrated by the smooth increase in the PSNR performance curve of the EFIF-Net model in Figure 6 as the number of input frames K increases. Table 2 presents additional quantitative results, including the PSNR and SSIM values, for all three test databases. The results demonstrate that the EFIF-Net model with $K = 60$ yields the best quantitative performance across all three databases.

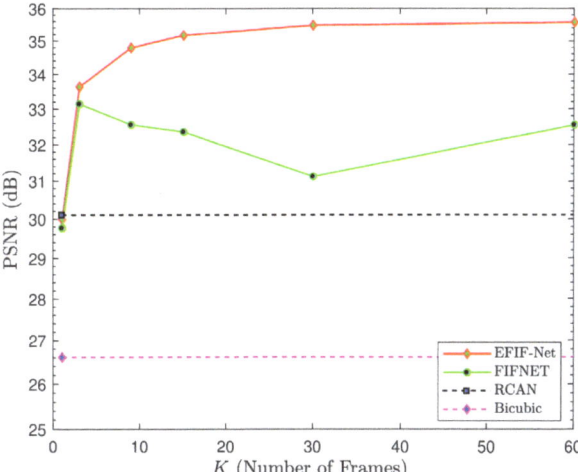

Figure 6. Quantitative performance comparison using simulated data from the DIV2K Validation dataset for $L = 3$ and $\sigma_\eta = 0.001$. The average PSNR is shown as a function of the number of input frames for the methods shown in the legend.

Table 2. Average PSNR(dB)/SSIM for $K = 1, 9, 30,$ and 60 using different methods and three different datasets: BSDS100, Set14, and DIV2K Validation. The numbers in bold font indicate the best performance for the corresponding metric and dataset category.

Dataset	K	PSNR(dB)/SSIM			
		Bicubic	RCAN	FIFNET	EFIF-Net
Set14	1	23.55/0.692	27.46/0.814	27.44/0.808	27.51/0.814
	9	-	-	31.45/0.909	32.67/0.925
	30	-	-	31.16/0.909	33.71/0.939
	60	-	-	31.03/0.903	**34.24/0.945**
BSDS100	1	24.31/0.720	26.36/0.781	26.29/0.775	26.36/0.781
	9	-	-	30.11/0.898	31.26/0.918
	30	-	-	29.98/0.897	32.56/0.937
	60	-	-	29.89/0.892	**33.18/0.945**
DIV2K	1	26.61/0.790	30.00/0.866	29.69/0.859	29.81/0.864
	9	-	-	33.10/0.919	34.48/0.940
	30	-	-	32.67/0.917	34.97/0.946
	60	-	-	32.41/0.9111	**35.16/0.948**

Figures 7–10 provide several processed images for subjective evaluation. Each figure includes the truth image in (a), and the processed ROIs in (b)–(e). It is important to note that the images in (b), and (c) in these figures are for single-frame methods, while the (d), and (e) are for multiframe methods with $K = 60$ frames. The captions of the figures provide the error metric values associated with the images. The image of an airplane in Figure 7 serves as a good example of the typical results. The bicubic interpolation method applied in Figure 7b results in significant blurring and aliasing artifacts in the form of jagged edges along the plane's wing. Conversely, the RCAN single-frame method applied in Figure 7c performs well in sharpening and reconstructing the edges of the wing. Nevertheless, the ill-posed nature of the inverse problem is evident in the numbers "0", "9" and "4" in this image. The multiframe methods are relatively successful in recovering these numbers. However, the EFIF-Net result in Figure 7e appears to offer the sharpest images with minimal aliasing artifacts. In addition, the EFIF-Net result in Figure 7e is 3.375 dB higher than the FIFNET model. Similar relative performance results are observed in Figures 8–10.

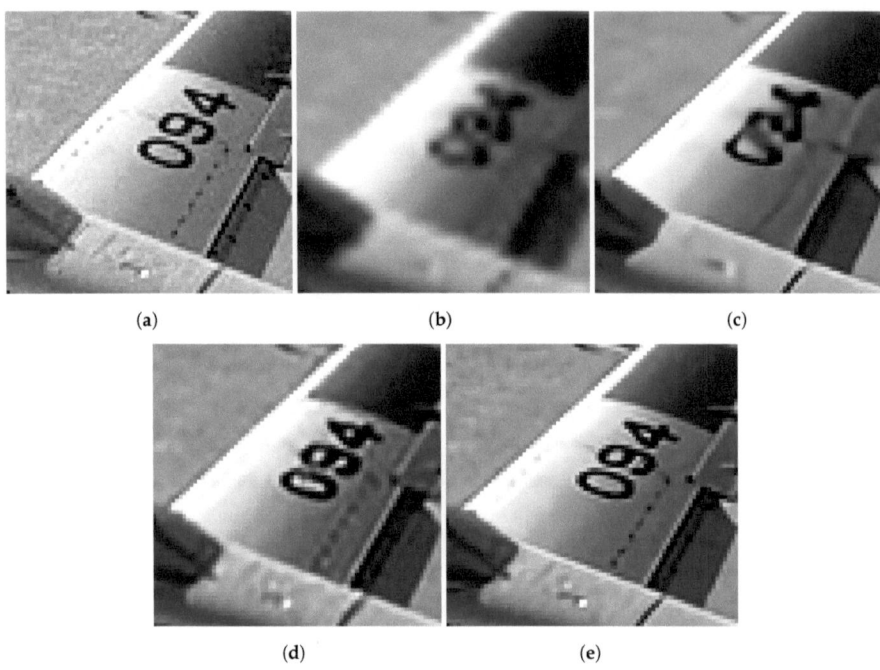

Figure 7. Results for image "071" in the BSDS100 dataset. The PSNR(dB)/SSIM values are (**b**) 26.755/0.726, (**c**) 30.895/0.839, (**d**) 33.183/0.886, and (**e**) 36.558/0.925. The noise has a standard deviation of $\sigma_\eta = 0.001$, and $K = 60$ frames are used in (**d**,**e**). (**a**) Truth, (**b**) Bicubic, (**c**) RCAN, (**d**) FIFNET, (**e**) EFIF-Net.

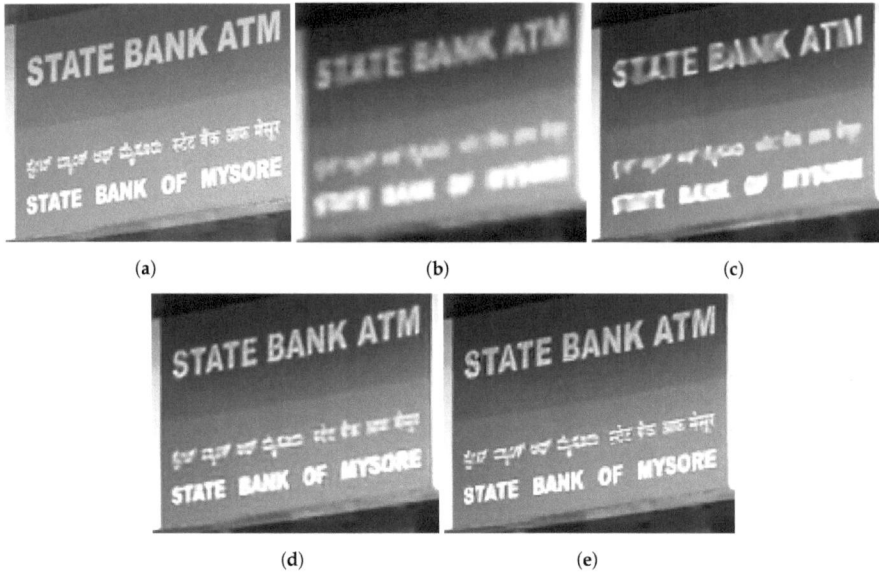

Figure 8. Results for image "91" in the DIV2K dataset. The PSNR(dB)/SSIM values are (**b**) 24.445/0.753, (**c**) 28.279/0.878, (**d**) 30.776/0.928, and (**e**) 34.440/0.961. The noise has a standard deviation of $\sigma_\eta = 0.001$, and $K = 60$ frames are used in (**d**,**e**). (**a**) Truth, (**b**) Bicubic, (**c**) RCAN, (**d**) FIFNET, (**e**) EFIF-Net.

Figure 9. Results for image "10" in the Set14 dataset. The PSNR(dB)/SSIM values are (**b**) 25.025/0.688, (**c**) 27.532/0.796, (**d**) 30.255/0.886, and (**e**) 33.168/0.934. The noise has a standard deviation of $\sigma_\eta = 0.001$, and $K = 60$ frames are used in (**d**,**e**). (**a**) Truth, (**b**) Bicubic, (**c**) RCAN, (**d**) FIFNET, (**e**) EFIF-Net.

Figure 10. *Cont.*

(d) (e)

Figure 10. Results for image "92" in the DSBS100 dataset. The PSNR(dB)/SSIM values are (**b**) 25.025/0.688, (**c**) 27.532/0.796, (**d**) 29.318.255/0.889, and (**e**) 34.753/0.960. The noise has a standard deviation of $\sigma_\eta = 0.001$, and $K = 60$ frames are used in (**d,e**). (**a**) Truth, (**b**) Bicubic, (**c**) RCAN, (**d**) FIFNET, (**e**) EFIF-Net.

3.2. Subjective Results with Real Camera Data

We present the results for two distinct real-camera multiframe datasets in Figures 11 and 12. For each dataset, the multiframe methods use $K = 60$ frames with $L = 3$. As mentioned above, these results showcase the potential of the algorithm in a real-world application (and without artificial degradation). Figure 11 illustrates the application of SR models on a bookshelf dataset captured indoors. Note that there is no actual ground truth available for such real images, but the scene content of books on bookshelf is familiar. We selected a value of $\sigma_\eta = 0.01$ for the training noise used for the network that processed these data. In Figure 11a, the bicubic interpolation image exhibits aliasing artifacts on the lettering. Although the RCAN method managed to decrease the aliasing artifacts on larger letters (Figure 11b), it performed more poorly on smaller lettering. On the other hand, the EFIF-Net outcome shown in Figure 11d offers better sharpness and superior noise suppression than that of FIFNET in Figure 11c.

The image in Figure 12 depicts a circularly symmetric chirp pattern that is specifically chosen to demonstrate the ability of the tested methods to reduce aliasing. For these data, we selected a value of $\sigma_\eta = 0.025$ for the training noise level. The bicubic interpolation image in Figure 12a shows aliasing in the form of a Moiré pattern on the high-frequency components of the chirp, causing the concentric ring pattern to appear inverted on the top and right of the image. The SISR RCAN method output shown in Figure 12b increases contrast but is unable to accurately reconstruct the true chirp structure from the undersampled imagery. The MFSR FIFNET method output shown in Figure 12c does show reduced aliasing and resolves more of the concentric rings than single-image bicubic or RCAN. However, the EFIF-Net output in Figure 12d provides the best sharpness and greater noise reduction.

Figure 11. Image results for the real camera data of a bookshelf. The images shown are (**a**) Bicubic, (**b**) RCAN, (**c**) FIFNET and (**d**) EFIF-Net. The noise has a standard deviation of $\sigma_\eta = 0.01$ and $K = 60$ frames for (**c**,**d**).

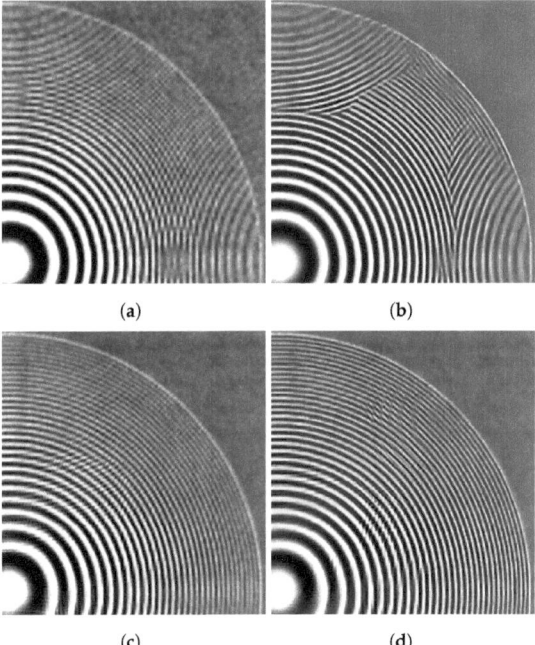

Figure 12. Image results for the real camera data of a chirp. The images shown are (**a**) Bicubic, (**b**) RCAN, (**c**) FIFNET and (**d**) EFIF-Net. The noise has a standard deviation of $\sigma_\eta = 0.025$ and $K = 60$ frames for (**c**,**d**).

4. Discussion

In this paper, we introduced the EFIF-Net approach to tackle the challenge of MFSR using CNNs. Diverging from many previous methodologies, our approach incorporates a realistic observation model in the creation of both the training and testing datasets. This model faithfully accounts for the genuine degradation effects introduced by the camera during image capture. Consequently, we developed a more robust CNN model equipped for real-world scenarios. Furthermore, we enhanced the previous FIFNET in several ways. First, we extended the camera motion model from translation to a more practical affine model. This affine model allows for rotation, scaling, and shear in addition to translation. Rotation, in particular, is important, as it is commonly introduced when acquiring an image sequence from a non-fixed platform. Another innovation in EFIF-Net, that is not in FIFNET, is the inclusion of the custom EWF layer for performing image fusion. Specifically, the EWF layer employs exponential weights to interpolated pixels so as to give more weight to pixels with less interpolation error. Finally, the current method provides superior restoration of the fused image by incorporating a modified RCAN network with many layers, whereas FIFNET uses a relative small CNN.

The experimental results presented in Section 3 demonstrate the performance benefits of the innovations described above. The results show that the MFSR methods provide higher PSNR values than the SISR benchmark method. This shows that the spatial sampling diversity provided by multiple frames, when properly exploited, is a significant benefit to the restoration quality given undersampled input frames. Furthermore, the EFIF-Net consistently outperforms the previous FIFNET in all quantitative and subjective performance analyses. The performance improvement of EFIF-Net is most pronounced with higher numbers of input frames. In fact, the performance of FIFNET initially went down in PSNR when more than four input frames were used. We attribute this drop in performance to the inability of the FIFNET architecture to efficiently fuse large numbers of input frames by simply concatenating them. The EWF layer in the EFIF-Net addresses this issue and reduces the number of learnable parameters associated with image fusion to just the $S = 7$ values in β, compared with a much larger number of learnable parameters associated with the full convolution layers in FIFNET. As a result, the EFIF-Net performance consistently improves with an increased number of input frames as one would hope from an MFSR method. We also observed, in the subjective image results with simulated and real data, that the EFIF-Net provided final results that appeared sharper and less noisy than the benchmark methods. We attribute this to the improved restoration provided by the modified RCAN processing layers, compared with the relatively small number of convolution layers used by the earlier FIFNET.

Our study had limitations. One limitation is that we considered only one camera model due to the significant network training time needed for multiple numbers of input frames and multiple noise levels. Additional performance analyses would be helpful in future work to better understand how the EFIF-Net performs with other cameras and optical parameters. Another important limitation to be aware of relates to the fact the MFSR methods presented here assume a static scene and affine camera motion. The presence of in-scene motion, like moving cars and people, is not addressed in the current study. However, it may be possible to incorporate some scene motion detection and alternative processing [52] in future work. Also, if there is no camera motion, there is generally no spatial sampling diversity. Consequently, the benefits of MFSR over SISR are more limited. Interestingly, the work in [46] demonstrated that mild atmospheric turbulence can provide sampling diversity to improve MFSR, even with a fully stationary camera and scene. Finally, we consider only grayscale images here so as to focus on the essential characteristics of MFSR. However, our approach can be readily extended to color in future work by employing some of the concepts described in [53].

Based on a subjective analysis, the results of the proposed EFIF-Net method using real camera data appear very consistent with the results with simulated data. In the real data, EFIF-Net produced clearly visible aliasing reduction, particularly in the chirp image (designed to illustrate aliasing). It also produced images with increased sharpness and increased noise reduction compared with all of the benchmark methods. Good results on real camera data are particularly notable since the training data were generated using our theoretical observation model and images with simulated degradation. The trained model generalized well from the simulated training data to the real camera testing data that had no artificial degradation, demonstrating robust real-world performance.

The main advantage of the proposed method is the ability to efficiently fuse multiple input frames using the EWF layer in a manner that scales to a virtually limitless number of input frames. In particular, after the EWF layer that produces $S = 7$ fusion images, the rest of the network does not grow in complexity with an increased number of input frames. One important area of future work is to investigate how to address scene motion within the proposed framework in such a way as to preserve moving objects and provide SR enhancement where possible.

Author Contributions: Conceptualization, H.E., D.F. and R.C.H.; Methodology, H.E., D.F. and R.C.H.; Software, H.E. and D.F.; Data curation, H.E.; Writing—original draft, H.E.; Writing—review & editing, D.F. and R.C.H.; Supervision, R.C.H. All authors have read and agreed to the published version of the manuscript.

Funding: This research received no external funding.

Institutional Review Board Statement: Not applicable

Informed Consent Statement: Not applicable

Data Availability Statement: Publicly available datasets were analyzed in this study [48,50].

Conflicts of Interest: The authors declare no conflict of interest.

References

1. Park, S.C.; Park, M.K.; Kang, M.G. Super-resolution image reconstruction: A technical overview. *IEEE Signal Process. Mag.* **2003**, *20*, 21–36. [CrossRef]
2. Bashir, S.M.A.; Wang, Y.; Khan, M.; Niu, Y. A comprehensive review of deep learning-based single image super-resolution. *PeerJ Comput. Sci.* **2021**, *7*, e621. [CrossRef] [PubMed]
3. Tan, R.; Yuan, Y.; Huang, R.; Luo, J. Video super-resolution with spatial-temporal transformer encoder. In Proceedings of the 2022 IEEE International Conference on Multimedia and Expo (ICME), Taipei, Taiwan, 18–22 July 2022; pp. 1–6.
4. Li, H.; Zhang, P. Spatio-temporal fusion network for video super-resolution. In Proceedings of the 2021 International Joint Conference on Neural Networks (IJCNN), Shenzhen, China, 18–22 July 2021; pp. 1–9.
5. Thawakar, O.; Patil, P.W.; Dudhane, A.; Murala, S.; Kulkarni, U. Image and video super resolution using recurrent generative adversarial network. In Proceedings of the 2019 16th IEEE International Conference on Advanced Video and Signal Based Surveillance (AVSS), Taipei, Taiwan, 18–21 September 2019; pp. 1–8.
6. Smith, E.; Fujimoto, S.; Meger, D. Multi-view silhouette and depth decomposition for high resolution 3d object representation. *Conf. on Neural Inf. Process. Syst.* **2018**, *32*, 6479–6489.
7. Li, B.; Li, X.; Lu, Y.; Liu, S.; Feng, R.; Chen, Z. Hst: Hierarchical swin transformer for compressed image super-resolution. In Proceedings of the European Conference on Computer Vision, Tel Aviv, Israel, 23–27 October 2022; pp. 651–668.
8. Li, H.; Yang, Y.; Chang, M.; Chen, S.; Feng, H.; Xu, Z.; Li, Q.; Chen, Y. Srdiff: Single image super-resolution with diffusion probabilistic models. *Neurocomputing* **2022**, *479*, 47–59. [CrossRef]
9. Ma, Z.; Liao, R.; Tao, X.; Xu, L.; Jia, J.; Wu, E. Handling motion blur in multi-frame super-resolution. In Proceedings of the IEEE Conference on Computer Vision and Pattern Recognition, Boston, MA, USA, 7–12 June 2015; pp. 5224–5232.
10. Johnson, J.; Alahi, A.; Fei-Fei, L. Perceptual losses for real-time style transfer and super-resolution. In Proceedings of the European Conference on Computer Vision, Amsterdam, The Netherlands, 11–14 October 2016; pp. 694–711.
11. Ledig, C.; Theis, L.; Huszár, F.; Caballero, J.; Cunningham, A.; Acosta, A.; Aitken, A.; Tejani, A.; Totz, J.; Wang, Z.; et al. Photo-Realistic Single Image Super-Resolution Using a Generative Adversarial Network. In Proceedings of the 2017 IEEE Conference on Computer Vision and Pattern Recognition (CVPR), Honolulu, HI, USA, 21–26 July 2017; pp. 105–114. [CrossRef]
12. Lai, W.S.; Huang, J.B.; Ahuja, N.; Yang, M.H. Deep Laplacian Pyramid Networks for Fast and Accurate Super-Resolution. In Proceedings of the 2017 IEEE Conference on Computer Vision and Pattern Recognition (CVPR), Honolulu, HI, USA, 21–26 July 2017; pp. 5835–5843. [CrossRef]
13. Lim, B.; Son, S.; Kim, H.; Nah, S.; Lee, K.M. Enhanced Deep Residual Networks for Single Image Super-Resolution. In Proceedings of the 2017 IEEE Conference on Computer Vision and Pattern Recognition Workshops (CVPRW), Honolulu, HI, USA, 21–26 July 2017; pp. 1132–1140. [CrossRef]

14. Sajjadi, M.S.M.; Schölkopf, B.; Hirsch, M. EnhanceNet: Single Image Super-Resolution Through Automated Texture Synthesis. In Proceedings of the 2017 IEEE International Conference on Computer Vision (ICCV), Venice, Italy, 22–29 October 2017; pp. 4501–4510. [CrossRef]
15. Bulat, A.; Tzimiropoulos, G. Super-FAN: Integrated Facial Landmark Localization and Super-Resolution of Real-World Low Resolution Faces in Arbitrary Poses with GANs. In Proceedings of the 2018 IEEE/CVF Conference on Computer Vision and Pattern Recognition, Salt Lake City, UT, USA, 18–23 June 2018; pp. 109–117. [CrossRef]
16. Chu, X.; Zhang, B.; Ma, H.; Xu, R.; Li, Q. Fast, Accurate and Lightweight Super-Resolution with Neural Architecture Search. In Proceedings of the 2020 25th International Conference on Pattern Recognition (ICPR), Milan, Italy, 10–15 January 2021; pp. 59–64. [CrossRef]
17. Ahn, N.; Kang, B.; Sohn, K.A. Image Super-Resolution via Progressive Cascading Residual Network. In Proceedings of the 2018 IEEE/CVF Conference on Computer Vision and Pattern Recognition Workshops (CVPRW), Salt Lake City, UT, USA, 18–22 June 2018; pp. 904–9048. [CrossRef]
18. Dong, C.; Loy, C.C.; He, K.; Tang, X. Image super-resolution using deep convolutional networks. *IEEE Trans. Pattern Anal. Mach. Intell.* **2015**, *38*, 295–307. [CrossRef]
19. Dong, C.; Loy, C.C.; Tang, X. Accelerating the super-resolution convolutional neural network. In Proceedings of the European Conference on Computer Vision, Amsterdam, The Netherlands, 11–14 October 2016; pp. 391–407.
20. Kim, J.; Lee, J.K.; Lee, K.M. Accurate Image Super-Resolution Using Very Deep Convolutional Networks. In Proceedings of the 2016 IEEE Conference on Computer Vision and Pattern Recognition (CVPR), Las Vegas, NV, USA, 27–30 June 2016; pp. 1646–1654. [CrossRef]
21. Kim, J.; Lee, J.K.; Lee, K.M. Deeply-recursive convolutional network for image super-resolution. In Proceedings of the IEEE Conference on Computer Vision and Pattern Recognition, Las Vegas, NV, USA, 27–30 June 2016; pp. 1637–1645.
22. Tai, Y.; Yang, J.; Liu, X. Image super-resolution via deep recursive residual network. In Proceedings of the IEEE Conference on Computer Vision and Pattern Recognition, Honolulu, HI, USA, 21–26 July 2017; pp. 3147–3155.
23. Tai, Y.; Yang, J.; Liu, X.; Xu, C. Memnet: A persistent memory network for image restoration. In Proceedings of the IEEE International Conference on Computer Vision, Venice, Italy, 22–29 October 2017; pp. 4539–4547.
24. He, K.; Zhang, X.; Ren, S.; Sun, J. Deep residual learning for image recognition. In Proceedings of the IEEE Conference on Computer Vision and Pattern Recognition, Las Vegas, NV, USA, 27–30 June 2016; pp. 770–778.
25. Yu, J.; Fan, Y.; Yang, J.; Xu, N.; Wang, Z.; Wang, X.; Huang, T. Wide activation for efficient and accurate image super-resolution. *arXiv* **2018**, arXiv:1808.08718.
26. Zhang, Y.; Li, K.; Li, K.; Wang, L.; Zhong, B.; Fu, Y. Image super-resolution using very deep residual channel attention networks. In Proceedings of the European Conference on Computer Vision (ECCV), Munich, Germany, 8–14 September 2018; pp. 286–301.
27. Hardie, R.C.; Droege, D.R.; Dapore, A.J.; Greiner, M.E. Impact of detector-element active-area shape and fill factor on super-resolution. *Front. Phys.* **2015**, *3*, 31. [CrossRef]
28. Milanfar, P. *Super-Resolution Imaging*; CRC Press: Boca Raton, FL, USA, 2017.
29. Deudon, M.; Kalaitzis, A.; Goytom, I.; Arefin, M.R.; Lin, Z.; Sankaran, K.; Michalski, V.; Kahou, S.E.; Cornebise, J.; Bengio, Y. Highres-net: Recursive fusion for multi-frame super-resolution of satellite imagery. *arXiv* **2020**, arXiv:2002.06460.
30. Arefin, M.R.; Michalski, V.; St-Charles, P.L.; Kalaitzis, A.; Kim, S.; Kahou, S.E.; Bengio, Y. Multi-image super-resolution for remote sensing using deep recurrent networks. In Proceedings of the IEEE/CVF Conference on Computer Vision and Pattern Recognition Workshops, Seattle, WA, USA, 14–19 June 2020; pp. 206–207.
31. Molini, A.B.; Valsesia, D.; Fracastoro, G.; Magli, E. Deepsum: Deep neural network for super-resolution of unregistered multitemporal images. *IEEE Trans. Geosci. Remote Sens.* **2019**, *58*, 3644–3656. [CrossRef]
32. Bajo, M. *Multi-Frame Super Resolution of Unregistered Temporal Images Using WDSR Nets*; 2020. Available online: https://zenodo.org/records/3733116 (accessed on 1 November 2023).
33. Dorr, F. Satellite image multi-frame super resolution using 3D wide-activation neural networks. *Remote Sens.* **2020**, *12*, 3812. [CrossRef]
34. Salvetti, F.; Mazzia, V.; Khaliq, A.; Chiaberge, M. Multi-image super resolution of remotely sensed images using residual attention deep neural networks. *Remote Sens.* **2020**, *12*, 2207. [CrossRef]
35. Bhat, G.; Danelljan, M.; Van Gool, L.; Timofte, R. Deep burst super-resolution. In Proceedings of the IEEE/CVF Conference on Computer Vision and Pattern Recognition, Nashville, TN, USA, 20–25 June 2021; pp. 9209–9218.
36. Tian, Y.; Zhang, Y.; Fu, Y.; Xu, C. Tdan: Temporally-deformable alignment network for video super-resolution. In Proceedings of the IEEE/CVF Conference on Computer Vision and Pattern Recognition, Seattle, WA, USA, 13–19 June 2020; pp. 3360–3369.
37. Ustinova, E.; Lempitsky, V. Deep multi-frame face super-resolution. *arXiv* **2017**, arXiv:1709.03196.
38. Wang, X.; Chan, K.C.; Yu, K.; Dong, C.; Loy, C.C. EDVR: Video Restoration With Enhanced Deformable Convolutional Networks. In Proceedings of the 2019 IEEE/CVF Conference on Computer Vision and Pattern Recognition Workshops (CVPRW), Long Beach, CA, USA, 16–17 June 2019; pp. 1954–1963. [CrossRef]
39. Zhu, X.; Hu, H.; Lin, S.; Dai, J. Deformable convnets v2: More deformable, better results. In Proceedings of the IEEE/CVF Conference on Computer Vision and Pattern Recognition, Long Beach, CA, USA, 15–20 June 2019; pp. 9308–9316.

40. Cao, F.; Su, M. Research on Face Recognition Algorithm Based on CNN and Image Super-resolution Reconstruction. In Proceedings of the 2022 IEEE 8th Intl Conference on Big Data Security on Cloud (BigDataSecurity), IEEE Intl Conference on High Performance and Smart Computing,(HPSC) and IEEE Intl Conference on Intelligent Data and Security (IDS), Jinan, China, 6–8 May 2022; pp. 157–161.
41. An, T.; Zhang, X.; Huo, C.; Xue, B.; Wang, L.; Pan, C. TR-MISR: Multiimage super-resolution based on feature fusion with transformers. *IEEE J. Sel. Top. Appl. Earth Obs. Remote Sens.* **2022**, *15*, 1373–1388. [CrossRef]
42. Gonbadani, M.M.A.; Abbasfar, A. Combined Single and Multi-frame Image Super-resolution. In Proceedings of the 2020 28th Iranian Conference on Electrical Engineering (ICEE), Tabriz, Iran, 4–6 August 2020; pp. 1–6.
43. Elwarfalli, H.; Hardie, R.C. Fifnet: A convolutional neural network for motion-based multiframe super-resolution using fusion of interpolated frames. *Comput. Vis. Image Underst.* **2021**, *202*, 103097. [CrossRef]
44. Hardie, R. A fast image super-resolution algorithm using an adaptive Wiener filter. *IEEE Trans. Image Process.* **2007**, *16*, 2953–2964. [CrossRef] [PubMed]
45. Hardie, R.C.; Barnard, K.J.; Ordonez, R. Fast super-resolution with affine motion using an adaptive Wiener filter and its application to airborne imaging. *Opt. Express* **2011**, *19*, 26208–26231. [CrossRef]
46. Hardie, R.C.; Rucci, M.; Karch, B.K.; Dapore, A.J.; Droege, D.R.; French, J.C. Fusion of interpolated frames superresolution in the presence of atmospheric optical turbulence. *Opt. Eng.* **2019**, *58*, 083103. [CrossRef]
47. Zhang, Y. RCAN: PyTorch Code for Our ECCV 2018 Paper "Image Super-Resolution Using Very Deep Residual Channel Attention Networks". Available online: https://github.com/yulunzhang/RCAN/tree/master (accessed on 1 November 2023).
48. Agustsson, E.; Timofte, R. NTIRE 2017 challenge on single image super-resolution: Dataset and study. In Proceedings of the IEEE Conference on Computer Vision and Pattern Recognition Workshops, Honolulu, HI, USA, 21–26 July 2017; pp. 1122–1131.
49. Kingma, D.P.; Ba, J. Adam: A Method for Stochastic Optimization. *arXiv* **2017**, arXiv:1412.6980.
50. Arbelaez, P.; Maire, M.; Fowlkes, C.; Malik, J. Contour detection and hierarchical image segmentation. *IEEE Trans. Pattern Anal. Mach. Intell.* **2010**, *33*, 898–916. [CrossRef]
51. Wang, Z.; Bovik, A.C.; Sheikh, H.R.; Simoncelli, E.P. Image quality assessment: From error visibility to structural similarity. *IEEE Trans. Image Process.* **2004**, *13*, 600–612. [CrossRef]
52. Hardie, R.C.; Barnard, K.J. Fast super-resolution using an adaptive Wiener filter with robustness to local motion. *Opt. Express* **2012**, *20*, 21053–21073. [CrossRef] [PubMed]
53. Karch, B.K.; Hardie, R.C. Robust super-resolution by fusion of interpolated frames for color and grayscale images. *Front. Phys.* **2015**, *3*, 28. [CrossRef]

Disclaimer/Publisher's Note: The statements, opinions and data contained in all publications are solely those of the individual author(s) and contributor(s) and not of MDPI and/or the editor(s). MDPI and/or the editor(s) disclaim responsibility for any injury to people or property resulting from any ideas, methods, instructions or products referred to in the content.

Article

SLMSF-Net: A Semantic Localization and Multi-Scale Fusion Network for RGB-D Salient Object Detection

Yanbin Peng *, Zhinian Zhai and Mingkun Feng

School of Information and Electronic Engineering, Zhejiang University of Science and Technology, Hangzhou 310023, China
* Correspondence: pyb@zust.edu.cn

Abstract: Salient Object Detection (SOD) in RGB-D images plays a crucial role in the field of computer vision, with its central aim being to identify and segment the most visually striking objects within a scene. However, optimizing the fusion of multi-modal and multi-scale features to enhance detection performance remains a challenge. To address this issue, we propose a network model based on semantic localization and multi-scale fusion (SLMSF-Net), specifically designed for RGB-D SOD. Firstly, we designed a Deep Attention Module (DAM), which extracts valuable depth feature information from both channel and spatial perspectives and efficiently merges it with RGB features. Subsequently, a Semantic Localization Module (SLM) is introduced to enhance the top-level modality fusion features, enabling the precise localization of salient objects. Finally, a Multi-Scale Fusion Module (MSF) is employed to perform inverse decoding on the modality fusion features, thus restoring the detailed information of the objects and generating high-precision saliency maps. Our approach has been validated across six RGB-D salient object detection datasets. The experimental results indicate an improvement of 0.20~1.80%, 0.09~1.46%, 0.19~1.05%, and 0.0002~0.0062, respectively in maxF, maxE, S, and MAE metrics, compared to the best competing methods (AFNet, DCMF, and C2DFNet).

Keywords: RGB-D; salient object detection; multi-modal and multi-scale features

Citation: Peng, Y.; Zhai, Z.; Feng, M. SLMSF-Net: A Semantic Localization and Multi-Scale Fusion Network for RGB-D Salient Object Detection. *Sensors* **2024**, *24*, 1117. https://doi.org/10.3390/s24041117

Academic Editors: Erik Blasch and Yufeng Zheng

Received: 9 January 2024
Revised: 2 February 2024
Accepted: 6 February 2024
Published: 8 February 2024

Copyright: © 2024 by the authors. Licensee MDPI, Basel, Switzerland. This article is an open access article distributed under the terms and conditions of the Creative Commons Attribution (CC BY) license (https://creativecommons.org/licenses/by/4.0/).

1. Introduction

Salient Object Detection (SOD) plays a crucial role in the field of computer vision, with its primary objective being the identification and accentuation of the most visually engaging objects within a scene [1,2]. These objects typically draw the majority of observer attention and play a vital role in image and video processing tasks, such as object tracking [3,4], image segmentation [5,6], and scene understanding [7,8]. With the rapid advancement of depth sensor technology, RGB-D salient object detection has elicited significant interest among researchers. Compared to using only RGB images, RGB-D datasets offer a richer array of information, including color and depth details, which are invaluable in enhancing the performance of salient object detection. However, the achievement of accurate salient object detection under complex scenarios, with multi-scale objects and noise interference, continues to present a substantial challenge. Current research is confronted with two main issues [9–37]:

1. The Modality Fusion Problem: Undoubtedly, depth information opens up significant possibilities for enhancing detection performance. The distance information it provides between objects aids in clearly distinguishing the foreground from the background, thereby endowing the algorithm with robustness when dealing with complex scenarios. However, an urgent challenge that remains to be solved is how to fully exploit this depth information and effectively integrate it with the color, texture, and other features of RGB images to extract richer and more discriminative features. This challenge becomes particularly pressing when dealing with issues of incomplete depth information and noise interference, which necessitate further exploration and research.

2. The Multi-level Feature Integration Problem: To more effectively integrate multi-level features, it's vital to fully consider the characteristics of both high-level and low-level features. High-level features contain discriminative semantic information, which aids in the localization of salient objects, while low-level features are rich in detailed information, beneficial for optimizing object edges. Traditional RGB-D salient object detection methods often fuse features from different levels directly, disregarding their inherent differences. This approach can lead to semantic information loss and make the method vulnerable to noise and background interference. Therefore, there is a need to explore more refined feature fusion techniques that fully take into account the characteristics of different levels of features, aiming to boost the performance of salient object detection.

To address the aforementioned challenges, we propose a Semantic Localization and Multi-Scale Fusion Network (SLMSF-Net) for RGB-D salient object detection. SLMSF-Net constitutes two stages: encoding and decoding. During the encoding phase, SLMSF-Net utilizes the ResNet50 network to separately extract features from RGB and depth images and employs a depth attention module for modal feature fusion. In the decoding phase, SLMSF-Net first accurately localizes salient objects through a semantic localization module, and then constructs a reverse decoder using a Multi-Scale Fusion Module to restore the detailed information of the salient objects. Our main contributions can be summarized as follows:

1. We propose a depth attention module that leverages channel and spatial attention mechanisms to fully explore the effective information of depth images and enhance the matching ability between RGB and depth feature maps.
2. We propose a semantic localization module that constructs a global view for the precise localization of salient objects.
3. We propose a reverse decoding network based on multi-scale fusion, which implements reverse decoding on modal fusion features and generates detailed information on salient objects through multi-scale feature fusion.

The design of the SLMSF-Net is poised to address key issues in the current RGB-D salient object detection domain and provide new research insights for other tasks within the field of computer vision. Extensive experimental results fully demonstrate that the SLMSF-Net exhibits excellent performance in RGB-D SOD tasks, enhancing the accuracy and effectiveness of salient object detection.

2. Related Works

In this section, we will review research works [17–37] related to the RGB-D salient object detection method that we propose. These related studies can be broadly divided into two categories: salient object detection based on RGB images and salient object detection based on RGB-D images.

2.1. Salient Object Detection Based on RGB Images

Salient object detection based on RGB images mainly focuses on visual cues such as color, texture, and contrast. Early saliency detection methods primarily depended on hand-crafted features and heuristic rules. For instance, Itti et al. [17] proposed a saliency detection model based on the biological visual system, which estimates saliency by calculating the local contrast of color, brightness, and directional features. Achanta et al. [18] introduced a frequency-tuned salient region detection method, which extracts global contrast features in the frequency domain of the image to detect salient regions. Tong et al. [19] combined global and local cues for salient object detection, using a variety of cues (such as color, texture, and contrast) to handle complex scenarios.

In recent years, deep learning technology has achieved significant success in the field of salient object detection. Models such as the deep learning saliency model proposed by Chen et al. [20] accomplish hierarchical representation of saliency features to realize end-to-end salient object detection. Cong et al. [21] proposed a salient object detection method based on a Fully Convolutional Network (FCN), which uses global contextual information and local detail information for saliency prediction. Hou et al. [22] developed a deeply supervised network for salient object detection, improving upon the Holistically Nested Edge Detector (HED) architecture. They introduced short connections between network layers, enhancing salient object detection by combining low-level and high-level features. Zhao et al. [23] proposed GateNet, a new network architecture for salient object detection. This model introduced multilevel gate units to balance encoder block contributions, suppressing non-salient features and contextualizing for the decoder. They also included Fold-ASPP to gather multiscale semantic information, enhancing atrous convolution for better feature extraction. Zhang et al. [24] combined neural network layer features to improve salient object detection accuracy in images. Their approach used both coarse and fine image details and incorporated edge-aware maps to enhance boundary detection. Wu et al. [25] proposed a cascaded partial decoder that discarded low-level features to reduce computational complexity while refining high-level features for accuracy.

Moreover, some researchers have applied attention mechanisms to RGB-based salient object detection models, such as [26–28]. These methods enable the models to concentrate their attention on the visually prominent regions of the image. Chen et al. [26] presented an approach for enhancing salient object detection through the use of reverse attention and side-output residual learning. This method aimed to refine saliency maps with a particular focus on improving resolution and reducing the model's size. Wang et al. [27] presented PAGE-Net, a model for salient object detection. The model utilized a pyramid attention module to enhance saliency representation by incorporating multi-scale information, thereby effectively boosting detection accuracy. Additionally, it featured a salient edge detection module, which sharpened the detection of salient object boundaries. Wang et al. [28] introduced PiNet, a salient object detection model designed for enhancing feature extraction and the progressive refinement of saliency. The model incorporated level-specific feature extraction mechanisms and employed a coarse-to-fine process for refining saliency features, which helped in overcoming common issues in existing methods like noise accumulation and spatial detail dilution. Although methods based on RGB images can achieve good performance in many situations, they lack the ability to handle depth information.

2.2. Salient Object Detection Based on RGB-D Images

With the advancement of depth sensors, RGB-D images (which contain both color and depth information) have been widely applied in salient object detection. For instance, Lang et al. [29] investigated the impact of depth cues on saliency detection, where they found that depth information holds significant value for salient object detection. Based on this, many researchers have begun to explore how to fully utilize depth information for salient object detection.

Peng et al. [30] proposed a multi-modal fusion framework that improves saliency detection performance by fusing local and global depth features with color and texture features. Zhang et al. [31] presented a new RGB-D salient object detection model, addressing challenges with depth image quality and foreground–background consistency. The model introduced a two-stage approach: firstly, an image generation stage that created high-quality, foreground-consistent pseudo-depth images, and secondly, a saliency reasoning stage that utilized these images for enhanced depth feature calibration and cross-modal fusion. Ikeda et al. [32] introduced a model for RGB-D salient object detection that integrated saliency and edge features with reverse attention. This approach effectively enhanced object boundary detection and saliency in complex scenes. The model also incorporated a Multi-Scale Interactive Module for improved global image information understanding and utilized supervised learning to enhance accuracy in salient object and boundary areas. Xu et al. [33] introduced a new approach to RGB-D salient object detection, addressing the object-part relationship dilemma in Salient Object Detection (SOD). The proposed CCNet model utilized a Convolutional Capsule Network based on Feature Extraction and

Integration (CCNet) to efficiently explore the object-part relationship in RGB-D SOD with reduced computational demand. Cong et al. [34] presented a comprehensive approach to RGB-D salient object detection, focusing on enhancing the interaction and integration of features from both RGB and depth modalities. It introduced a new network architecture that efficiently combined these modalities, addressing challenges in feature representation and fusion. However, these methods overlook the feature differences between different modalities, resulting in insufficient information fusion.

To address this issue, Qu et al. [35] introduced a simple yet effective deep learning model, which learns the interaction mechanism between RGB and depth-induced saliency features. Yi et al. [36] proposed a Cross-stage Multi-scale Interaction Network (CMINet), which intertwines features at different stages with the use of a Multi-scale Spatial Pooling (MSP) module and a Cross-stage Pyramid Interaction (CPI) module. They then designed an Adaptive Weight Fusion (AWF) module for balancing the importance of multi-modal features and fusing them. Liu et al. [37] proposed a cross-modal edge-guided salient object detection model for RGB-D images. This model extracts edge information from cross-modal color and depth information and integrates the edge information into cross-modal color and depth features, generating a saliency map with clear boundaries. Sun et al. [38] introduced an RGB-D salient object detection method that combined cross-modal interactive fusion with global awareness. This method embedded a transformer network within a U-Net structure to merge global attention mechanisms with local convolution, aiming for enhanced feature extraction. It utilized a U-shaped structure for extracting dual-stream features from RGB and depth images, employing a multi-level information reconstruction approach to suppress lower-layer disturbances and minimize redundant details. Peng et al. [39] introduced MFCG-Net, an RGB-D salient object detection method that leveraged multimodal fusion and contour guidance to improve detection accuracy. It incorporated attention mechanisms for feature optimization and designed an interactive feature fusion module to effectively integrate RGB and depth image features. Additionally, the method utilized contour features to guide the detection process, achieving clearer boundaries for salient objects. Sun et al. [40] introduced a new approach for RGB-D salient object detection, leveraging a cascaded and aggregated Transformer Network structure to enhance feature extraction and fusion. They employed three key modules: the Attention Feature Enhancement Module (AFEM) for multi-scale semantic information, the Cross-Modal Fusion Module (CMFM) to address depth map quality issues, and the Cascaded Correction Decoder (CCD) to refine feature scale differences and suppress noise. Although some significant results have been achieved in existing research, it remains a formidable challenge to achieve accurate salient object detection in complex scenes through cross-modal and cross-level feature fusion.

3. Proposed Method

In this section, we first provide an overview of our method in Section 3.1. Following that, in Section 3.2, we elaborate on the depth attention module we propose, which is used to mine valuable depth information. In Sections 3.3 and 3.4, we introduce the semantic localization module and the reverse decoding network based on the Multi-Scale Fusion Module, respectively. Finally, in Section 3.5, we discuss the loss function.

3.1. Overview of SLMSF-Net

Figure 1 displays the overall network structure of SLMSF-Net. Without loss of generality, we adopt Resnet50 [41] as the backbone network to extract features from both RGB images and depth images separately. Resnet50 encompasses five convolution stages; we removed the final pooling layer and the fully connected layer, resulting in a fully convolutional neural network, and use the outputs of the intermediate five convolution blocks as feature outputs. These output feature maps are denoted as M1, M2, M3, M4, and M5, with their sizes being 1/2, 1/4, 1/8, 1/16, and 1/32 of the original image, respectively.

1. Modal Feature Fusion: As shown in Figure 1, we proposed a Depth Attention Module. This module performs a modal fusion of RGB image features and depth image features, forming the modal fusion features F_1^{Fuse}, F_2^{Fuse}, F_3^{Fuse}, F_4^{Fuse} and F_5^{Fuse}.
2. Semantic Localization: We proposed a Semantic Localization Module. This module first downsamples the top-level modal fusion feature to compute a global view. It then performs coordinate localization on the global view and ultimately fuses the localization information with the global view, thereby precisely locating the salient object. Assuming the semantic localization module is represented as the SLM function, its output result can be written as: $F^{of} = \text{SLM}(F_5^{Fuse})$.
3. Multi-Scale Fusion Decoding: After performing semantic localization, we predicted the clear boundaries of the salient object through reverse multi-level feature integration from front to back. To accomplish this multi-level feature integration, we constructed a Multi-Scale Fusion Module, which effectively fuses features at all levels.

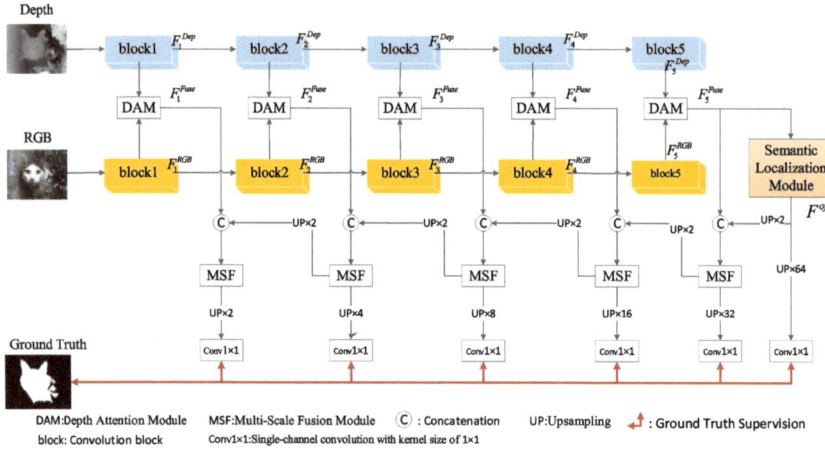

Figure 1. The overall network architecture of SLMSF-Net.

3.2. Depth Attention Module

In the process of fusing RGB and depth features, we need to address two main issues. The first one is the modal mismatch problem, which requires us to resolve the modal differences between the two types of features. The second one is the information complementarity problem; since RGB and depth features often capture different aspects of object information, we need to consider how to let these two types of features complement each other's information, aiming to enhance the accuracy and robustness of object detection. Inspired by [42], we designed a depth attention module to improve the matching and complementarity of multi-modal features.

Specifically, F_i^{RGB} represents the ith RGB image feature and F_i^{Dep} represents the ith depth image feature, where i is a natural number from 1 to 5. As shown in Figure 2, the depth attention module first enhances the depth image feature through channel attention. The enhanced result is then multiplied element-wise with the RGB image feature to obtain the channel-enhanced fusion feature. Following this, the channel-enhanced fusion feature undergoes spatial attention enhancement, and the enhanced result is multiplied element-wise with the RGB image feature, thus obtaining the modal fusion feature. To enhance the matching of depth features, we stacked a depth attention module behind each depth feature branch. By introducing attention units, we can enhance the saliency representation ability of depth features. The fusion process of the two modal features can be expressed as follows:

$$F_i^{\text{Fuse}} = F_i^{\text{RGB}} \times \text{SA}(F_i^{\text{RGB}} \times \text{CA}(F_i^{\text{Dep}}))\tag{1}$$

Herein, $\text{CA}(\cdot)$ symbolizes the channel attention operation, $\text{SA}(\cdot)$ indicates the spatial attention operation, and $\times$ represents the element-wise multiplication operation.

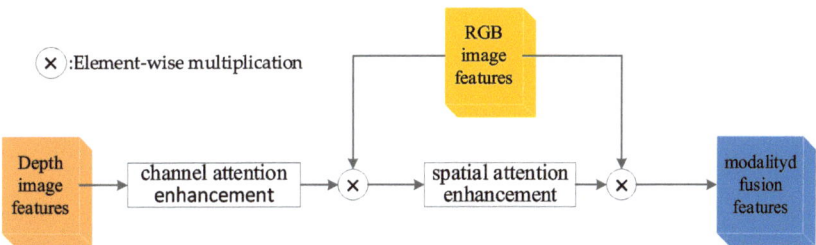

Figure 2. Depth attention module.

3.3. Semantic Localization Module

In the process of salient object localization, high-level features play a crucial role. Compared to low-level features, high-level features are capable of capturing more abstract information, which aids in highlighting the location of salient objects. Therefore, we introduced a semantic localization module designed to effectively learn the global view of the entire image, thereby achieving more precise salient object localization. As depicted in Figure 3, the semantic localization process is divided into three stages: initially, the first stage downsamples the top-level modal fusion features to compute a global view; subsequently, the second stage carries out coordinate localization on the global view; finally, the third stage fuses the localization information with the global view.

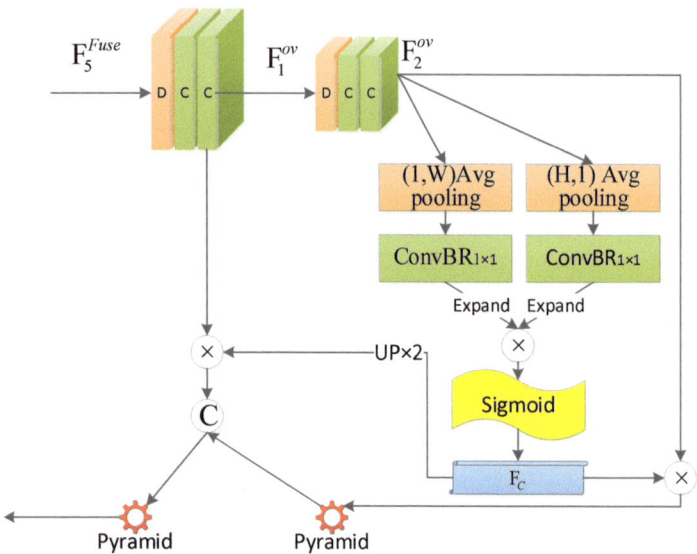

Figure 3. Semantic localization module.

In the first stage, we implement a 1/2 scale downsample operation on the top-level modal fusion features F_5^{Fuse}, followed by two ConvBR$_{3\times3}$ operations, thereby obtaining the first layer of the global feature map F_1^{ov}. Subsequently, we perform the same 1/2 scale downsample and two ConvBR$_{3\times3}$ operations on the first layer of the global feature map, resulting in the second layer of the global feature map F_2^{ov}. As observed, these two global feature maps possess a significantly large receptive field, enabling them to serve as the global view of the entire image. The computation process for the global view can be described as follows:

$$F_1^{\text{ov}} = \text{ConvBR}_{3\times3}(\text{ConvBR}_{3\times3}(\text{DownS}_{1/2}(F_5^{\text{Fuse}}))) \qquad (2)$$

$$F_2^{\text{ov}} = \text{ConvBR}_{3\times3}(\text{ConvBR}_{3\times3}(\text{DownS}_{1/2}(F_1^{\text{ov}}))) \qquad (3)$$

Herein, DownS$_{1/2}(\cdot)$ denotes a 1/2 scale downsample operation on the input feature map. ConvBR$_{3\times3}(\cdot)$ represents a convolution operation performed on the input feature map using a kernel size of 3×3, followed by batch normalization and activation operations, where the activation function is Relu. This can be expressed as:

$$\text{ConvBR}_{3\times3}(X) = \text{Relu}(\text{BN}(\text{Conv}_{3\times3}(X))) \qquad (4)$$

Herein, Conv$(\cdot)$ symbolizes the convolution operation, BN$(\cdot)$ denotes the batch normalization operation, and Relu$(\cdot)$ represents the Relu activation function.

In the second stage, for the second layer of the global feature map F_2^{ov}, we utilize a pooling kernel of size (1, W) to perform average pooling along the vertical coordinate of the feature map, followed by a convolution operation with a kernel size of 1×1, resulting in the height-oriented feature map T_H. Simultaneously, we use a pooling kernel of size (H, 1) to conduct average pooling along the horizontal coordinate of the feature map F_2^{ov}, then perform a convolution operation with a kernel size of 1×1, yielding the width-oriented feature map T_W. This can be described as:

$$T_H = \text{ConvBR}_{1 \times 1}(\frac{1}{W} \sum_{0 \leq i < W} F_2^{ov}(H, i)) \tag{5}$$

$$T_W = \text{ConvBR}_{1 \times 1}(\frac{1}{H} \sum_{0 \leq j < H} F_2^{ov}(j, W)) \tag{6}$$

Herein, feature map T_H extends in the width direction, while feature map T_W expands in the height direction. The two expanded feature maps undergo pixel-wise multiplication, and then through a Sigmoid activation function, a coordinate localization feature map F_C is formed. This can be described as:

$$F_C = \text{Sigmoid}(K(T_H) \times K(T_W)) \tag{7}$$

Herein, the $K(\cdot)$ operation refers to expanding the input feature map in the width or height direction to match the size of feature map A, while $\text{Sigmoid}(\cdot)$ signifies the Sigmoid activation function.

In the third stage, we view the localization feature map as a self-attention mechanism for calibrating the global view. Specifically, we perform a pixel-wise multiplication operation between the localization feature map F_C and the second layer of the global feature map F_2^{ov}, followed by a pyramid feature fusion operation on the multiplication results, yielding feature map F_*^{of}. Subsequently, we upscale the localization feature map F_C twice and perform a pixel-wise multiplication operation with the first layer of the global feature map F_1^{ov}. The result is stacked with F_*^{of}, and then the stacked result is subjected to a pyramid feature fusion operation to finally obtain the global localization fusion feature F^{of}. This can be described as:

$$F_*^{of} = \text{Pyramid}(F_C \times F_2^{ov}) \tag{8}$$

$$F^{of} = \text{Pyramid}(\text{concat}(\text{UP}(F_C) \times F_1^{ov}, F_*^{of})) \tag{9}$$

Herein, $\text{Pyramid}(\cdot)$ represents the pyramid feature fusion operation, $\text{concat}(\cdot)$ signifies the stacking operation along the channel, and $\text{UP}(\cdot)$ denotes the operation of upscaling by a factor of two.

The pyramid feature fusion operation is depicted in Figure 4. Initially, we conduct a convolution operation with a kernel size 1×1, adjusting the number of channels in the input feature map X to 32, which yields the feature map Y. Following this, we execute feature extraction on Y, with the specific extraction method detailed as follows:

$$Y = \text{ConvBR}_{1 \times 1}(X) \tag{10}$$

$$P_1 = \text{ConvBR}_{3 \times 3}(\text{ConvBR}_{3 \times 3}(Y)) \tag{11}$$

$$P_i = \text{ConvBR}_{3 \times 3}^{2i-1}(\text{ConvBR}_{3 \times 3}(Y) + P_{i-1})), i(i \in \{2, 3\}) \tag{12}$$

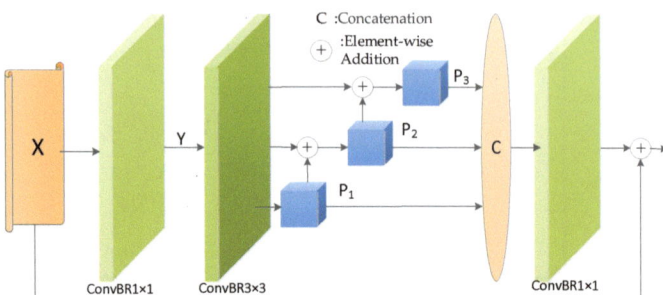

Figure 4. Pyramid feature fusion operation.

Herein, $\text{ConvBR}_{3\times3}^{2i-1}(\cdot)$ represents a dilated convolution with a kernel size of 3×3 and a dilation rate of $2i - 1$. We perform a concatenation operation along the channel with the three extracted features. Subsequently, we conduct a convolution operation on the concatenation result with a kernel size of 1×1, adjusting the channel count to match that of the input feature map. Finally, a residual connection is established with the input feature map. This process can be described as follows:

$$\text{Pyramid}(X) = \text{ConvBR}_{1\times1}(\text{concat}(P_1, P_2, P_3, P_4)) + X \tag{13}$$

3.4. Multi-Scale Fusion Module and the Reverse Decoding Process

Following semantic localization, we integrate multi-layer features in a forward-to-backward manner to delineate intricate details of the salient object. To achieve this multi-layer feature integration, we designed and constructed a Multi-Scale Fusion Module. The reverse decoder operates in five stages, each accepting the output from the preceding stage for reverse multi-scale fusion decoding. Importantly, the input for the fifth stage of the decoder is the global localization fusion feature F^{of}. The process of the reverse decoder can be described as follows:

$$Decode_5^* = \text{UP}(\text{ConvBR}_{1\times1}(F^{\text{of}})) \tag{14}$$

$$Decode_5 = \text{MSF}(\text{concat}(\text{ConvBR}_{1\times1}(F_5^{\text{Fuse}}), Decode_5^*)) \tag{15}$$

$$Decode_i^* = \text{UP}(\text{ConvBR}_{1\times1}(Decode_{i+1})) \tag{16}$$

$$Decode_i = \text{MSF}(\text{concat}(\text{ConvBR}_{1\times1}(F_i^{\text{Fuse}}), Decode_i^*)), i(i \in \{1,2,3,4\}) \tag{17}$$

Herein, $\text{MSF}(\cdot)$ stands for the Multi-Scale Fusion Module. We upscale the output from the first stage of the decoder to the size of the input image, thereby obtaining the final saliency prediction map. The specific formula used to generate the saliency prediction map is as follows:

$$S = \text{Sigmoid}(\text{Conv}_{1\times1}(\text{UP}_{\text{in}}(Decode_1))) \tag{18}$$

Herein, S represents the saliency prediction map, $\text{UP}_{\text{in}}(\cdot)$ denotes the upscaling of the feature map to the size of the input image, while $\text{Conv}_{1\times1}(\cdot)$ signifies a single-channel convolution with a kernel size of 1×1. The primary purpose of $\text{Conv}_{1\times1}(\cdot)$ is to adjust the channel count of the feature map to 1.

As illustrated in Figure 5, the multi-scale feature fusion module comprises four parallel branches and a residual connection. Initially, we employ a convolutional operation with a kernel of size 1×1 to reduce the number of channels in the input feature map to 64. Following this, in the first branch, we sequentially execute a convolution with a kernel also of size 1×1, followed by another with a kernel of size 3×3. For the i-th($i \in \{2,3,4\}$) branch of the module, the procedure commences with a convolution involving a kernel of size $(2i-1) \times 1$, proceeded by another convolution with a kernel of size $1 \times (2i-1)$. Finally, a dilated convolution operation with a kernel of size 3×3 and a dilation rate of $2i-1$ is applied. This design strategy is aimed at extracting multi-scale information from the multi-modal fusion features, thereby enriching the representational power of the model. Next, the outputs from the four branches are stacked along the channel dimension, and the channel count of the stacked output is adjusted to match the input feature map's channel count, using a convolution operation with a kernel size of 1×1. Finally, the adjusted result is connected to the input feature map via a residual connection. The entire fusion process can be described as follows:

$$branch_1(x) = \text{ConvBR}_{3\times3}(\text{ConvBR}_{1\times1}(x)) \tag{19}$$

$$branch_i(x) = \text{ConvBR}_{3\times3}^{2i-1}(\text{ConvBR}_{1\times(2i-1)}(\text{ConvBR}_{(2i-1)\times1}(\text{ConvBR}_{1\times1}(x)))), \\ i \in \{2,3,4\} \tag{20}$$

$$\text{MSF}(x) = \text{ConvBR}_{1\times1}(\text{concat}(branch_1(x), branch_2(x), branch_3(x), branch_4(x))) + x \tag{21}$$

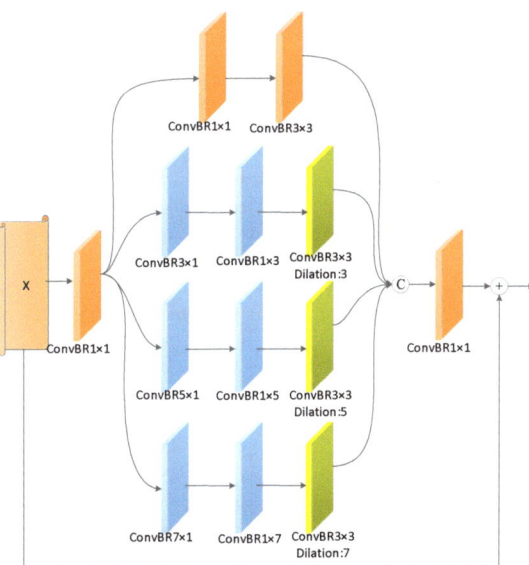

Figure 5. Multi-Scale Fusion Module.

Herein, $branch_i(x)$ denotes the ith parallel branch, while x symbolizes the input feature map.

3.5. Loss Function

As depicted in Figure 1, at each stage of the decoder, the decoded output is upsampled to the size of the input image. Following this, a convolution operation with a single-channel convolution kernel of 1×1 is performed, and then a prediction saliency map is generated through a sigmoid activation function. The saliency maps predicted at each of the five stages are denoted as O_i ($i = 1, 2, \cdots, 5$). Following the same process, we can also generate the predicted saliency map O^{of} corresponding to the output of the semantic localization module. This process can be described as follows:

$$O_i = \text{Sigmoid}(\text{Conv}_{1\times 1}(\text{UP}_{in}(Decode_i))) \tag{22}$$

$$O^{of} = \text{Sigmoid}(\text{Conv}_{1\times 1}(\text{UP}_{in}(F^{of}))) \tag{23}$$

Assuming the predicted saliency map is denoted as O, and the real saliency map is denoted as GT, the formula for calculating the loss value of the prediction results is as follows:

$$\text{Loss}(O, GT) = \text{Bce}(O, GT) + \text{Dice}(O, GT) \tag{24}$$

$$\text{Bce}(O, GT) = GT \cdot \log O + (1 - GT) \cdot \log(1 - O) \tag{25}$$

$$\text{Dice}(O, GT) = 1 - \frac{2 \cdot GT \cdot O}{||GT|| + ||O||} \tag{26}$$

Herein, $\text{Bce}(\cdot)$ represents the binary cross-entropy loss function, $\text{Dice}(\cdot)$ denotes the Dice loss function [43], and $||\cdot||$ represents the L_1 norm. The total loss function during the training phase is described as follows:

$$L = \alpha \cdot \sum_{i=1}^{5} \text{Loss}(O_i, GT) + (1 - \alpha) \cdot \text{Loss}(O^{of}, GT) \tag{27}$$

wherein, α represents the weight coefficients. During the testing phase, O_1 is the final prediction result of the model.

4. Experiments

Section 4.1 provides a detailed description of the implementation details, Section 4.2 discusses the sources of the datasets used, Section 4.3 introduces the setup of the evaluation metrics, Section 4.4 presents the comparison with the current state-of-the-art (SOTA) methods, and Section 4.5 is dedicated to the discussion of the ablation experiments. Together, these sections form the experimental analysis and evaluation part of the paper, comprehensively demonstrating the effectiveness and reliability of the research method.

4.1. Implementation Details

The salient object detection method proposed in this paper is implemented based on the Pytorch framework [44,45], and all experimental procedures were carried out on a single NVIDIA RTX A6000 GPU(NVIDIA, Santa Clara City, CA, USA). The initialization parameters of the backbone model, ResNet50, are derived from a pre-trained model on ImageNet [46]. Specifically, both the RGB image branch and the depth image branch use a ResNet50 model for feature extraction, with the only difference being that the input channel number for the depth image branch is 1. To enhance the model's generalization capability, various augmentation strategies, such as random flipping, rotation, and boundary cropping, were applied to all training images. Throughout the training process, the Adam optimizer was employed, with parameters set to $\beta 1 = 0.9$ and $\beta 2 = 0.999$, and a batch size of 10. The initial learning rate was set to 1×10^{-4} and was divided by 10 every 50 rounds. The dimensions of the input images were all adjusted to 768×768. The model converged within 200 rounds. In order to show the training process of our model more clearly, we report the training and validation loss curve of our network in Figure 6.

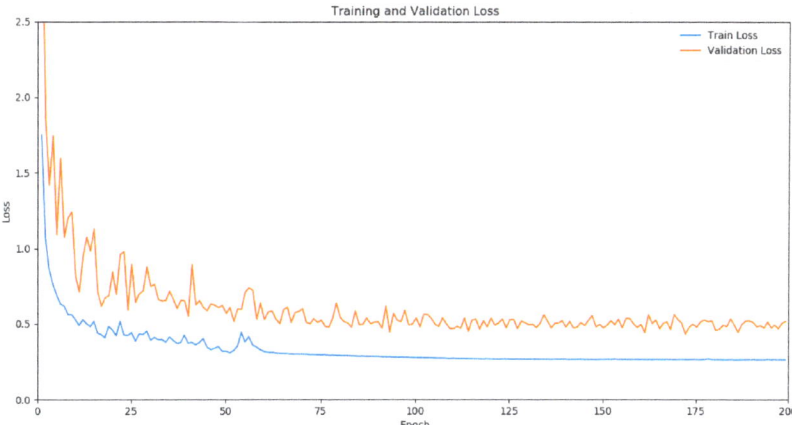

Figure 6. Training and validation loss curve.

4.2. Datasets

In this study, SLMSF-Net was extensively evaluated across six widely used datasets, including NJU2K [47], NLPR [30], STERE [48], SSD [49], SIP [50], and DES [51]. These datasets contain 1985, 1000, 1000, 80, 929, and 135 images, respectively. For the training phase, we utilized 1485 images from the NJU2K dataset and 700 images from the NLPR dataset. During the testing phase, the remaining images from the NJU2K and NLPR datasets, as well as the entire STERE, SSD, SIP, and DES datasets were used.

4.3. Evaluation Metrics

We employed four widely used evaluation metrics to compare SLMSF-Net with previous state-of-the-art methods, namely E-Measure, F-measure, S-measure, and MAE.

E-Measure (E_ξ) is a saliency map evaluation method based on cognitive vision, capable of integrating statistical information at both the image level and local pixel level. This measurement strategy was proposed by [52] and is defined as follows:

$$E_\xi = \frac{1}{W \times H} \sum_{i=1}^{W} \sum_{j=1}^{H} \xi(i,j) \tag{28}$$

Here, W and H represent the width and height of the saliency map, respectively, while ξ signifies the enhanced alignment matrix. E-measure has three different variants: maximum E-measure, adaptive E-measure, and average E-measure. In our experiments, we used the maximum E-measure (maxE) as the evaluation criterion.

F-measure (F_β) serves as a weighted harmonic mean of precision and recall. It is defined as follows:

$$F_\beta = \frac{(1+\beta^2)\text{Precision} \times \text{Recall}}{\beta^2 \times \text{Precision} + \text{Recall}} \tag{29}$$

Here, β is a parameter used to balance Precision and Recall. In this study, we set β^2 to 0.3. Similar to E-measure, F-measure also has three different variants: maximum F-measure, adaptive F-measure, and average F-measure. In our experiments, we reported the results of the maximum F-measure (maxF).

S-measure (S_α) is a method for evaluating structural similarity. It assesses from two perspectives: region awareness (S_r) and object awareness (S_o). It is defined as follows:

$$S_\alpha = \alpha \times S_o + (1-\alpha) S_r \tag{30}$$

Here, $\alpha \in [0,1]$ is a hyperparameter used to balance between S_o and S_r. In our experiments, α is set to 0.5.

MAE (Mean Absolute Error) represents the average per-pixel absolute error between the predicted saliency map S and the ground truth map GT. It is defined as follows:

$$\text{MAE} = \frac{1}{W \times H} \sum_{i=1}^{W} \sum_{j=1}^{H} |S(i,j) - \text{GT}(i,j)| \tag{31}$$

Here, W and H, respectively, denote the width and height of the saliency map. The MAE is normalized to a value in the [0, 1] interval.

4.4. Comparison with SOTA Methods

We compared the SLMSF-Net model proposed in this study with ten deep learning-based RGB-D saliency detection methods, including AFNet [53], HINet [54], C2DFNet [55], DCMF [56], CFIDNet [57], CIR-Net [58], DCF [59], DASNet [60], D3Net [50], and ICNet [61]. To ensure a fair comparison, we used the saliency maps provided by the authors. If the saliency maps were not provided, we computed them using the source code and model files provided by the authors.

4.4.1. Quantitative Comparison

Figure 7 presents the comparison results of PR curves from different methods, while Table 1 presents the quantitative comparison results for four evaluation metrics. As shown in the figure and table, our PR curve outperforms all other comparison methods, whether on the NJU2K, NLPR, DES, SIP, SSD, or STERE datasets. This advantage is largely attributed to our designed semantic localization and multi-scale fusion strategies, which, respectively, achieve precise localization of salient objects and capture of detailed boundary information. Additionally, our designed depth attention module can effectively utilize depth information to enhance the model's segmentation performance. Concurrently, the table data reflects the same conclusion, i.e., our method outperforms all comparison methods in performance on the NJU2K, NLPR, DES, SIP, SSD, and STERE datasets. Compared with the best comparison methods (AFNet, C2DFNet, and DCMF), we have improved the MAE, maxF$_\beta$, maxE$_\xi$, S$_\alpha$ evaluation metrics by 0.0002~0.0062, 0.2~1.8%, 0.09~1.46%, and 0.19~1.05%, respectively. Therefore, both the PR curves and evaluation metrics affirm the effectiveness and superiority of our method proposed for the RGB-D SOD task.

Table 1. Comparison of results for four evaluation metrics—mean absolute error (MAE), maximum F-measure (maxF), maximum E-measure (maxE), and S-measure (S)—across six datasets. The symbol "↑" indicates that a higher value is better for the metric, while "↓" indicates that a lower value is better. The best performance in each row is highlighted in bold.

Datasets	Evaluation Metrics	Deep Learning-Based RGB-D Saliency Detection Methods									
		DASNet ICMM2020	D3Net TNNLS 2020	ICNet TIP2020	DCF CVPR2021	CIRNet TIP2022	CFIDNet NCA2022	DCMF TIP2022	C2DFNet TMM2022	HINet PR2023	AFNet NC2023	Ours
NJU2K [47]	MAE↓	0.0418	0.0467	0.0519	0.0357	0.0350	0.0378	0.0357	0.0387	0.0385	0.0317	**0.0315**
	maxF↑	0.9015	0.8993	0.8905	0.9147	0.9281	0.9148	0.9252	0.9089	0.9138	0.9282	**0.9352**
	maxE↑	0.9393	0.9381	0.9264	0.9504	0.9547	0.9464	0.9582	0.9425	0.9447	0.9578	**0.9615**
	S↑	0.9025	0.9	0.8941	0.9116	0.9252	0.9142	0.9247	0.9082	0.9153	0.9262	**0.9306**

Table 1. Cont.

Datasets	Evaluation Metrics	Deep Learning-Based RGB-D Saliency Detection Methods										
		DASNet ICMM2020	D3Net TNNLS 2020	ICNet TIP2020	DCF CVPR2021	CIRNet TIP2022	CFIDNet NCA2022	DCMF TIP2022	C2DFNet TMM2022	HINet PR2023	AFNet NC2023	Ours
NLPR [30]	MAE↓	0.0212	0.0298	0.0281	0.0217	0.0280	0.0256	0.0290	0.0217	0.0257	0.0201	**0.0199**
	maxF↑	0.9218	0.8968	0.9079	0.9118	0.9071	0.9054	0.9057	0.9166	0.9062	0.9249	**0.9298**
	maxE↑	0.9641	0.9529	0.9524	0.9628	0.9554	0.9553	0.9541	0.9605	0.9565	0.9684	**0.9693**
	S↑	0.9294	0.9118	0.9227	0.9239	0.9208	0.9219	0.9220	0.9279	0.9223	0.9362	**0.9388**
DES [51]	MAE↓	0.0246	0.0314	0.0266	0.0241	0.0287	0.0233	0.0232	0.0199	0.0215	0.0221	**0.0176**
	maxF↑	0.9025	0.8842	0.9132	0.8935	0.8917	0.9108	0.9239	0.9159	0.9220	0.9225	**0.9307**
	maxE↑	0.9390	0.9451	0.9598	0.9514	0.9407	0.9396	0.9679	0.9590	0.9670	0.9529	**0.9739**
	S↑	0.9047	0.8973	0.9201	0.9049	0.9067	0.9169	0.9324	0.9217	0.9274	0.9252	**0.9403**
SIP [50]	MAE↓	0.0508	0.0632	0.0695	0.0518	0.0685	0.0601	0.0623	0.0529	0.0656	0.0434	**0.0422**
	maxF↑	0.8864	0.861	0.8571	0.8844	0.8662	0.8699	0.8719	0.8770	0.8550	0.9089	**0.9114**
	maxE↑	0.9247	0.9085	0.9033	0.9217	0.9047	0.9088	0.9111	0.9160	0.8993	0.9389	**0.9408**
	S↑	0.8767	0.8603	0.8538	0.8756	0.8615	0.8638	0.8700	0.8715	0.8561	0.8959	**0.9045**
SSD [49]	MAE↓	0.0423	0.0585	0.0637	0.0498	0.0523	0.0504	0.0731	0.0478	0.0488	0.0383	**0.0321**
	maxF↑	0.8725	0.834	0.8414	0.8509	0.8547	0.8707	0.8108	0.8598	0.8524	0.8848	**0.9007**
	maxE↑	0.9298	0.9105	0.9025	0.9090	0.9119	0.9261	0.8970	0.9171	0.9160	0.9427	**0.9565**
	S↑	0.8846	0.8566	0.8484	0.8644	0.8725	0.8791	0.8382	0.8718	0.8652	0.8968	**0.9062**
STERE [48]	MAE↓	0.0368	0.0462	0.0446	0.0389	0.0457	0.0426	0.0433	0.0385	0.0490	0.0336	**0.0331**
	maxF↑	0.9043	0.8911	0.8978	0.9009	0.8966	0.8971	0.9061	0.8973	0.8828	0.9177	**0.9195**
	maxE↑	0.9436	0.9382	0.9415	0.9447	0.9388	0.9420	0.9463	0.9429	0.9325	0.9572	**0.9584**
	S↑	0.9104	0.8985	0.9025	0.9022	0.9013	0.9012	0.9097	0.9023	0.8919	0.9184	**0.9201**

4.4.2. Qualitative Comparison

For a qualitative comparison, we present a selection of representative visual examples in Figure 8. Upon observation, our method demonstrates superior performance in several challenging scenarios compared to other methods. Examples of these scenarios include situations where the foreground and background colors are similar (rows 1–2), in complex environments (rows 3–4), in scenes with multiple objects present (rows 5–6), for small object detection (rows 7–8), and under conditions of low-quality depth images (rows 9–10). These visual examples show that our method can more precisely locate salient objects and generate more accurate saliency maps.

4.5. Ablation Studies

As shown in Table 2, we conducted an in-depth ablation analysis to verify the effectiveness of each module. DAM represents the Deep Attention Module, SLM is the Semantic Localization Module, and MSFM stands for Multi-Scale Fusion Module. "Without DAM", "without SLM", and "without MSFM" refer to the models obtained after removing the DAM, SLM, and MSFM modules from the SLMSF-Net model, respectively. By comparing the data in the third column with the sixth column, we can clearly see that the introduction of the DAM module significantly improves the performance of the model. Similarly, by comparing the data in the fourth and sixth columns, we can see that the introduction of the SLM module can significantly enhance the performance of the model. Comparing the data in the fifth and sixth columns, we can see that adding the MSFM module will enhance the model's performance. These results prove the importance of the three modules: the DAM module introduces depth image information, the SLM module realizes the precise semantic location of salient objects, and the MSFM module can fuse multi-scale features to refine the boundaries of salient objects. Each of these three functional modules resulted in a significant increase in model performance. In the last column, we can see that the SLMSF-Net model that incorporates these three modules achieved the best results.

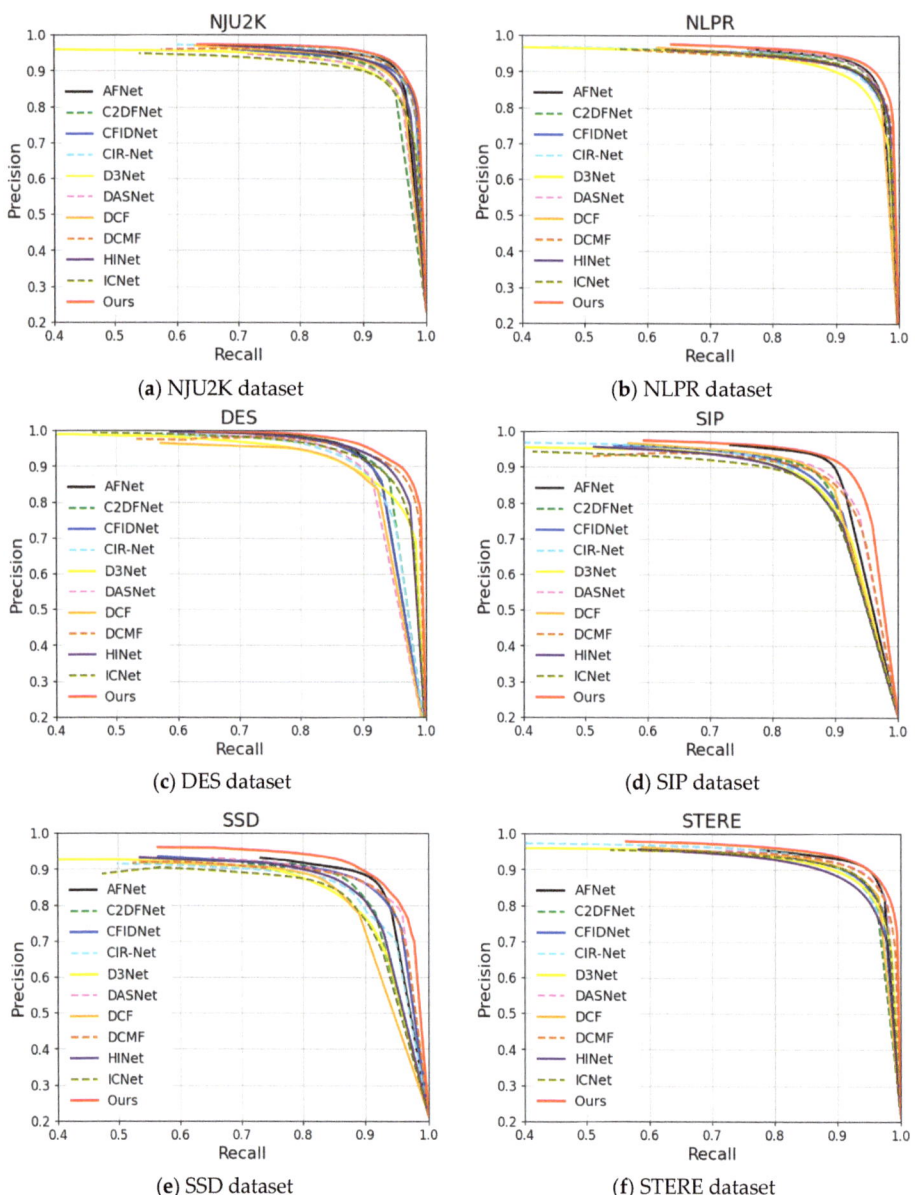

Figure 7. Comparison of precision-recall (P-R) curves for different methods across six RGB-D datasets. Our SLMSF-Net method is represented by a solid red line.

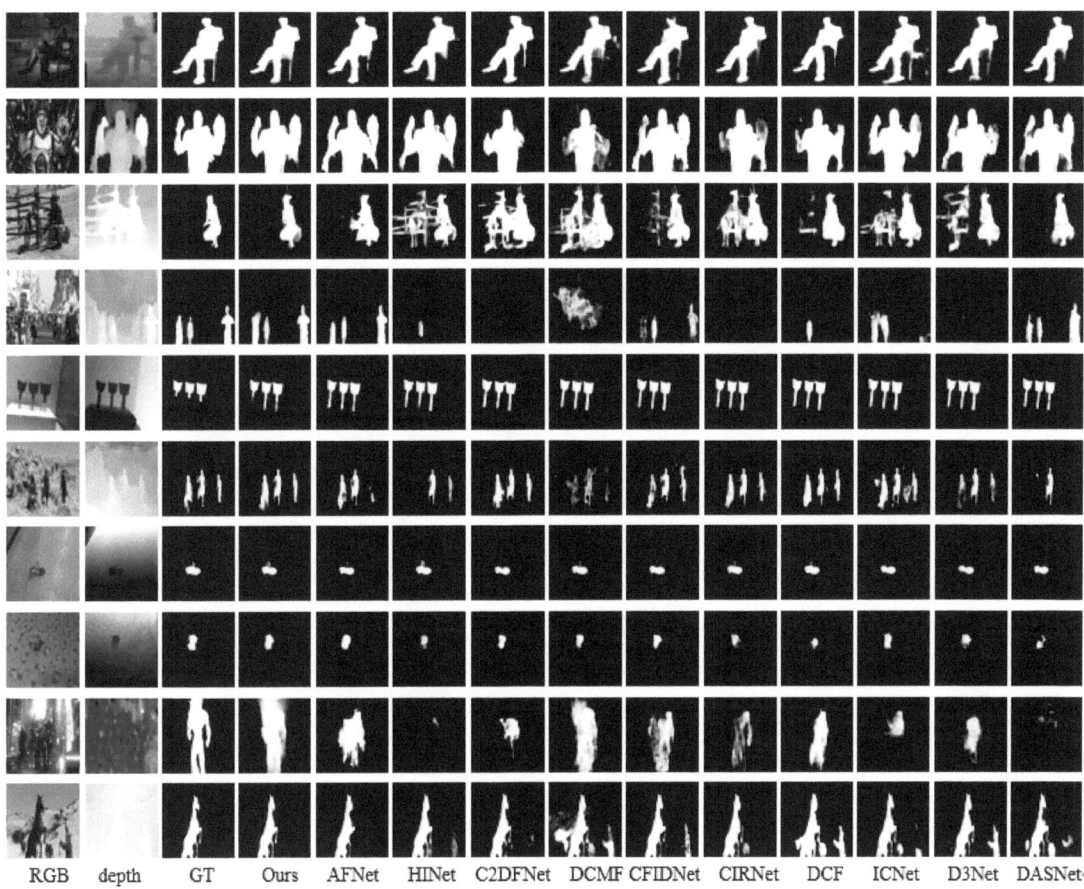

Figure 8. Visual comparison between SLMSF-Net and state-of-the-art RGB-D models.

Table 2. Comparison of ablation study results. The symbol "↑" indicates that a higher value is better for the metric, while "↓" indicates that a lower value is better. The best performance in each row is highlighted in bold.

Datasets	Evaluation Metrics	Without DAM	Without SLM	Without MSFM	SLMSF-Net
NJU2K [47]	MAE↓	0.0362	0.0352	0.0393	**0.0315**
	maxF↑	0.9214	0.9215	0.9165	**0.9352**
	maxE↑	0.9516	0.9508	0.9478	**0.9615**
	S↑	0.921	0.9225	0.9185	**0.9306**
NLPR [30]	MAE↓	0.0235	0.0234	0.0284	**0.0199**
	maxF↑	0.9226	0.9193	0.911	**0.9298**
	maxE↑	0.9627	0.964	0.9593	**0.9693**
	S↑	0.9328	0.9327	0.9238	**0.9388**

Table 2. *Cont.*

Datasets	Evaluation Metrics	Without DAM	Without SLM	Without MSFM	SLMSF-Net
DES [51]	MAE↓	0.0191	0.0228	0.0231	**0.0176**
	maxF↑	0.9284	0.9174	0.9263	**0.9307**
	maxE↑	0.9704	0.9618	0.9687	**0.9739**
	S↑	0.9342	0.9260	0.9317	**0.9403**
SIP [50]	MAE↓	0.0569	0.0528	0.0600	**0.0422**
	maxF↑	0.8827	0.8916	0.8748	**0.9114**
	maxE↑	0.9154	0.9202	0.9119	**0.9408**
	S↑	0.8776	0.8830	0.8739	**0.9045**
SSD [49]	MAE↓	0.0537	0.0534	0.0548	**0.0321**
	maxF↑	0.8378	0.8381	0.8395	**0.9007**
	maxE↑	0.9093	0.9042	0.9045	**0.9565**
	S↑	0.8661	0.865	0.8658	**0.9062**
STERE [48]	MAE↓	0.0443	0.0376	0.0508	**0.0331**
	maxF↑	0.8919	0.9100	0.8906	**0.9195**
	maxE↑	0.9381	0.9479	0.9330	**0.9584**
	S↑	0.9014	0.9143	0.8986	**0.9201**

5. Conclusions

In complex scenarios, achieving precise RGB-D salient object detection against multiple scales of objects and noisy backgrounds remains a daunting task. Current research primarily faces two major challenges: modality fusion and multi-level feature integration. To address these challenges, we propose an innovative RGB-D salient object detection network, the Semantic Localization and Multi-Scale Fusion Network (SLMSF-Net). This network comprises two main stages: encoding and decoding. In the encoding stage, SLMSF-Net utilizes ResNet50 to extract features from RGB and depth images and employs a depth attention module for the effective fusion of modal features. In the decoding stage, the network precisely locates salient objects through the semantic localization module and restores the detailed information of salient objects in the reverse decoder via the Multi-Scale Fusion Module. Rigorous experimental validation shows that SLMSF-Net exhibits superior accuracy and robustness on multiple RGB-D salient object detection datasets, outperforming existing technologies. In the future, we plan to further optimize the model, improve the attention mechanism, delve into refining edge details, and explore its application in RGB-T salient object detection tasks.

Author Contributions: Conceptualization, Y.P.; Data curation, Z.Z.; Formal analysis, M.F.; Funding acquisition, Y.P.; Investigation, M.F.; Methodology, Y.P.; Project administration, Y.P.; Resources, Z.Z.; Software, Y.P. and Z.Z.; Supervision, Y.P.; Validation, Y.P., Z.Z. and M.F.; Visualization, Z.Z.; Writing—original draft, Y.P.; Writing—review and editing, Z.Z. All authors have read and agreed to the published version of the manuscript.

Funding: This work was supported by the National Natural Science Foundation of China (No. 61972357), the basic public welfare research program of Zhejiang Province (No. LGF22F020017 and No. GG21F010013), and the Natural Science Foundation of Zhejiang Province (No. Y21F020030).

Institutional Review Board Statement: Not applicable.

Informed Consent Statement: Not applicable.

Data Availability Statement: Data are contained within the article.

Conflicts of Interest: The authors declare no conflicts of interest.

References

1. Liu, J.J.; Hou, Q.; Liu, Z.A.; Cheng, M.M. Poolnet+: Exploring the potential of pooling for salient object detection. *IEEE Trans. Pattern Anal. Mach. Intell.* **2022**, *45*, 887–904. [CrossRef] [PubMed]
2. Liang, Y.; Qin, G.; Sun, M.; Qin, J.; Yan, J.; Zhang, Z.H. Multi-modal interactive attention and dual progressive decoding network for RGB-D/T salient object detection. *Neurocomputing* **2022**, *490*, 132–145. [CrossRef]
3. Zakharov, I.; Ma, Y.; Henschel, M.D.; Bennett, J.; Parsons, G. Object Tracking and Anomaly Detection in Full Motion Video. In Proceedings of the IGARSS 2022, 2022 IEEE International Geoscience and Remote Sensing Symposium, Kuala Lumpur, Malaysia, 17–22 July 2022; pp. 7910–7913.
4. Zhang, Y.; Sun, P.; Jiang, Y.; Yu, D.D.; Weng, F.C.; Yuan, Z.H.; Luo, P.; Liu, W.Y.; Wang, X.G. Bytetrack: Multi-object tracking by associating every detection box. In Proceedings of the Computer Vision–ECCV 2022: 17th European Conference, Tel Aviv, Israel, 23–27 October 2022; Springer Nature: Cham, Switzerland, 2022; pp. 1–21.
5. Wang, R.; Lei, T.; Cui, R.; Zhang, B.T.; Meng, H.Y.; Nandi, A.K. Medical image segmentation using deep learning: A survey. *IET Image Process.* **2022**, *16*, 1243–1267. [CrossRef]
6. He, B.; Hu, W.; Zhang, K.; Yuan, S.D.; Han, X.L.; Su, C.; Zhao, J.M.; Wang, G.Z.; Wang, G.X.; Zhang, L.Y. Image segmentation algorithm of lung cancer based on neural network model. *Expert Syst.* **2022**, *39*, e12822. [CrossRef]
7. Fan, J.; Zheng, P.; Li, S. Vision-based holistic scene understanding towards proactive human–robot collaboration. *Robot. Comput.-Integr. Manuf.* **2022**, *75*, 102304. [CrossRef]
8. Gong, T.; Zhou, W.; Qian, X.; Lei, J.S.; Yu, L. Global contextually guided lightweight network for RGB-thermal urban scene understanding. *Eng. Appl. Artif. Intell.* **2023**, *117*, 105510. [CrossRef]
9. Chen, G.; Shao, F.; Chai, X.; Chen, H.; Jiang, Q.; Meng, X.; Ho, Y.S. Modality-Induced Transfer-Fusion Network for RGB-D and RGB-T Salient Object Detection. *IEEE Trans. Circuits Syst. Video Technol.* **2022**, *33*, 1787–1801. [CrossRef]
10. Gao, L.; Fu, P.; Xu, M.; Wang, T.; Liu, B. UMINet: A unified multi-modality interaction network for RGB-D and RGB-T salient object detection. *Vis. Comput.* **2023**, 1–18. [CrossRef]
11. Wu, Y.H.; Liu, Y.; Xu, J.; Bian, J.W.; Gu, Y.C.; Cheng, M.M. MobileSal: Extremely efficient RGB-D salient object detection. *IEEE Trans. Pattern Anal. Mach. Intell.* **2021**, *44*, 10261–10269. [CrossRef]
12. Liu, S.; Huang, D. Receptive field block net for accurate and fast object detection. In Proceedings of the European Conference on Computer Vision (ECCV), Munich, Germany, 8–14 September 2018; pp. 385–400.
13. Zhang, N.; Han, J.; Liu, N. Learning implicit class knowledge for rgb-d co-salient object detection with transformers. *IEEE Trans. Image Process.* **2022**, *31*, 4556–4570. [CrossRef]
14. Wu, Y.H.; Liu, Y.; Zhang, L.; Cheng, M.M.; Ren, B. EDN: Salient object detection via extremely-downsampled network. *IEEE Trans. Image Process.* **2022**, *31*, 3125–3136. [CrossRef]
15. Wu, Z.; Li, S.; Chen, C.; Hao, A.; Qin, H. Recursive multi-model complementary deep fusion for robust salient object detection via parallel sub-networks. *Pattern Recognit.* **2022**, *121*, 108212. [CrossRef]
16. Fan, D.P.; Zhai, Y.; Borji, A.; Yang, J.; Shao, L. BBS-Net: RGB-D salient object detection with a bifurcated backbone strategy network. In Proceedings of the Computer Vision–ECCV 2020: 16th European Conference, Glasgow, UK, 23–28 August 2020; Springer International Publishing: Cham, Switzerland, 2020; pp. 275–292.
17. Itti, L.; Koch, C.; Niebur, E. A model of saliency-based visual attention for rapid scene analysis. *IEEE Trans. Pattern Anal. Mach. Intell.* **1998**, *20*, 1254–1259. [CrossRef]
18. Achanta, R.; Hemami, S.; Estrada, F.; Susstrunk, S. Frequency-tuned salient region detection. In Proceedings of the 2009 IEEE Conference on Computer vision And Pattern Recognition, Miami, FL, USA, 20–25 June 2009; pp. 1597–1604.
19. Tong, N.; Lu, H.; Zhang, Y.; Ruan, X. Salient object detection via global and local cues. *Pattern Recognit.* **2015**, *48*, 3258–3267. [CrossRef]
20. Chen, C.; Wei, J.; Peng, C.; Qin, H. Depth-quality-aware salient object detection. *IEEE Trans. Image Process.* **2021**, *30*, 2350–2363. [CrossRef] [PubMed]
21. Cong, R.; Yang, N.; Li, C.; Fu, H.; Zhao, Y.; Huang, Q.; Kwong, S. Global-and-local collaborative learning for co-salient object detection. *IEEE Trans. Cybern.* **2022**, *53*, 1920–1931. [CrossRef] [PubMed]
22. Hou, Q.; Cheng, M.M.; Hu, X.; Borji, A.; Tu, Z.; Torr, P. Deeply supervised salient object detection with short connections. In Proceedings of the IEEE Conference on Computer Vision and Pattern Recognition, Honolulu, HI, USA, 21–26 July 2017.
23. Zhao, X.; Pang, Y.; Zhang, L.; Lu, H. Suppress and balance: A simple gated network for salient object detection. In Proceedings of the Computer Vision–ECCV, Glasgow, UK, 23–28 August 2020.
24. Zhang, P.; Wang, D.; Lu, H.; Wang, H.; Ruan, X. Amulet: Aggregating multi-level convolutional features for salient object detection. In Proceedings of the IEEE International Conference on Computer Vision, Venice, Italy, 22–29 October 2017.
25. Wu, Z.; Su, L.; Huang, Q. Cascaded partial decoder for fast and accurate salient object detection. In Proceedings of the IEEE/CVF Conference on Computer Vision and Pattern Recognition, Long Beach, CA, USA, 16–20 July 2019.
26. Chen, S.; Tan, X.L.; Wang, B.; Hu, X.L. Reverse attention for salient object detection. In Proceedings of the European Conference on Computer Vision ECCV, Munich, Germany, 8–14 September 2018.
27. Wang, W.; Zhao, S.Y.; Shen, J.B.; Hoi, S.C.H.; Borji, A. Salient object detection with pyramid attention and salient edges. In Proceedings of the IEEE/CVF Conference on Computer Vision and Pattern Recognition, Long Beach, CA, USA, 16–20 June 2019.

28. Wang, X.; Liu, Z.; Liesaputra, V.; Huang, Z. Feature specific progressive improvement for salient object detection. *Pattern Recognit.* **2024**, *147*, 110085. [CrossRef]
29. Lang, C.; Nguyen, T.V.; Katti, H.; Yadati, K.; Kankanhalli, M.; Yan, S. Depth matters: Influence of depth cues on visual saliency. In Proceedings of the Computer Vision–ECCV 2012: 12th European Conference on Computer Vision, Florence, Italy, 7–13 October 2012; Springer: Berlin/Heidelberg, Germany, 2012; pp. 101–115.
30. Peng, H.; Li, B.; Xiong, W.; Hu, W.; Ji, R. RGBD salient object detection: A benchmark and algorithms. In Proceedings of the Computer Vision–ECCV 2014: 13th European Conference, Zurich, Switzerland, 6–12 September 2014; Springer International Publishing: Berlin/Heidelberg, Germany, 2014; pp. 92–109.
31. Zhang, Q.; Qin, Q.; Yang, Y.; Jiao, Q.; Han, J. Feature Calibrating and Fusing Network for RGB-D Salient Object Detection. *IEEE Trans. Circuits Syst. Video Technol.* **2023**, 1–15. [CrossRef]
32. Ikeda, T.; Masaaki, I. RGB-D Salient Object Detection Using Saliency and Edge Reverse Attention. *IEEE Access* **2023**, *11*, 68818–68825. [CrossRef]
33. Xu, K.; Guo, J. RGB-D salient object detection via convolutional capsule network based on feature extraction and integration. *Sci. Rep.* **2023**, *13*, 17652. [CrossRef]
34. Cong, R.; Liu, H.; Zhang, C.; Zhang, W.; Zheng, F.; Song, R.; Kwong, S. Point-aware interaction and cnn-induced refinement network for RGB-D salient object detection. In Proceedings of the 31st ACM International Conference on Multimedia, Ottawa, ON, Canada, 29 October–3 November 2023.
35. Qu, L.; He, S.; Zhang, J.; Tang, Y. RGBD salient object detection via deep fusion. *IEEE Trans. Image Process.* **2017**, *26*, 2274–2285. [CrossRef]
36. Yi, K.; Zhu, J.; Guo, F.; Xu, J. Cross-Stage Multi-Scale Interaction Network for RGB-D Salient Object Detection. *IEEE Signal Process. Lett.* **2022**, *29*, 2402–2406. [CrossRef]
37. Liu, Z.; Wang, K.; Dong, H.; Wang, Y. A cross-modal edge-guided salient object detection for RGB-D image. *Neurocomputing* **2021**, *454*, 168–177. [CrossRef]
38. Sun, F.; Hu, X.H.; Wu, J.Y.; Sun, J.; Wang, F.S. RGB-D Salient Object Detection Based on Cross-modal Interactive Fusion and Global Awareness. *J. Softw.* **2023**, 1–15. [CrossRef]
39. Peng, Y.; Feng, M.; Zheng, Z. RGB-D Salient Object Detection Method Based on Multi-modal Fusion and Contour Guidance. *IEEE Access* **2023**, *11*, 145217–145230. [CrossRef]
40. Sun, F.; Ren, P.; Yin, B.; Wang, F.; Li, H. CATNet: A cascaded and aggregated transformer network for RGB-D salient object detection. *IEEE Trans. Multimed.* **2023**, *26*, 2249–2262. [CrossRef]
41. Theckedath, D.; Sedamkar, R.R. Detecting affect states using VGG16, ResNet50 and SE-ResNet50 networks. *SN Comput. Sci.* **2020**, *1*, 79. [CrossRef]
42. Li, H.; Wu, X.J.; Durrani, T. NestFuse: An infrared and visible image fusion architecture based on nest connection and spatial/channel attention models. *IEEE Trans. Instrum. Meas.* **2020**, *69*, 9645–9656. [CrossRef]
43. Milletari, F.; Navab, N.; Ahmadi, S.A. V-net: Fully convolutional neural networks for volumetric medical image segmentation. In Proceedings of the 2016 Fourth International Conference on 3D Vision (3DV), Stanford, CA, USA, 25–28 October 2016; pp. 565–571.
44. Paszke, A.; Gross, S.; Massa, F.; Lerer, A.; Bradbury, J.; Chanan, G.; Killeen, T.; Lin, Z.; Gimelshein, N. Pytorch: An imperative style, high-performance deep learning library. *Adv. Neural Inf. Process. Syst.* **2019**, *32*, 1–12.
45. Ketkar, N.; Moolayil, J. Introduction to pytorch. In *Deep Learning with Python: Learn Best Practices of Deep Learning Models with PyTorch*; Apress: New York, NY, USA, 2021; pp. 27–91.
46. Krizhevsky, A.; Sutskever, I.; Hinton, G.E. ImageNet classification with deep convolutional neural networks. *Commun. ACM* **2017**, *60*, 84–90. [CrossRef]
47. Ju, R.; Ge, L.; Geng, W.; Ren, T.; Wu, G. Depth saliency based on anisotropic center-surround difference. In Proceedings of the 2014 IEEE International Conference on Image Processing (ICIP), Paris, France, 27–30 October 2014; pp. 1115–1119.
48. Niu, Y.; Geng, Y.; Li, X.; Liu, F. Leveraging stereopsis for saliency analysis. In Proceedings of the 2012 IEEE Conference on Computer Vision and Pattern Recognition, Providence, RI, USA, 16–21 June 2012; pp. 454–461.
49. Zhu, C.; Li, G. A three-pathway psychobiological framework of salient object detection using stereoscopic technology. In Proceedings of the IEEE International Conference on Computer Vision Workshops, Venice, Italy, 22–29 October 2017; pp. 3008–3014.
50. Fan, D.P.; Lin, Z.; Zhang, Z.; Zhu, M.; Cheng, M.M. Rethinking RGB-D salient object detection: Models, data sets, and large-scale benchmarks. *IEEE Trans. Neural Netw. Learn. Syst.* **2020**, *32*, 2075–2089. [CrossRef]
51. Cheng, Y.; Fu, H.; Wei, X.; Xiao, J.; Cao, X. Depth enhanced saliency detection method. In Proceedings of the International Conference on Internet Multimedia Computing and Service, Xiamen, China, 10–12 July 2014; pp. 23–27.
52. Fan, D.P.; Gong, C.; Cao, Y.; Ren, B.; Cheng, M.M.; Borji, A. Enhanced-alignment measure for binary foreground map evaluation. In Proceedings of the 27th International Joint Conference on Artificial Intelligence, Stockholm, Sweden, 16 July 2018; pp. 698–704.
53. Chen, T.; Xiao, J.; Hu, X.; Zhang, G.; Wang, S. Adaptive fusion network for RGB-D salient object detection. *Neurocomputing* **2023**, *522*, 152–164. [CrossRef]
54. Bi, H.; Wu, R.; Liu, Z.; Zhu, H.; Zhang, C.; Xiang, T.Z. Cross-modal hierarchical interaction network for RGB-D salient object detection. *Pattern Recognit.* **2023**, *136*, 109194. [CrossRef]
55. Zhang, M.; Yao, S.; Hu, B.; Piao, Y. C2DFNet: Criss-Cross Dynamic Filter Network for RGB-D Salient Object Detection. *IEEE Trans. Multimed.* **2022**, *25*, 5142–5154. [CrossRef]

56. Wang, F.; Pan, J.; Xu, S.; Tang, J. Learning discriminative cross-modality features for RGB-D saliency detection. *IEEE Trans. Image Process.* **2022**, *31*, 1285–1297. [CrossRef]
57. Chen, T.; Hu, X.; Xiao, J.; Zhang, G.; Wang, S. CFIDNet: Cascaded feature interaction decoder for RGB-D salient object detection. *Neural Comput. Appl.* **2022**, *34*, 7547–7563. [CrossRef]
58. Cong, R.; Lin, Q.; Zhang, C.; Li, C.; Cao, X.; Huang, Q.; Zhao, Y. CIR-Net: Cross-modality interaction and refinement for RGB-D salient object detection. *IEEE Trans. Image Process.* **2022**, *31*, 6800–6815. [CrossRef] [PubMed]
59. Ji, W.; Li, J.; Yu, S.; Zhang, M.; Piao, Y.; Yao, S.; Bi, Q.; Ma, K.; Zheng, Y.; Lu, H.; et al. Calibrated RGB-D salient object detection. In Proceedings of the IEEE/CVF Conference on Computer Vision and Pattern Recognition, Virtual, 19–25 June 2021; pp. 9471–9481.
60. Zhao, J.; Zhao, Y.; Li, J.; Chen, X. Is depth really necessary for salient object detection? In Proceedings of the 28th ACM International Conference on Multimedia, Seattle, WA, USA, 12–16 October 2020; pp. 1745–1754.
61. Li, G.; Liu, Z.; Ling, H. ICNet: Information conversion network for RGB-D based salient object detection. *IEEE Trans. Image Process.* **2020**, *29*, 4873–4884. [CrossRef] [PubMed]

Disclaimer/Publisher's Note: The statements, opinions and data contained in all publications are solely those of the individual author(s) and contributor(s) and not of MDPI and/or the editor(s). MDPI and/or the editor(s) disclaim responsibility for any injury to people or property resulting from any ideas, methods, instructions or products referred to in the content.

Article

Enhancing Mask Transformer with Auxiliary Convolution Layers for Semantic Segmentation

Zhengyu Xia and Joohee Kim *

Department of Electrical and Computer Engineering, Illinois Institute of Technology, Chicago, IL 60616, USA
* Correspondence: joohee@ece.iit.edu

Abstract: Transformer-based semantic segmentation methods have achieved excellent performance in recent years. Mask2Former is one of the well-known transformer-based methods which unifies common image segmentation into a universal model. However, it performs relatively poorly in obtaining local features and segmenting small objects due to relying heavily on transformers. To this end, we propose a simple yet effective architecture that introduces auxiliary branches to Mask2Former during training to capture dense local features on the encoder side. The obtained features help improve the performance of learning local information and segmenting small objects. Since the proposed auxiliary convolution layers are required only for training and can be removed during inference, the performance gain can be obtained without additional computation at inference. Experimental results show that our model can achieve state-of-the-art performance (57.6% mIoU) on the ADE20K and (84.8% mIoU) on the Cityscapes datasets.

Keywords: deep learning; semantic segmentation; image segmentation; transformer; convolutional neural networks

1. Introduction

Citation: Xia, Z.; Kim, J. Enhancing Mask Transformer with Auxiliary Convolution Layers for Semantic Segmentation. *Sensors* **2023**, *23*, 581. https://doi.org/10.3390/s23020581

Academic Editors: Erik Blasch and Yufeng Zheng

Received: 6 December 2022
Revised: 30 December 2022
Accepted: 31 December 2022
Published: 4 Janauary 2023

Copyright: © 2023 by the authors. Licensee MDPI, Basel, Switzerland. This article is an open access article distributed under the terms and conditions of the Creative Commons Attribution (CC BY) license (https:// creativecommons.org/licenses/by/ 4.0/).

Transformer, a type of a deep learning model based on self-attention [1], was first applied to natural language processing (NLP) tasks and achieved significant improvements. Inspired by the huge success of Transformer architectures in NLP, extensive research has been recently performed to apply Transformer to various computer vision tasks [2–4]. The basic idea for vision transformers is to break down images into sequential patches and learn self-attention features without using convolutional layers. Unlike traditional convolutional neural network (CNN) models [5,6], transformer-based ones can better capture global attention and broader range relations throughout the entire layers.

Recently, several semantic segmentation approaches [7–11] based on vision transformers have been proposed to exploit the benefits of transformer models for improving semantic segmentation. One way to improve semantic segmentation is to adopt a feature pyramid network (FPN) [12] in a transformer model to obtain multi-scale feature maps. For example, SETR [7] designs a top-down feature aggregation at the decoder side. It generates the final predictions by collecting the feature maps from the transformer backbone. SegFormer [8] proposes a hierarchical transformer at the encoder side. The feature outputs are then fused into a multilayer perceptron (MLP) decoder to aggregate information. Another way is to replace per-pixel classification with mask classification to predict the final outputs. Segmenter [9] utilizes a transformer-based decoder to generate class masks by computing the scalar product between the patch embeddings and the class embeddings. MaskFormer [10] observes that mask classification is sufficiently general to solve both semantic- and instance-level segmentation tasks. It converts per-pixel classification into a mask classification model using a set prediction mechanism. Mask2Former [11] improves the performance on top of [10] and presents a universal segmentation model using the same mask classification mechanism.

However, we observe that these segmentation approaches rely heavily on transformer models and therefore lose local information at a certain level. Even Mask2Former [11], a powerful unified segmentation model, still faces the issue of learning local features and segmenting small objects. In contrast, convolutional layers can better capture local features since most CNN models adopt a small window-sized learning manner. In addition, the optimization with CNN models is easier and more robust compared to transformer models. Therefore, many researchers consider a hybrid model which combines the benefits of CNNs and transformers. For example, ref. [13] replaces the ViT patching module with a convolutional stem to achieve faster convergence and more stable training. BotNet [14] incorporates multi-head self-attention modules on top of the ResNet. It provides a backbone architecture that uses transformer-like blocks for downstream tasks. Visformer [15] offers an empirical study by transforming a transformer-based model to a CNN model and then proposes a new hybrid architecture by absorbing the advantages and discarding the disadvantages.

Inspired by these hybrid approaches, we propose a simple yet effective method on top of [11] to boost semantic segmentation performance. In this work, we introduce an auxiliary CNN on the encoder side. It encourages the model to learn dense local features compared to a pure transformer-based backbone. Additionally, unlike the existing hybrid models, our proposed auxiliary convolution layers can be removed. Therefore, it enhances the semantic segmentation performance without any additional computational cost at inference. Since [11] is a universal segmentation model, we will also show that our proposed method can improve the semantic segmentation performance using a single panoptic model. The contributions of this work can be summarized as follows:

(1) We design an auxiliary CNN on top of Mask2Former [11] to help improve semantic segmentation performance. The proposed network consists of simple convolutional layers without bells and whistles. We demonstrate that the proposed method improves the semantic segmentation performance quantitatively and qualitatively. Specifically, we show that the proposed method is effective in learning local features and segmenting small objects more accurately.

(2) Since the proposed auxiliary convolution layers are required during the training stage only, the proposed method incurs no additional computation overhead at inference. This is one of the important properties of the proposed method because enhancing the performance while maintaining the complexity at inference is crucial for real-world applications.

(3) The proposed auxiliary convolution layers are effective for both semantic and panoptic segmentation. Since Mask2Former is a universal architecture for different segmentation tasks and our proposed method is designed to enhance Mask2Former, we show that the proposed method achieves state-of-the-art performance for semantic and panoptic segmentation on the ADE20K [16] and Cityscapes [17] datasets.

The rest of the paper is organized as follows: In Section 2, the related work is discussed. In Section 3, the proposed method is explained in detail. Section 4 introduces the dataset, implement details, ablation study and experimental results. Section 5 is the conclusion and future work.

2. Related Work
2.1. Semantic Segmentation

Semantic segmentation aims to assign a category label to each pixel. Ref. [18] is the first work to train a fully convolutional network (FCN) end-to-end for semantic segmentation. SegNet [19] and UNet [20] extend the segmentation model with a symmetric encoder-decoder architecture to gradually recover image resolutions. ParseNet [21] augments the features with the average feature for each layer to exploit global context information. PSP-Net [22] and DeepLab [23–25] follow the ideas of Spatial Pyramid Pooling (SPP) [26] to capture dense contextual information at multiple levels. DANet [27] appends two separate attention modules on top of FCN to obtain global dependencies in spatial and channel dimensions, respectively. CCNet [28] proposes a criss-cross attention module on the decoder side to harvest contextual information along the criss-cross path. OCNet [29] presents an object context aggregation scheme with an interlaced spare self-attention to address the semantic segmentation task. These well-known models are all based on convolutional neural networks to learn image features. With the advent of vision transformers [2–4], many semantic segmentation approaches are proposed based on transformers. SETR [7] reformulates semantic segmentation as a sequence-to-sequence learning problem and deploys a pure encoder-decoder transformer model for semantic segmentation. SegFormer [8] designs a hierarchical transformer encoder with a lightweight MLP decoder to generate segmentation results without heavy computational cost. Segmenter [9] refers to ViT [2] and extends it to semantic segmentation. It adopts a mask transformer on the decoder side to generate class masks.

2.2. Panoptic Segmentation

Panoptic segmentation [30] aims to combine semantic and instance segmentation into a general unified output. Panoptic-Deeplab [31] and TASCNet [32] build one shared backbone with two segmentation heads to learn semantic and instance features individually. UPSNet [33] designs a parameter-free panoptic head using pixel-wise classification to resolve the conflicts between semantic and instance features. BGRNet [34] adopts a graph structure on top of a panoptic network to mine intra- and inter-modular relations between foreground and background classes. Auto-Panoptic [35] proposes an automated framework to search for main components simultaneously in a panoptic network, achieving a reciprocal relation between things and stuff classes. Panoptic-FCN [36] represents things and stuff uniformly using a proposed kernel head, which generates unique weights for both classes. MaskFormer [10] demonstrates that mask classification is sufficient to be used for both semantic- and instance-level segmentation tasks. It shows that a simple mask classification can outperform state-of-the-art per-pixel classification models. Mask2Former [11] is an improved version of [10] and utilizes masked attention to extract localized features. It is a universal image segmentation model that outperforms specialized segmentation models across different tasks.

2.3. Hybrid Models Using Convolutions and Transformers

Recently, numerous approaches that combine both convolutions and transformers have been proposed. DETR [26] adopts a CNN backbone with a transformer decoder for object detection. ViLBERT [37] builds a multimodal two-steam model to process visual and textual inputs through co-attentional transformer layers. It utilizes a BERT [4] architecture for the linguistic stream and a Faster-RCNN [38] to capture image regions. PVT [39,40] borrows the pyramid structure concept in CNNs and designs a pyramid vision transformer for learning multi-scale features with high resolutions. P2T [41] implements a pooling-based self-attention module with depthwise convolutional operations for multi-scale feature learning. Ref. [13] demonstrates that the optimization challenges in ViT [2] are related to the patchify stem and shows that the use of convolutional stem enables a much faster convergence in training. BotNet [14] and Visformer [15] analyze the behaviors in convolution- and transformer-based models. Both methods incorporate Multi-Head Self-Attention (MHSA) modules on top of the ResNet-like models to improve the performance of the baseline models. In this work, we propose a simple yet efficient method that introduces an auxiliary CNN on top of the Mask2Fomer [11]. It helps increase the semantic segmentation performance, especially for the local features and small objects. Unlike the existing hybrid models, the proposed method can be removed at the inference stage and therefore does not incur any additional computation overhead at inference.

3. Proposed Method

3.1. Overall Architecture

Our proposed method is integrated with the transformer-based model to improve semantic segmentation. Figure 1 illustrates the overall architecture, where the proposed auxiliary CNN is jointly trained with the main segmentation network. First, the input image is fed to a Swin [42] backbone to generate feature embeddings F_l, where $l \in \{1, \ldots, L\}$ and L is the total number of the stages represented in the Swin backbone. Then, feature embeddings F_l are shared between two separate branches: the main segmentation head and the proposed auxiliary CNN. We use Mask2Former [11] which adopts a pixel decoder and a transformer decoder to generate mask predictions as the main segmentation head. In the auxiliary CNN, feature embeddings F_l are first fed to a simple CNN-based network, aiming to learn local features with different resolutions. Then, an auxiliary loss is calculated based on the auxiliary outputs and added to the main loss to compute the total loss.

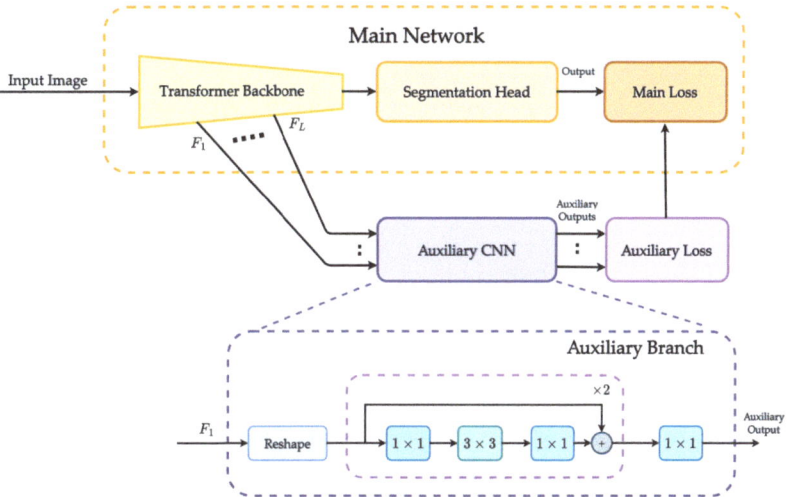

Figure 1. Architecture overview. The proposed method is instantiated on top of Mask2Former [11], which uses Swin Transformer [42] as the backbone network to extract feature embeddings $\{F_1, \ldots, F_L\}$. The proposed auxiliary CNN consists of several simple convolutional layers to learn more accurate local features by using the feature embedding produced by the Transformer backbone as input. An auxiliary loss is used along with the main loss to compute the total loss for segmentation. The auxiliary CNN is used for training only and will be removed at inference.

3.2. Auxiliary CNN

We design an auxiliary branch with convolutional layers to generate multi-scale local features from the backbone, as illustrated in Figure 1.

For a feature embedding $F_l \in \mathbb{R}^{\frac{HW}{r_l^2} \times C_l}$, we first reshape it into a feature map F_l' with a size of $C_l \times \frac{H}{r_l} \times \frac{W}{r_l}$, where C_l is the channel dimension of the feature map at the lth stage in the Swin backbone. H and W are the height and width of the input image, respectively. r_l is the resolution factor equal to 4, 8, 16, and 32 for stages 1, 2, 3, and 4, respectively. Then, the reshaped feature map F_l' is applied to a series of residual blocks for local feature learning. The residual block consists of a stack of three convolutional layers. The three layers are 1×1, 3×3, and 1×1 convolutions, where the 1×1 layers are responsible for downsampling and upsampling the channel dimensions and 3×3 filters are used for feature learning. A skip connection and an element-wise summation are included in the residual block to refine the optimization processing during the training phase. Then, the output from the residual block is fed to a 1×1 convolutional layer to reduce the feature dimension from $C_l \times \frac{H}{r_l} \times \frac{W}{r_l}$ to $N \times \frac{H}{r_l} \times \frac{W}{r_l}$, where N is the number of categories in the dataset. Finally, we adopt a cross-entropy function for auxiliary loss calculation. Note that the reshaping operation in the proposed method is not mandatory. Depending on the shape of the output obtained from the Transformer backbone, the proposed auxiliary CNN can be used without reshaping the feature embeddings.

3.3. Auxiliary Loss

We define the loss function for auxiliary CNN as a cross-entropy loss. Specifically, the loss function for auxiliary branch at the lth stage is computed as:

$$L_{aux}^l = \sum_{w=1}^{W^l} \sum_{h=1}^{H^l} CE(y^l(w,h), gt^l(w,h)), \qquad (1)$$

where $W^l = W/r_l$ and $H^l = W/r_l$. y^l is the auxiliary prediction at the l stage, and gt^l is the corresponding ground truth for semantic segmentation. CE is the cross-entropy loss function. The total auxiliary segmentation loss L_{aux} is the normalized sum of the cross-entropy loss L_{aux}^l over all L stages and is defined as:

$$L_{aux} = \sum_{n=1}^{L} norm(L_{aux}^l). \qquad (2)$$

When training with the auxiliary CNN, the total loss function is defined as:

$$L_{total} = L_{mask-cls} + \beta L_{aux}, \qquad (3)$$

where $L_{mask-cls}$ is the mask classification loss defined in [11], β is the weight for auxiliary segmentation loss. In our ablation study, the best β is selected as 0.1.

4. Experimental Results

4.1. Dataset

We conduct experiments on the ADE20K [16] and Cityscapes [17] datasets. The ADE20K dataset is a densely annotated dataset for scene parsing with 150 categories. The training set contains 20K images, and the validation set contains 2K images. The Cityscapes dataset is a street-view dataset with 19 classes, focusing on a semantic understanding of urban street scenes. It contains 5K images with fine annotations and 20K images with coarse annotations. The fine-annotated dataset contains 2975, 500, and 1525 images for training, validation, and testing, respectively. The ADE20K validation dataset is used for the ablation study to compare the performance with our baseline Mask2Former [11] and other setups.

We use the mean Intersection-over-Union (mIoU) metric for semantic segmentation and the standard Panoptic Quality (PQ) metric for panoptic segmentation. PQ metric [30] evaluates the performance of both stuff and things in a unified manner. Additionally, we use the same metric settings for semantic and instance segmentation based on a single panoptic model as in [11]. Specifically, we report mIoU$_{pan}$ for semantic segmentation by merging instance masks with the same category, and AP$_{pan}$ for instance segmentation, evaluated on the "thing" categories with instance segmentation annotations.

4.2. Implementation Details

Our implementation is based on PyTorch [43] framework with Detectron2 [44]. We use the AdamW [45] optimizer and the step learning rate schedule, where the base learning rate is initialized to 0.0001. All the training has a weight decay of 0.05 and a momentum of 0.9. The input image is resized to 640 × 640 and 512 × 1024 for ADE20K and Cityscapes, respectively. Data augmentation includes random crop, random flip, and large-scale jittering (LSJ) [46]. Following the default settings in [11], we adopt batch normalization for the Cityscapes dataset only. The query number is 100 for all training except that we set 200 queries for the panoptic model with the Swin-L backbone.

Due to the GPU memory limitation, we use smaller batch sizes with higher numbers of training iterations so that we can have similar training settings as in Mask2Former. Specifically, for the ADE20K dataset, we set the batch sizes to 16, 16, 12, and 8 for the Swin-T, Swin-S, Swin-B, and Swin-L transformer backbones, respectively. The corresponding training iterations for these Swin transformer backbones are set to 160K, 160K, 240K, and 360K, respectively. For the Cityscapes dataset, we assign the batch sizes to 12, 8, and 6 for the Swin-S, Swin-B, and Swin-L, respectively. The corresponding training iterations are set to 120K, 180K, and 240K, respectively. By doing so, the number of training epochs is the same as [11]. We also represent the reproduced Mask2Former with our settings, marked as Mask2Former(ours), for a fair comparison.

4.3. Ablation Study

We conduct our ablation study on the ADE20K validation dataset. To evaluate our proposed method fairly, we use the same experimental environments to compare the performance with different settings. We use the Mask2Former with Swin-B backbone as the base network. The cropping size of the input data is set to 640×640.

Effectiveness of auxiliary CNN: To determine the best architecture of the proposed auxiliary CNN for local feature learning, we first use different combinations of the multi-scale feature maps obtained from the Swin transformer backbone as input and evaluate the performance. Specifically, we set various setups by using $\{F_1\}$, $\{F_2\}$, $\{F_3\}$, $\{F_1, F_2\}$, $\{F_1, F_3\}$, $\{F_2, F_3\}$, and $\{F_1, F_2, F_3\}$ as the feature inputs for our auxiliary branches. The subscript in F_l indicates the stage number in the Swin backbone. Table 1 shows that the use of auxiliary CNN with any feature map generated from the transformer backbone improves the performance compared to the baseline method. Among all settings, the best performance is obtained when $\{F_1, F_2, F_3\}$ is used as input to the proposed auxiliary CNN. The experimental results verify that the proposed auxiliary CNN is effective in learning additional local features and achieves better performance when multi-scale features are used. Since the proposed auxiliary CNN will be removed at inference, we use $\{F_1, F_2, F_3\}$ as input to the proposed auxiliary CNN for the remaining experiments to achieve the best performance.

Table 1. Performance comparison of different auxiliary CNN setups using the ADE20K validation set. baseline: Mask2Former with Swin-B backbone for semantic segmentation. F_l: feature embeddings extracted at the lth stage from the Swin-B backbone. ss: single-scale. ms: multi-scale.

Setups	Baseline	F_1	F_2	F_3	mIoU (ss)	mIoU (ms)	#params
Setup 1	√				53.9	55.1	107.0M
Setup 2	√	√			54.2 (↑0.3)	55.3 (↑0.2)	107.1M
Setup 3	√		√		54.3 (↑0.4)	55.3 (↑0.2)	107.3M
Setup 4	√			√	54.0 (↑0.1)	55.1 (-)	109.2M
Setup 5	√	√	√		54.3 (↑0.4)	55.3 (↑0.2)	107.4M
Setup 6	√	√		√	54.2 (↑0.3)	55.2 (↑0.1)	109.3M
Setup 7	√		√	√	54.3 (↑0.4)	55.3 (↑0.2)	109.5M
Setup 8	√	√	√	√	54.5 (↑0.6)	55.5 (↑0.4)	109.6M

Architecture of auxiliary CNN: One of the main design criteria for the proposed auxiliary CNN is to learn some useful local information based on the feature maps generated from the transformer backbone network using simple architectures. We consider four different simple CNN architectures: a 1×1 convolutional layer, a 3×3 convolutional layer, a residual block (a stack of 1×1, 3×3, and 1×1 convolutional layers with a skip connection), and a stack of two residual blocks. Table 2 shows the comparison of performance gain in semantic segmentation obtained by using these different auxiliary CNN architectures for the ADE20K validation dataset. Among the simple architectures we considered, a stack of two residual blocks achieved the best performance improvement. Since stacking more than two residual blocks does not improve the performance gain significantly, we build our proposed auxiliary CNN by using a stack of two residual blocks.

Table 2. Performance comparison of various auxiliary CNNs using the ADE20K validation set. ss: single-scale. ms: multi-scale.

Auxiliary Structure	mIoU (ss)	mIoU (ms)	#params
-	53.9	55.1	-
1×1 conv.	53.6 (↓0.3)	54.7 (↓0.4)	1.4M
3×3 conv.	54.0 (↑0.1)	55.0 (↓0.1)	12.4M
one residual block	54.2 (↑0.3)	55.2 (↑0.1)	1.3M
two residual blocks	54.5 (↑0.6)	55.5 (↑0.4)	2.6M

Weighting parameter of auxiliary CNN: A weighting parameter β is introduced in Equation (3) to balance the loss between the main and the auxiliary tasks. The auxiliary CNN is trained along with the main segmentation network to enhance local features and improve segmenting small objects. However, while achieving this objective, the auxiliary task should not dominate the overall segmentation task. Table 3 shows the performance comparison when four different weighting parameters are used to adjust the contribution of the auxiliary loss. To maximize the overall performance by balancing the main and the auxiliary tasks, we set the weighting parameter β to 0.1.

Table 3. Performance comparison of different weighting parameters β using the ADE20K validation set. ss: single-scale. ms: multi-scale.

Weighting Parameter β	mIoU (ss)	mIoU (ms)
-	53.9	55.1
0.1	54.5 (↑0.6)	55.5 (↑0.4)
0.2	54.4 (↑0.5)	55.3 (↑0.2)
0.3	54.1 (↑0.2)	55.2 (↑0.1)
0.05	54.2 (↑0.3)	55.3 (↑0.2)

4.4. Experimental Results for Semantic Segmentation

We compare the semantic segmentation performance of the proposed method with the recent transformer-based semantic segmentation models on the ADE20K and Cityscapes validation datasets. Since the performance of each model can be different from the one presented in the original paper depending on the hardware environment, we also include the performance of the baseline model Mask2Former obtained by our reproduced experiments.

For the ADE20K dataset, we can observe in Table 4 that our proposed method improves the performance of Mask2Former for all Swin Transformer backbones. Specifically, the proposed auxiliary CNN with Swin-T transformer backbone improves the baseline Mask2former by 0.9% and achieves 48.8% in mIoU (ss). With Swin-S, Swin-B[†], and Swin-L[†] Transformer backbones, the proposed method improves the mIoU by 0.9%, 0.4%, and 0.4%, respectively.

For the Cityscapes dataset, it can be seen from Table 5 that the proposed auxiliary CNN can enhance the Mask2Former's semantic segmentation performance by 0.5%, 0.6% and 0.3% when Swin-S, Swin-B[†], and Swin-L[†] transformer backbones are used, respectively. Both experimental results show that our proposed method consistently outperforms Mask2Former with different Swin Transformer-based backbones. We observe that the performance with Swin-B is slightly better than with Swin-L. Two possible explanations for these results are: the use of smaller batch size for Swin-L in our experimental settings and the better multi-scale inference performance of Swin-B compared to Swin-L from the baseline.

Table 4. Performance comparison of semantic segmentation on the ADE20K validation dataset with 150 categories. †: backbone pretrained on ImageNet-22K. ss: single-scale. ms: multi-scale.

Method	Backbone	Crop Size	mIoU (ss)	mIoU (ms)
PVTv1 [39]	PVTv1-L	512 × 512	44.8	-
PVTv2 [40]	PVTv2-B5	512 × 512	48.7	-
P2T [41]	P2T-L	512 × 512	49.4	-
Swin-UperNet [42,47]	Swin-L †	640 × 640	-	53.5
FaPN-MaskFormer [10,48]	Swin-L †	640 × 640	55.2	56.7
BEiT-UperNet [4,47]	BEiT-L †	640 × 640	-	57.0
MaskFormer [10]	Swin-T	512 × 512	46.7	48.8
	Swin-S	512 × 512	49.8	51.0
	Swin-B †	640 × 640	52.7	53.9
	Swin-L †	640 × 640	54.1	55.6
Mask2Former [11]	Swin-T	512 × 512	47.7	49.6
	Swin-S	512 × 512	51.3	52.4
	Swin-B †	640 × 640	53.9	55.1
	Swin-L †	640 × 640	56.1	57.3
Mask2Former (Ours)	Swin-T	512 × 512	47.9	49.7
	Swin-S	512 × 512	51.3	52.5
	Swin-B †	640 × 640	54.1	54.9
	Swin-L †	640 × 640	56.0	57.1
Ours	Swin-T	512 × 512	48.8	50.3
	Swin-S	512 × 512	52.2	53.1
	Swin-B †	640 × 640	54.5	55.5
	Swin-L †	640 × 640	56.4	57.6

Table 5. Performance comparison of semantic segmentation on the Cityscapes validation dataset with 19 categories. †: backbone pretrained on ImageNet-22K. ss: single-scale. ms: multi-scale.

Method	Backbone	Crop Size	mIoU (ss)	mIoU (ms)
Segmenter [9]	ViT-L †	768 × 768	-	81.3
SETR [7]	ViT-L †	768 × 768	-	82.2
SegFormer [8]	MiT-B5	768 × 768	-	84.0
Mask2Former [11]	Swin-S	512 × 1024	82.6	83.6
	Swin-B †	512 × 1024	83.3	84.5
	Swin-L †	512 × 1024	83.3	84.3
Mask2Former (Ours)	Swin-S	512 × 1024	82.4	83.5
	Swin-B †	512 × 1024	83.2	84.3
	Swin-L †	512 × 1024	83.3	84.3
Ours	Swin-S	512 × 1024	82.9	83.8
	Swin-B †	512 × 1024	83.8	84.8
	Swin-L †	512 × 1024	83.6	84.5

Since one of the main objectives of the proposed auxiliary CNN is to improve the segmentation performance in complex scenes which include small objects and require detailed local information for accurate segmentation, we show several qualitative results for the ADE20K and Cityscapes datasets. Figure 2 presents the qualitative results of the ADE20K validation dataset. In the first row, the category "light" on the ceiling is misclassified as a pillar by the baseline. Our proposed method can label the small object with the correct category. In the second row, the category "bread" labeled with khaki color is not segmented correctly using the baseline approach. However, our method can accurately segment most of them. In the third row, the baseline model fails to segment the category "plant" in the middle, while ours can detect and fully segment it.

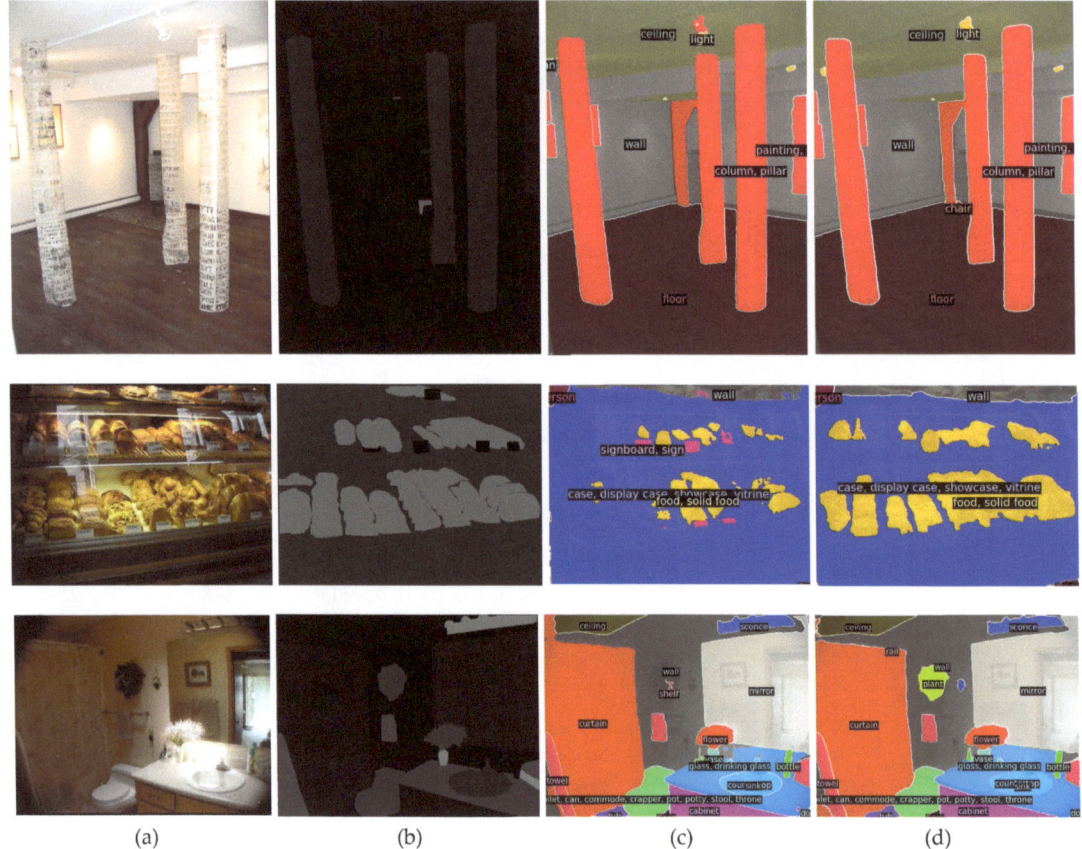

Figure 2. Semantic segmentation results based on the ADE20K validation dataset. (**a**) RGB input, (**b**) ground truth, (**c**) baseline method, and (**d**) our proposed method. The proposed method using the auxiliary CNN improves the detection of local information and small objects compared with the baseline method Mask2Former.

Figure 3 shows the qualitative results of the Cityscapes validation dataset. In the first column, we can observe that the results generated by the baseline mislabeled the category "road" (labeled with purple) on the right-middle side as the category "sidewalk" (labeled with pink). Our proposed method can well distinguish both categories and segment them accurately. In the second column, the baseline approach cannot tell the difference between the category "terrain" (labeled with cyan) and "sidewalk" (labeled with pink), shown on the left side. As a result, the baseline erroneously merges both categories into one, while ours can correctly detect and segment these two categories. In the third column, we can observe that the baseline has difficulty detecting objects with similar textures on the left side. It recognizes the category "terrain" (labeled with cyan) and "road" (labeled with purple) as "sidewalk" (labeled with pink). Our proposed method can distinguish them clearly and accurately. The qualitative results prove that our proposed method can effectively learn local features and identify small objects much better than its baseline method.

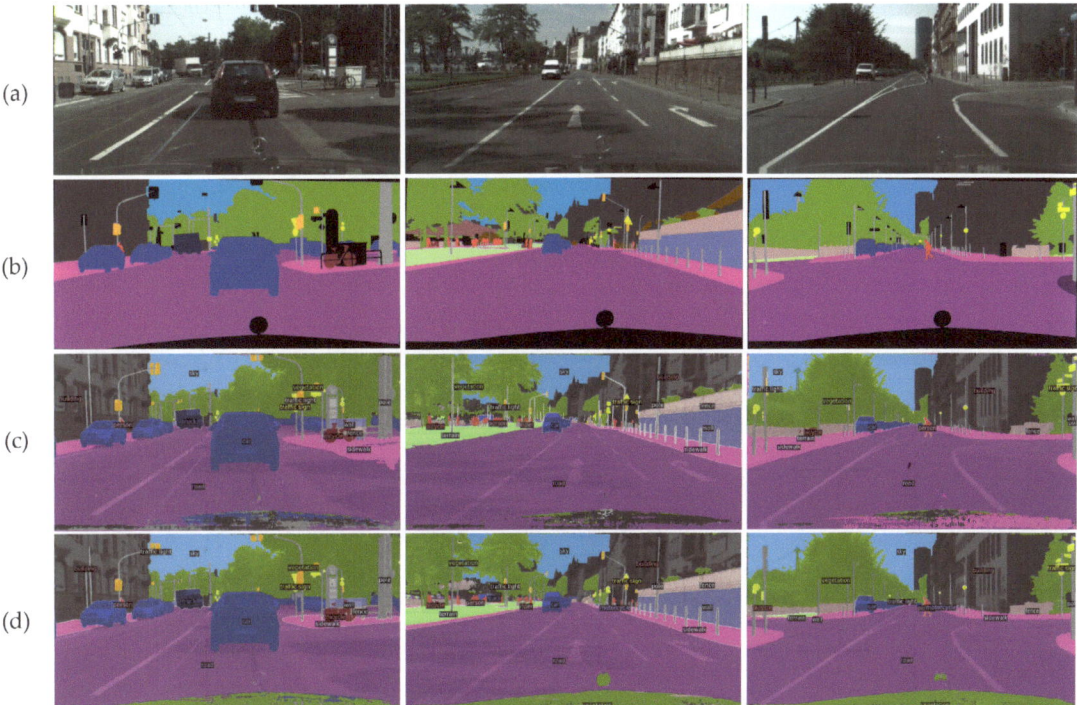

Figure 3. Semantic segmentation results based on the Cityscapes validation dataset. (**a**) RGB input, (**b**) ground truth, (**c**) baseline method, and (**d**) our proposed method. The proposed method using the auxiliary CNN improves the detection of local information and small objects compared with the baseline method Mask2Former.

4.5. Experimental Results for Panoptic Segmentation

Since our baseline method Mask2Former is a well-known universal segmentation model, we evaluate our proposed method using a single panoptic model. Again, since the hardware's difference, we marked "Mask2Former(Ours)" as our reproduced results for the baseline method. Following the baseline's settings, we set 100 queries for Swin-B backbone and 200 queries for Swin-L backbone.

The experimental results in Tables 6 and 7 show that our proposed method can improve all segmentation performance. Specifically, we enhance the ADE20K's panoptic, instance, and semantic segmentation performance with Swin-L backbone by 0.5%, 1.1%, and 0.3%, respectively. We also improve the panoptic, instance, and semantic segmentation performance for the Cityscapes dataset by 0.3%, 1.6%, and 0.3%, respectively. It proves that our proposed method can also improve all segmentation performance, even using a single panoptic model.

4.6. Limitations

The proposed method aims to adopt a simple auxiliary CNN on top of a transformer backbone to increase the overall segmentation performance. In Tables 4 and 5, we can observe that the performance gain gradually decreases when the scale size of a Swin transformer backbone increases. It indicates that a fixed-size auxiliary CNN has less impact on a larger transformer. In our future work, we hope to design an auxiliary CNN that can be adaptive to the transformer backbones with different scales.

Table 6. Performance comparison of panoptic segmentation on the ADE20K validation dataset. Single-scale (ss) inference is adopted by default. Multi-scale results are marked with *. †: backbone pretrained on ImageNet-22K.

Method	Backbone	Panoptic Model		
		PQ (ss)	AP_{pan}	$mIoU_{pan}$
BGRNet [34]	R50	31.8	-	-
Auto-Panoptic [35]	ShuffleNetV2 [49]	32.4	-	-
MaskFormer [10]	R50	34.7	-	-
Kirillov et al. [30]	R50	35.6 *	-	-
Panoptic-DeepLab [31]	SWideRNet [50]	37.9 *	-	50.0 *
Mask2Former [11]	Swin-L †	48.1	34.2	54.5
Mask2Former (Ours)	Swin-L †	48.3	34.0	54.4
Ours	Swin-L †	48.8	35.1	54.7

Table 7. Performance comparison of panoptic segmentation on the Cityscapes validation dataset. Single-scale (ss) inference is adopted by default. Multi-scale results are marked with *. †: backbone pretrained on ImageNet-22K. ‡: backbone pretrained on ImageNet-1K and COCO.

Method	Backbone	Panoptic Model		
		PQ (ss)	AP_{pan}	$mIoU_{pan}$
TASCNet [32]	R50 ‡	59.2	-	-
Kirillov et al. [30]	R50	61.2 *	36.4 *	80.9 *
UPSNet [33]	R101 ‡	61.8 *	39.0 *	79.2 *
Panoptic-DeepLab [31]	SWideRNet [50]	66.4	40.1	82.2
Panoptic-FCN [36]	Swin-L †	65.9	-	-
Mask2Former [11]	Swin-B †	66.1	42.8	82.7
	Swin-L †	66.6	43.6	82.9
Mask2Former (Ours)	Swin-B †	65.7	42.8	82.1
	Swin-L †	66.4	43.0	82.9
Ours	Swin-B †	66.6	43.8	82.9
	Swin-L †	66.7	44.6	83.2

5. Conclusions

In this paper, we propose a simple yet effective auxiliary CNN architecture that introduces auxiliary convolutional layers to Mask2Former during training to learn dense local features. Since the proposed auxiliary CNN is required only for training and can be removed at inference, the segmentation performance can be improved without additional computation overhead at inference. Experimental results show that our proposed method achieves an mIoU of 57.6% on the ADE20K validation dataset and an mIoU of 84.8% on the Cityscapes validation dataset. In the future, we hope to develop a model that can be adaptive to the transformer backbones with different scales to improve the segmentation performance.

Author Contributions: Conceptualization , Z.X.; methodology, Z.X.; software, Z.X.; validation, Z.X. and J.K.; formal analysis, Z.X.; investigation, Z.X.; resources, J.K.; data curation, Z.X.; writing—original draft preparation, Z.X.; writing—review and editing, Z.X. and J.K.; visualization, Z.X.; supervision, J.K.; project administration, J.K.; funding acquisition, J.K. All authors have read and agreed to the published version of the manuscript.

Funding: This work was supported by the Technology Innovation Program of the Ministry of Trade, Industry & Energy (MOTIE, Republic of Korea). [#1415181272, Software and Hardware Development of cooperative autonomous driving control platform for commercial special and work-assist vehicles].

Institutional Review Board Statement: Not applicable.

Informed Consent Statement: Not applicable.

Data Availability Statement: The data presented in this study are openly available in ADE20K at 10.1109/CVPR.2017.544, reference number [16] and Cityscapes at 10.48550/arXiv.1604.01685, reference number [17].

Conflicts of Interest: The authors declare no conflict of interest. The funders had no role in the design of the study; in the collection, analyses, or interpretation of data; in the writing of the manuscript, or in the decision to publish the results.

References

1. Vaswani, A.; Shazeer, N.; Parmar, N.; Uszkoreit, J.; Jones, L.; Gomez, A.N.; Kaiser, L.; Polosukhin, I. Attention Is All You Need. In Proceedings of the Conference Neural Information Processing Systems (NeurIPS), Long Beach, CA, USA, 4–9 December 2017.
2. Dosovitskiy, A.; Beyer, L.; Kolesnikov, A.; Weissenborn, D.; Zhai, X.; Unterthiner, T.; Dehghani, M.; Minderer, M.; Heigold, G.; Gelly, S.; et al. An Image is Worth 16 × 16 Words: Transformers for Image Recognition at Scale. In Proceedings of the International Conference on Learning Representations (ICLR), Addis Ababa, Ethiopia, 26–30 April 2020.
3. Touvron, H.; Cord, M.; Douze, M.; Massa, F.; Sablayrolles, A.; Jégou, H. Training Data-efficient Image Transformers & Distillation through Attention. In Proceedings of the International Conference on Machine Learning (ICML), Virtual, 18–24 July 2021.
4. Bao, H.; Dong, L.; Piao, S.; Wei, F. BEiT: BERT Pre-Training of Image Transformers. *arXiv* **2021**, arXiv:2106.08254.
5. Krizhevsky, A.; Sutskever, I.; Hinton, G.E. ImageNet Classification with Deep Convolutional Neural Networks. In Proceedings of the Conference Neural Information Processing Systems (NeurIPS), Lake Tahoe, Nevada, USA, 3–6 December 2012.
6. He, K.; Zhang, X.; Ren, S.; Sun, J. Deep Residual Learning for Image Recognition. In Proceedings of the IEEE Conference on Computer Vision and Pattern Recognition (CVPR), Las Vegas, NV, USA, 27–30 June 2016.
7. Zheng, S.; Lu, J.; Zhao, H.; Zhu, X.; Luo, X.Z.; Wang, Y.; Fu, Y.; Feng, J.; Xing, T.; Torr, P.H.S.; et al. Rethinking Semantic Segmentation from a Sequence-to-Sequence Perspective with Transformers. In Proceedings of the IEEE Conference on Computer Vision and Pattern Recognition (CVPR), Nashville, TN, USA, 20–25 June 2021.
8. Xie, E.; Wang, W.; Yu, Z.; Anadkumar, A.; Alvarez, J.M.; Luo, P. SegFormer: Simple and Efficient Design for Semantic Segmentation with Transformers. In Proceedings of the Conference Neural Information Processing Systems (NeurIPS), Virtual, 6–14 December 2021.
9. Strudel, R.; Garcia, R.; Laptev, I.; Schmid, C. Segmenter: Transformer for Semantic Segmentation. In Proceedings of the IEEE International Conference on Computer Vision (ICCV), Virtual, 11–17 October 2021.
10. Cheng, B.; Schwing, A.G.; Kirillov, A. Per-Pixel Classification is Not All You Need for Semantic Segmentation. In Proceedings of the Conference Neural Information Processing Systems (NeurIPS), 6–14 December 2021.
11. Cheng, B.; Misra, I.; Schwing, A.G.; Kirillov, A.; Girdhar, R. Masked-attention Mask Transformer for Universal Image Segmentation. In Proceedings of the IEEE Conference on Computer Vision and Pattern Recognition (CVPR), New Orleans, LA, USA, 19–24 June 2022.
12. Lin, T.Y.; Dollár, P.; Girshick, R.; He, K.; Hariharan, B.; Belongie, S. Feature Pyramid Networks for Object Detection. In Proceedings of the IEEE Conference on Computer Vision and Pattern Recognition (CVPR), Honolulu, HI, USA, 21–26 July 2017.
13. Xiao, T.; Singh, M.; Mintun, E.; Darrell, T.; Dollár, P.; Girshick, R. Early Convolutions Help Transformers See Better. In Proceedings of the Conference Neural Information Processing Systems (NeurIPS), Virtual, 6–14 December 2021.
14. Srinivas, A.; Lin, T.Y.; Parmar, N.; Shlens, J.; Abbeel, P.; Vaswani, A. Bottleneck Transformers for Visual Recognition. In Proceedings of the IEEE Conference on Computer Vision and Pattern Recognition (CVPR), Nashville, TN, USA, 20–25 June 2021.
15. Chen, Z.; Xie, L.; Niu, J.; Liu, X.; Wei, L.; Tian, Q. Visformer: The Vision-friendly Transformer. In Proceedings of the IEEE International Conference on Computer Vision (ICCV), Montreal, BC, Canada, 11–17 October 2021.
16. Zhou, B.; Zhao, H.; Puig, X.; Fidler, S.; Barriuso, A.; Torralba, A. Scene Parsing through ADE20K Dataset. In Proceedings of the IEEE Conference on Computer Vision and Pattern Recognition (CVPR), Honolulu, HI, USA, 21–26 July 2017.
17. Cordts, M.; Omran, M.; Ramos, S.; Rehfeld, T.; Enzweiler, M.; Benenson, R.; Franke, U.; Roth, S.; Schiele, B. The Cityscapes Dataset for Semantic Urban Scene Understanding. In Proceedings of the IEEE Conference on Computer Vision and Pattern Recognition (CVPR), Las Vegas, NV, USA, 27–30 June 2016.
18. Long, J.; Shelhamer, E.; Darrell, T. Fully Convolutional Networks for Semantic Segmentation. In Proceedings of the IEEE Conference on Computer Vision and Pattern Recognition (CVPR), Boston, MA, USA, 7–12 June 2015.
19. Badrinarayanan, V.; Kendall, A.; Cipolla, R. SegNet: A Deep Convolutional Encoder-Decoder Architecture for Image Segmentation. In Proceedings of the IEEE Conference on Computer Vision and Pattern Recognition (CVPR), Boston, MA, USA, 7–12 June 2015.
20. Ronneberger, O.; Fischer, P.; Brox, T. U-Net: Convolutional Networks for Biomedical Image Segmentation. In Proceedings of the International Conference on Medical Image Computing and Computer Assisted Intervention (MICCAI), Munich, Germany, 5–9 October 2015.
21. Liu, W.; Rabinovich, A.; Berg, A.C. ParseNet: Looking Wider to See Better. In Proceedings of the International Conference on Learning Representations (ICLR), San Juan, Puerto Rico, 2–4 May 2016.
22. Zhao, H.; Shi, J.; Qi, X.; Wang, X.; Jia, J. Pyramid Scene Parsing Network. In Proceedings of the IEEE Conference on Computer Vision and Pattern Recognition (CVPR), Honolulu, HI, USA, 21–26 July 2017.

23. Chen, L.C.; Papandreou, G.; Kokkinos, I.; Murphy, K.; Yuille, A.L. DeepLab: Semantic Image Segmentation with Deep Convolutional Nets, Atrous Convolution, and Fully Connected CRFs. *TPAMI* **2017**, *40*, 834–848. [CrossRef]
24. Chen, L.C.; Papandreou, G.; Schroff, F.; Adam, H. Rethinking Atrous Convolution for Semantic Image Segmentation. *arXiv* **2017**, arXiv:1706.05587.
25. Chen, L.C.; Zhu, Y.; Papandreou, G.; Schroff, F.; Adam, H. Encoder-Decoder with Atrous Separable Convolution for Semantic Image Segmentation. In Proceedings of the European Conference on Computer Vision (ECCV), Munich, Germany, 8–14 September 2018.
26. Carion, N.; Massa, F.; Synnaeve, G.; Usunier, N.; Kirillov, A.; Zagoruyko, S. End-to-End Object Detection with Transformers. In Proceedings of the European Conference on Computer Vision (ECCV), Glasgow, UK, 23–28 August 2020.
27. Fu, J.; Liu, J.; Tian, H.; Li, Y.; Bao, Y.; Fang, Z.; Lu, H. Dual Attention Network for Scene Segmentation. In Proceedings of the IEEE Conference on Computer Vision and Pattern Recognition (CVPR), Long Beach, CA, USA, 15–20 June 2019.
28. Huang, Z.; Wang, X.; Wei, Y.; Huang, L.; Shi, H.; Liu, W.; Huang, T.S. CCNet: Criss-Cross Attention for Semantic Segmentation. In Proceedings of the IEEE International Conference on Computer Vision (ICCV), Seoul, Republic of Korea, October 27–2 November 2019.
29. Yuan, Y.; Huang, L.; Guo, J.; Zhang, C.; Chen, X.; Wang, J. OCNet: Object Context for Semantic Segmentation. *IJCV* **2021**, *129*, 2375–2398. [CrossRef]
30. Kirillov, A.; He, K.; Girshick, R.; Rother, C.; Dollár, P. Panoptic Segmentation. In Proceedings of the IEEE Conference on Computer Vision and Pattern Recognition (CVPR), Long Beach, CA, USA, 15–20 June 2019.
31. Cheng, B.; Collins, M.D.; Zhu, Y.; Liu, T.; Huang, T.S.; Adam, H.; Chen, L.C. Panoptic-DeepLab: A Simple, Strong, and Fast Baseline for Bottom-Up Panoptic Segmentation. In Proceedings of the IEEE Conference on Computer Vision and Pattern Recognition (CVPR), Long Beach, CA, USA, 15–20 June 2019.
32. Li, J.; Raventos, A.; Bhargava, A.; Tagawa, T.; Gaidon, A. Learning to Fuse Things and Stuff. *arXiv* **2018**, arXiv:1812.01192.
33. Xiong, Y.; Liao, R.; Zhao, H.; Hu, R.; Bai, M.; Yumer, E.; Urtasun, R. UPSNet: A Unified Panoptic Segmentation Network. In Proceedings of the IEEE Conference on Computer Vision and Pattern Recognition (CVPR), Long Beach, CA, USA, 15–20 June 2019.
34. Wu, Y.; Zhang, G.; Gao, Y.; Deng, X.; Gong, K.; Liang, X.; Lin, L. Bidirectional Graph Reasoning Network for Panoptic Segmentation. In Proceedings of the IEEE Conference on Computer Vision and Pattern Recognition (CVPR), Seattle, WA, USA, 13–19 June 2020.
35. Wu, Y.; Zhang, G.; Xu, H.; Liang, X.; Lin, L. Auto-Panoptic: Cooperative Multi-Component Architecture Search for Panoptic Segmentation. In Proceedings of the Conference Neural Information Processing Systems (NeurIPS), Vancouver, Canada, 6–12 December 2020.
36. Li, Y.; Zhao, H.; Qi, X.; Chen, Y.; Qi, L.; Wang, L.; Li, Z.; Sun, J.; Jia, J. Fully Convolutional Networks for Panoptic Segmentation with Point-based Supervision. *arXiv* **2021**, arXiv:2012.00720.
37. Lu, J.; Batra, D.; Parikh, D.; Lee, S. ViLBERT: Pretraining Task-Agnostic Visiolinguistic Representations for Vision-and-Language Tasks. In Proceedings of the Conference Neural Information Processing Systems (NeurIPS), Vancouver, Canada, 8–14 December 2019.
38. Ren, S.; He, K.; Girshick, R.; Sun, J. Faster. R-CNN: Towards Real-Time Object Detection with Region Proposal Networks. In Proceedings of the Conference Neural Information Processing Systems (NeurIPS), Montréal, Canada, 11–12 December 2015.
39. Wang, W.; Xie, E.; Li, X.; Fan, D.P.; Song, K.; Liang, D.; Lu, T.; Luo, P.; Shao, L. Pyramid Vision Transformer: A Versatile Backbone for Dense Prediction without Convolutions. In Proceedings of the IEEE International Conference on Computer Vision (ICCV), Montreal, BC, Canada, 11–17 October 2021.
40. Wang, W.; Xie, E.; Li, X.; Fan, D.P.; Song, K.; Liang, D.; Lu, T.; Luo, P.; Shao, L. PVT v2: Improved Baselines with Pyramid Vision Transformer. *CVMJ* **2022**, *8*, 1–10. [CrossRef]
41. Wu, Y.H.; Liu, Y.; Zhan, X.; Cheng, M.M. P2T: Pyramid Pooling Transformer for Scene Understanding. *TPAMI* **2022**, *99*, 1–12. [CrossRef] [PubMed]
42. Liu, Z.; Lin, Y.; Cao, Y.; Hu, H.; Wei, Y.; Zhang, Z.; Lin, S.; Guo, B. Swin Transformer: Hierarchical Vision Transformer using Shifted Windows. In Proceedings of the IEEE International Conference on Computer Vision (ICCV), Montreal, BC, Canada, 11–17 October 2021.
43. Paszke, A.; Gross, S.; Massa, F.; Lerer, A.; Bradbury, J.; Chanan, G.; Killeen, T.; Lin, Z.; Gimelshein, N.; Antiga, L.; et al. PyTorch: An Imperative Style, High-Performance Deep Learning Library. In Proceedings of the Conference Neural Information Processing Systems (NeurIPS), Vancouver, BC, Canada, 8–14 December 2019.
44. Wu, Y.; Kirillov, A.; Massa, F.; Lo, W.Y.; Girshick, R. Detectron2. 2019. Available online: https://github.com/facebookresearch/detectron2 (accessed on 6 February 2020).
45. Loshchilov, I.; Hutter, F. Decoupled Weight Decay Regularization. In Proceedings of the International Conference on Learning Representations (ICLR), New Orleans, LA, USA, 6–9 May 2019.
46. Ghiasi, G.; Cui, Y.; Srinivas, A.; Qian, R.; Lin, T.; Cubuk, E.D.; Le, Q.V.; Zoph, B. Simple Copy-paste is A Strong Data Augmentation Method for Instance Segmentation. In Proceedings of the IEEE Conference on Computer Vision and Pattern Recognition (CVPR), Nashville, TN, USA, 20–25 June 2021.
47. Xiao, T.; Liu, Y.; Zhou, B.; Jiang, Y.; Sun, J. Unified Perceptual Parsing for Scene Understanding. In Proceedings of the European Conference on Computer Vision (ECCV), Munich, Germany, 8–14 September 2018.

48. Huang, S.; Lu, Z.; Cheng, R.; He, C. Fapn: Feature-aligned Pyramid Network for Dense Image Prediction. In Proceedings of the IEEE International Conference on Computer Vision (ICCV), Montreal, BC, Canada, 11–17 October 2021.
49. Ma, N.; Zhang, X.; Zheng, H.T.; Sun, J. ShuffleNet V2: Practical Guidelines for Efficient CNN Architecture Design. In Proceedings of the European Conference on Computer Vision (ECCV), Munich, Germany, 8–14 September 2018.
50. Chen, L.; Wang, H.; Qiao, S. Scaling Wide Residual Networks for Panoptic Segmentation. *arXiv* **2020**, arXiv:2011.11675.

Disclaimer/Publisher's Note: The statements, opinions and data contained in all publications are solely those of the individual author(s) and contributor(s) and not of MDPI and/or the editor(s). MDPI and/or the editor(s) disclaim responsibility for any injury to people or property resulting from any ideas, methods, instructions or products referred to in the content.

Article

Fusion of Multi-Modal Features to Enhance Dense Video Caption

Xuefei Huang [1], Ka-Hou Chan [1,2], Weifan Wu [1], Hao Sheng [1,3,4] and Wei Ke [1,2,*]

1. Faculty of Applied Sciences, Macao Polytechnic University, Macau 999078, China; xuefei.huang@mpu.edu.mo (X.H.); chankahou@mpu.edu.mo (K.-H.C.); weifan.wu@mpu.edu.mo (W.W.); shenghao@buaa.edu.cn (H.S.)
2. Engineering Research Centre of Applied Technology on Machine Translation and Artificial Intelligence of Ministry of Education, Macao Polytechnic University, Macau 999078, China
3. State Key Laboratory of Virtual Reality Technology and Systems, School of Computer Science and Engineering, Beihang University, Beijing 100191, China
4. Beihang Hangzhou Innovation Institute Yuhang, Yuhang District, Hangzhou 310023, China
* Correspondence: wke@mpu.edu.mo

Abstract: Dense video caption is a task that aims to help computers analyze the content of a video by generating abstract captions for a sequence of video frames. However, most of the existing methods only use visual features in the video and ignore the audio features that are also essential for understanding the video. In this paper, we propose a fusion model that combines the Transformer framework to integrate both visual and audio features in the video for captioning. We use multi-head attention to deal with the variations in sequence lengths between the models involved in our approach. We also introduce a Common Pool to store the generated features and align them with the time steps, thus filtering the information and eliminating redundancy based on the confidence scores. Moreover, we use LSTM as a decoder to generate the description sentences, which reduces the memory size of the entire network. Experiments show that our method is competitive on the ActivityNet Captions dataset.

Keywords: dense video caption; video captioning; multi-modal feature fusion; feature extraction; neural network

Citation: Huang, X.; Chan, K.-H.; Wu, W.; Sheng, H.; Ke, W. Fusion of Multi-Modal Features to Enhance Dense Video Caption. *Sensors* **2025**, *23*, 5565. https://doi.org/10.3390/s23125565

Academic Editors: Erik Blasch and Yufeng Zheng

Received: 11 May 2023
Revised: 30 May 2023
Accepted: 7 June 2023
Published: 14 June 2023

Copyright: © 2025 by the authors. Licensee MDPI, Basel, Switzerland. This article is an open access article distributed under the terms and conditions of the Creative Commons Attribution (CC BY) license (https://creativecommons.org/licenses/by/4.0/).

1. Introduction

A dense video caption is an abstract representation of the important events in unedited videos that may contain different scenes. Thanks to the popularity of Internet resources and mobile devices, the amount of video data is increasing, and the types of which are becoming very rich. As an important way to disseminate information, videos have become inseparable from people's daily lives, and the use of live videos has become a trend. The tasks of dense video captions are to effectively label different types of content, and can be applied to various scenarios in videos towards extraction of higher-level semantics. For instance, monitoring traffic safety, finding target people, assisting in reviewing video content, and improving network security [1–3]. In contrast to tasks such as object recognition and tracking, video captioning requires a combination of computer vision [4–7] and Natural Language Processing (NLP) techniques [8]. In addition to spatial and temporal information, there is also the contextual information contained in the video, including sound effects and speeches. It makes dense captioning much more challenging. Video captioning requires the computer to not only unambiguously recognize objects in the video, but also to understand the relationships between the objects. Finally, it requires the computer to express the contents of the video in logical terms of human languages [9–11].

The emergence of deep learning [12] first achieved breakthroughs in image caption tasks, and the encoder–decoder framework was quickly transferred to video caption domains by researchers who have obtained various results [13,14]. Video caption tasks aim to generate a natural language sentence that summarizes the main content of a given short video. However, most videos do not contain only one event, but are composed of multiple scenarios. The events in a given video are usually related to each other, and most events are action-oriented and can even overlap [15,16]. These characteristics make it difficult for a single sentence to fully express the complex content of the video. To solve this problem, a dense video caption task was proposed to generate a natural language paragraph that describes all of the important events and details in a given video. The dense video caption task accurately expresses the complex content of the video through multiple compound sentences, which is more in line with human needs for artificial intelligence [17].

Figure 1 shows one of the successive frames of an unedited video containing different scenes. It can be seen that this video converts four different scenes, and it is difficult to fully capture all of the content of the video without the help of a dense video caption task. As a result, the generated text is incomplete and incoherent. Furthermore, Figure 1 indicates the importance of combining audio patterns to generate text descriptions. As we can see, the first segment of the video determines the caption as "a woman in a red top talking to the camera" from a purely visual perspective. In fact, the woman in this video is a broadcaster providing the news via audio, and talking about a surfer who accidentally got lost while attempting a huge wave. From the above example, we can conclude that the audio enables our model to produce captions that match the full content of the video more closely. Therefore, it is important to integrate audio features into dense video caption tasks, which help computers to comprehend relatively abstract videos and express rich scenes in text.

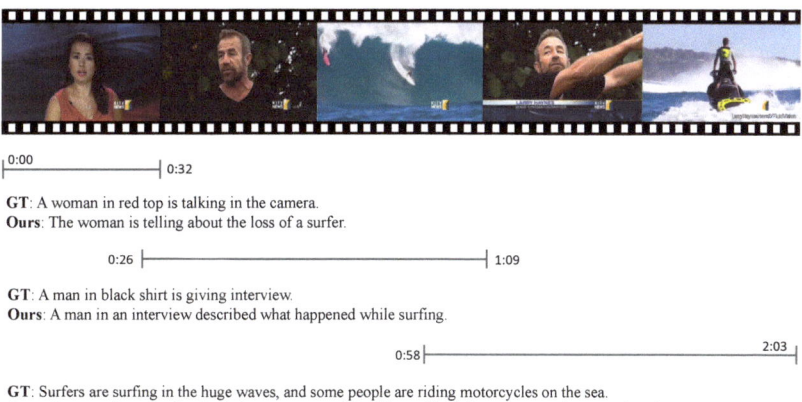

GT: A woman in red top is talking in the camera.
Ours: The woman is telling about the loss of a surfer.

GT: A man in black shirt is giving interview.
Ours: A man in an interview described what happened while surfing.

GT: Surfers are surfing in the huge waves, and some people are riding motorcycles on the sea.
Ours: Two people are surfing in the huge waves, and the lifeguard is riding a motorcycle to search and rescue.

Figure 1. Example video with the predictions of our model alongside the ground truth.

Building on previous work, we focus on how to fully integrate features into video and audio sequences as vectors of different lengths. The main contributions of this paper are as follows:

(1) We introduce a new framework for dense video caption generation. Such framework makes use of the Transformer's multi-head attention module to efficiently fuse video and audio features in video sequences, thus improving the accuracy and richness of the model-generated captions.

(2) We propose a confidence module to select major events, which addresses the problem of unequal recall and precision after using fused video–audio features, making the fused audiovisual features more effective in generating descriptive texts.

(3) We employ LSTM as a decoder for sentence representation, which has the advantage of long-term memory to meet the requirements of text description generation, and also enhances the overall computational efficiency of the framework.

(4) We show that our framework is competitive with existing methods on the ActivityNet Captions dataset.

We arrange the following content as below. Section 2 gives an overview of the related work in video captioning and the deep learning approaches. Section 3 describes the structure of our multi-modal approach and the technical details. The setup of experiments, the discussions of results, and the comparisons with other methods are presented in Section 4. Finally, we conclude the paper in Section 5.

2. Related Work

Video captioning is an introduction to what the video contains in logical sentences. A typical video sequence is formed by playing more than a dozen frames (images) per second quickly. Therefore, the initial methods for dense video captioning were largely inspired by the image caption field, especially the encoder–decoder structure based on deep learning, which can encode the visual features and decode them into natural language sentences [18–20]. The sequence-to-sequence–video-to-text [21] (S2VT) model follows this idea, where a certain number of frames are extracted from the video as images. The encoding part uses the VGG [22] network to process the characteristics of the input data, and adds the optical flow method as an auxiliary. Then, the extracted features are averaged, and a text description is generated in the decoding part using LSTM [23]. However, due to the particularity of video, this method does not take into account the timing information contained in the video, and the generated text description is not detailed enough.

The emergence of convolutional three-dimension networks (C3D) has solved the above problems to a certain extent [24]. It adds the time dimension to the original structure of a 2D CNN, which is more conducive to processing complex video data, and the extracted video features are more comprehensive. Therefore, C3D gradually occupies a major position in the field of video captioning [25–28], and many other projects use it as an encoder for the feature extraction of videos. The inflated 3D convNet (I3D) adds optical flow features on the basis of C3D [29]. The weight of the 2D CNN model pre-trained on ImageNet is used as the initial parameter to train the model, which further improves the performance of video feature extraction. The pseudo-3D residual network (P3D) decomposes 3D convolutions into two-dimensional space convolutions and one-dimensional time convolutions, and adds the concept of residual connection to increase the overall depth of the network, and obtains good results [30].

Dense video caption [31] has raised the video caption task to a new level. On the basis of the S2VT model, the original short description text is extended to the problem of caption generation based on regional sequences, which improves the comprehensiveness and diversity of descriptions and maintains the accuracy. For this task, the ActivityNet Captions dataset is also proposed, which has a high position in the field of dense video captioning [32,33]. Yu et al. proposed the concept of converting a video into an article, using multiple sentences to form a long text paragraph to summarize the video substance [34]. The decoding part of the model is divided into two modules: sentence generation and paragraph composition. The description text as referenced in the dataset is used as a part of the input to train the model to learn the correlation information between sentences. The effective results of this work have inspired the research on video caption tasks to a certain extent. The single-stream temporary action proposal (SST) [35] model obtains the timing information of the video by filtering the threshold, and uses the attention mechanism to analyze the information; the output video features are input to the decoder as the initial state. The Meteor [36] score of this model on the ActivityNet Captions dataset is 9.65, which is higher than that of other models at the same stage.

Because of the outstanding performance, recurrent neural networks (RNN) in the NLP field are applied to video caption tasks. Most models use the characteristics of LSTM that can remember long sequences as a decoder to generate description text. Pan et al. proposed the LSTM-E model [37], which is on the basis of traditional cross-entropy loss. A correlation loss is added to allow the model to learn both semantic relationships and visual content, fully associated sentences used as references with visual features, and improve the accuracy of the output. The boundary-aware encoder [38] model uses LSTM as the encoding part, and proposes a recurrent video coding scheme, which can better explore and use the hierarchical structure in the video, and enhance the matching degree with the timing information in the video.

More importantly, researchers have sought to apply the attention mechanism to the field of video captioning and achieved good results. Yao et al. proposed to introduce attention weight α on the basis of an S2VT model to calculate features of time series, paying high attention to important information in the video, and ignoring some interference or unimportant information [39]. The evaluation index of this behavior is higher than other models in the same period. The spatio-temporal and temporo-spatial attention (STaTS) model [40] takes the language state as the premise, complements the spatial and temporal information of a video through two different attention combinations, and proposes an LSTM-based time sequence function (sorting attention), which can be used to capture actions in the video.

In order to solve the problem that the LSTM structure cannot be trained in parallel, the Transformer frame builds a global relationship on the semantic information of the reference description statements in the dataset based on the attention mechanism, and has achieved good results in dealing with the problem of missing details of video caption tasks [41–45]. Wang et al. proposed a training model based on a Transformer (EEDVC) [46], with one encoder corresponding to two decoders. The video is divided according to different events, and each extracted event is described separately. However, this approach relies too much on the quality of video feature extraction, and the time information captured in the video is insufficient. Wang et al. proposed the parallel decoding method (PDVC) [47] on the basis of EEVDC, which enhanced the model's learning of video features and semantic relationships in reference sentences through two different parallel Transformer methods. The diverse paragraph captioning for untrimmed videos (TDPC) [48] uses the Transformer framework, and adds a dynamic video memory module to interpret the global features of video in stages, taking into account the accuracy and diversity of text descriptions.

The description text generated only for the visual modal information in the video cannot cover all of its content [49,50]. The enhanced topic-guided system [51] introduces the Mel frequency cepstrum coefficient (MFCC) [52] to extract the audio feature information in the video, and fully integrates the visual and audio features, to achieve the purpose of an all-around description. Iashin et al. continued this theory, using I3D and VGGish [53] to extract features of visual and audio modes in the video, and introduced a bi-Transformer framework to abstract video expression [54]. The multi-modal dense video caption (MDVC) model [55] builds upon the Transformer architecture, where the visual, audio, and speech in the video are used as input data, and finally converted into text descriptions. Chang et al. proposed an EMVC method [56], using visual-audio cues to generate event proposals, and developed an attention gate that dynamically fused and adjusted the multi-modal information control mechanism. Hao et al. proposed three different depth fusion strategies for multi-modal information in videos, trying to maximize the advantages of audio-visual resonance [57]. Park et al. combined the human face information in the video, extracted the spatio-temporal features in the video using I3D, and combined it with the Transformer architecture to predict the relationship between different IDs and objects [58].

Based on the above analysis, we summarize some methods for video captioning in Table 1. In addition, we can clearly understand that the video caption field has achieved significant performance improvement in recent years, but it still faces some challenges and problems. On the one hand, video captioning needs to fully utilize the multimodal data features in videos, such as visual, audio, and text, but most current methods only focus on visual features and ignore the importance of other features for generating accurate and rich descriptions. On the other hand, video captions need to generate natural language descriptions that are highly relevant and grammatically correct for the video content, but most of the current methods only use template-based or sequence learning-based language models, lacking the modeling of complex relationships and logical reasoning abilities between video and text.

Table 1. Summary of selected video caption methods.

Method	CNN	RNN	Attention	Transformer	Visual	Audio	Others
S2VT [21], LSTM-E [37]	✓	✓			✓		
DCE [31], SST [35], STS [39], STaTS [40]	✓	✓	✓		✓		
AMT [42], SwinBERT [43], PDVC [47], TDPC [48]				✓	✓		
ETGS [51], VGA [57]	✓	✓	✓		✓	✓	
DVMF [50]	✓	✓	✓		✓	✓	✓
MDVC [55], BMT [54], EMVC [56], FiI [58]				✓	✓	✓	

Therefore, how to better utilize the multi-modal data features in videos, and how to more effectively model the complex relationships and logical reasoning abilities between video and text, are still challenges that need to be focused on and solved in the future research of this field.

3. Methodology

Considering the importance of audio patterns in a video and combining the above research points, a dense video caption generation model that fully integrates visual and audio patterns is proposed. This model applies the I3D and VGGish approaches to extract visual and audio features, respectively. Moreover, the output features produced by these approaches are always presented in different sizes, we thus introduce a multi-head attention module to integrate the extracted visual and audio features. Finally, the LSTM is used as a decoder to implement a descriptive textual representation of the video content.

3.1. Model Overview

The framework of the proposed model consists of three parts: feature extraction, multi-modal feature fusion, and caption generation. Figure 2 presents the entire framework and the data flow between the three parts in a schematic diagram.

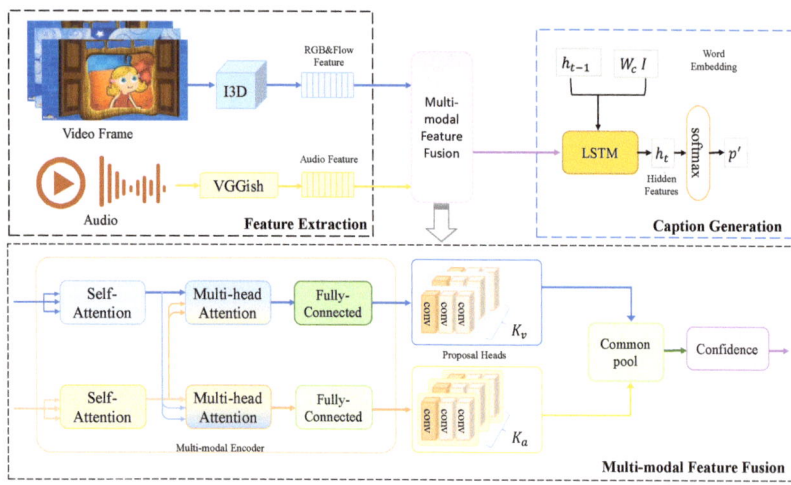

Figure 2. Overall framework of the proposed model.

Feature Extraction Since there are size differences between visual and audio features, they need to be extracted separately to remove noise and redundancy. For the visual pattern features, the I3D network is applied to achieve the extraction of spatial features present in the video, while optical flow features are also added to further improve the performance. Next, VGGish is used to extract a selection of audio features that can effectively convert the audio stream into a feature vector corresponding to natural language elements.

Multi-Modal Feature Fusion The features extracted from visual and audio modalities produce vectors of different dimensions that cannot be directly fused. Therefore, a multi-model attention fusion module is proposed as an encoder based on the Transformer framework, aiming to fully fuse the audio and visual features for information resonance. Furthermore, a confidence module is added to filter the major information in this part.

Caption Generation We employ LSTM to retain the attributes of lengthy sequences as a decoder. The proposals evaluated by the confidence module serve as the initial state input of the decoder, which simulates the distribution in the vocabulary encoded by the embedded position. Finally, a detailed textual description for the video is generated automatically.

3.2. Feature Extraction

Currently, there is no method to extract both visual and audio features from a video simultaneously. These features can only be extracted from different modalities. As noted in Section 2, CNN is a highly regarded method in the field of computer vision that outperforms in dense video caption. Since a video is essentially a collection of still images that contain temporal information, relying solely on 2D CNN networks to extract information from video frames ignores the correlation between frames and thus fails to fully extract the rich information in the video. Therefore, we recommend applying the I3D pre-trained on the Kinetics dataset [59] as the backbone for extracting visual features from the video. This approach adds optical flow features to the spatio-temporal features of the video that can be learned by the C3D approach. We also train the RGB and optical flow networks separately as the input features, then average them during testing. Additionally, I3D contains a deeper network structure and a multi-branch structure, allowing for reduced parameters and increased efficiency. In practice, we apply I3D to extract RGB and optical flow features from each video frame, then combine these two features and encode them using linear layers to achieve a simple, compact, and well-suited network model for intensive video annotation tasks.

We also use the VGGish network to extract audio features from videos. Numerous studies have demonstrated that VGGish outperformed traditional methods for audio extraction. This is achieved by combining the deep structure of the VGG network with log-mel features and training on a large amount of audio data from the Audioset dataset [60]. The pre-trained parameters show strong generalization capabilities. In our framework, the audio clips are represented as log-mel-scale spectrograms of size 96×64, obtained by a short-time Fourier transform. The VGGish network converts the audio into 128-dimensional feature vectors with semantic information, where high-level feature vectors have more expressive power.

3.3. Multi-Modal Feature Fusion

Most dense video caption tasks rely on visual features, and a few integrate multi-modal features by concatenating or sharing the weights of different features rather than fully fusing them. As a result, multi-modal features are not fully functional in nature. To address this problem, we propose a multi-modal feature fusion approach that includes a multi-modal encoder, proposal heads, and confidence module. The multi-modal encoder stacks visual (V_n) and audio (A_n) features into N multi-modal encoder blocks to enable full fusion of the two features. Each multi-modal encoder block consists of self-attentive, multi-head attention, and fully connected layers. The process is described as follows. First, the self-attentive layer of the Transformer processes variable-length information sequences and dynamically generates different weights for the extracted visual and auditory features.

$$\text{self Attention}(Q, K, V) = \text{Softmax}\left(\frac{QK^T}{\sqrt{d}}\right)V, \tag{1}$$

where Q, K, and V denote query, key, and value, respectively. The $\sqrt{d}$ as a training parameter controls the gradient of *Softmax* with the purpose of enhancing the attention weights and distinguishing the differences between these features.

$$V_n^{self} = \text{self Attention}\left(V_{n-1}^{fc}, V_{n-1}^{fc}, V_{n-1}^{fc}\right), \tag{2}$$

$$A_n^{self} = \text{self Attention}\left(A_{n-1}^{fc}, A_{n-1}^{fc}, A_{n-1}^{fc}\right). \tag{3}$$

Then, multiple queries (Q) of multi-head attention are obtained to compute and produce a score for determination. These heads are concatenated as an output feature after the multi-head attentions are determined.

$$\text{head}_h(Q, K, V) = \text{self Attention}\left(QW_h^Q, KW_h^K, VW_h^V\right), \quad h \in [1, H]. \tag{4}$$

All head_h will be concatenated and input to the multi-head attention:

$$\text{MultiHeadAttention}(Q, K, V) = [\text{head}_1(Q, K, V), \ldots, \text{head}_H(Q, K, V)]W^{out}. \tag{5}$$

Note that there are two different dimensions of multi-head attention weights produced by the two modalities, so concatenation processing is required to fuse them together.

$$V_n^A = \text{MultiHeadAttention}\left(V_n^{self}, A_n^{self}, A_n^{self}\right), \tag{6}$$

$$A_n^V = \text{MultiHeadAttention}\left(A_n^{self}, V_n^{self}, V_n^{self}\right). \tag{7}$$

At this point, the module will produce two fully merged new feature sequences: the visual feature V_n^{fc} and audio feature A_n^{fc}, which also contain the most interesting information for the visual and audio components in this part, respectively.

$$V_n^{fc} = \text{FullyConnected}\left(V_n^A\right), \tag{8}$$

$$A_n^{fc} = \text{FullyConnected}\left(A_n^V\right). \tag{9}$$

Once the visual and audio features have been corrected, they are passed to the proposal heads to predict a set of proposal tags to initialize the video. This process can help select features that match the video content and improve the accuracy of the captioning module. However, due to differences in sequence lengths between the video and audio modalities, the feature sequences of the two modalities cannot match every proposal in the video at every time step. To alleviate this problem, a Common Pool is introduced to store the video or audio modality proposal corresponding to each time stamp. This allows filtering out noise and redundant information based on confidence scores.

As shown in Figure 3, there are two proposal heads K_v (video) and K_a (audio) as input features, which are passed to the proposal model in parallel. Each proposal head is a fully-CNN model consisting of three convolutional layers with different kernel sizes, and their paddings are used to unify the sequence length between each layer. The kernel size of the first convolutional layer is configured as $k \times k$, which is used to scale-down the input size while extracting the most important features. Next, the kernel size of the other two convolutional layers is set to 1, with the goal of performing trainable weight batching for tensor learning. Each layer is separated by an activation function *ReLU*, and a *Dropout* layer is connected to the end in order to avoid overfitting.

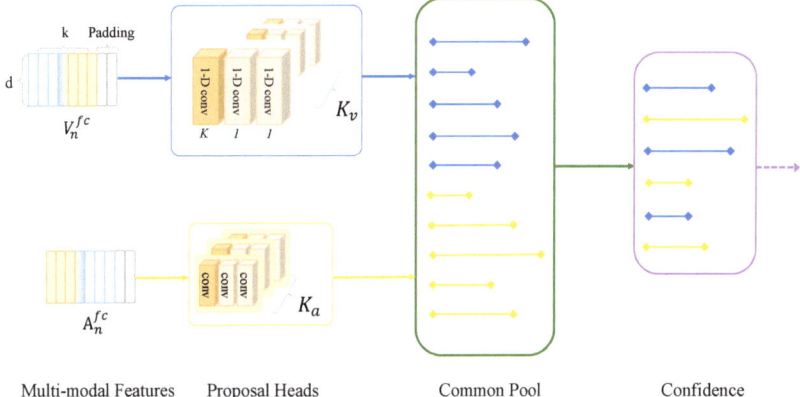

Figure 3. After the multi-modal encoder, the output features are marked in the proposal heads. The Common Pool is used to store the proposals predicted for each mode at each time step, and extract more important proposals by confidence.

The purpose of the Common Pool is to store proposals from different modalities, which are feature vectors extracted from video or audio modalities that reflect the content and semantics at each timestamp. The Common Pool puts these feature vectors into a shared space, allowing for comparison and communication between different modalities. Then, by using contrastive learning methods, the Common Pool can learn a unified representation that enables alignment and interaction between cross-modal proposals, thus providing a basis for subsequent processing and fusion.

We aim to select the most accurate proposals from the Common Pool based on their confidence score. We use the top-100 accuracy as a metric to evaluate the quality of the proposals and filter out the ones that are redundant or irrelevant. To further refine the proposals, we apply the K-means clustering algorithm with a Euclidean distance metric to estimate the optimal size of the kernel for each proposal. The kernel size is determined by predicting the threshold and receptive range that correspond to various high probability events in the feature space. We then scale the feature time span according to the clustering centroids and use them to obtain the values in grid cell coordinates. This way, we can generate more precise and compact proposals for different modalities. The details are as follows,

$$c_p = p_i + \theta(l_c), \tag{10}$$

where $p_i = (\text{start}, \text{end}, \text{confidence})$, $i \in [1, 100]$, and l_c is the length. This $\theta(l_c)$ represents the c_p (centre) position relative to position p in the sequences. The *Sigmoid* function θ ensures that the range must be $[0, 1]$.

$$\text{start} = c_p - \frac{l_c}{2}, \tag{11}$$

$$\text{end} = c_p + \frac{l_c}{2}, \tag{12}$$

$$\text{confidence} = \theta(l_o), \tag{13}$$

where the proposal p_i can be judged by the time bounds of "begin" and "end", as well as the confidence scores, as in (13). Finally, the accuracy of the top 100 is obtained and the encoder is used to process the pruned features. Overall, the feature vectors with rich event relationships can be obtained after completing the multi-modal feature fusion module, which is important for generating comprehensive and logical text descriptions in this work.

3.4. Caption Generation

The decoder part of the model takes as input the feature vector output by the multimodal feature fusion module and the previously generated word embedding representation. These inputs are then passed to the decoder layer for processing. We recommend using the LSTM as the RNN decoder part because it can effectively alleviate gradient explosion and remember long sequences. The LSTM iterates the next predicted word and generates the description text. Compared with the Transformer, LSTM also has good performance and can greatly reduce the model training time. Moreover, the output features of the last layer of the decoder can be used in the generator to predict the next caption word. The LSTM-based headline generation module is implemented by the following,

$$I_t = \left\{ \sum_{i=1}^{N} \exp(\theta(l_c)) \left(V_n^{fc}, A_n^{fc} \right), W_c^t \right\}, \quad (14)$$

where I_t represents the initial state of the current caption generation module, V_n^{fc} and A_n^{fc} represent the video feature, including part of the audio and video feature, respectively. The $\exp(\cdot)$ function is used to determine the weight of the input feature, and W_c^t denotes the subtitle word embedding feature. An input rule of LSTM is expressed in (15),

$$h_t = \text{LSTM}(y_{t-1}, h_{t-1}, I_{t-1}), \quad (15)$$

which is used to produce the current hidden state h_t. It receives the caption word embedding feature y, and h_{t-1} and l_{t-1} denote the two inter-hidden states that must be considered in each recurrent. The *Softmax* layer is used to compute the probability distribution p for the word prediction of the video caption,

$$p(y_t \mid y_{t-1}) = \text{Softmax}(y_{t-1}, h_t, I_t). \quad (16)$$

According to the predicted index of the vocabulary, the determined word can be continuously embedded and pass the net recurrent for the next word prediction until the end-of-sentence (<EOS>) symbol is received. Finally, the predictions are measured against the ground truth results, the difference is calculated using the cross-entropy function $L_{CE}(\mu)$, and the gradient is contributed to the training energy. Given g, the reference caption in the dataset is the ground truth, μ is the training parameter in the proposed model, and a complete cross-entropy function is thus expressed as follows,

$$L_{CE}(\mu) = -\sum_{t=1}^{T} \log\big(p_\mu(g_t \mid g_{t-1})\big). \quad (17)$$

4. Experiment

To evaluate the effectiveness of the proposed model, we performed experiments on the public ActivityNet Captions dataset and compared it with state-of-the-art methods.

4.1. Dataset and Data Pre-Processing

In our experiments, MSVD [61], ActivityNet Captions [31], and YouCook2 [62] are commonly used public datasets in the field of video caption. ActivityNet is mainly designed for dense video captioning tasks and covers a wide range of domains, which is exactly what our method requires. The ActivityNet Captions dataset contains 20,000 video clips with an average duration of 2 min. Each video is labeled with the events it contains, and the start and end times of each event are clearly marked, along with a manually created description of the event content. Note that some videos have been removed or altered by the original author and are no longer directly downloadable from the online resource. Following the approach used by most scholars, there are 10,024 videos in the training set, 4926 in the validation set, and 5044 in the test set. However, the labels in the test set are not yet publicly available, so we use the validation set for experiments and comparisons.

For the training preprocessing, the truecase, tokenization, and cleaning symbols must be completed, and the start mark <BOS> and end mark <EOS> must be inserted at the beginning and end of a sentence, respectively. Due to the limited size of the vocabulary and the misleading description of low-frequency word pairs, words with a frequency of less than 5 in the text are uniformly replaced by <UNK>, whose semantics are discarded and considered to be out of the vocabulary. To prevent no input from the decoder at the beginning, the <BOS> is also inserted as the first token, and the caption will be generated verbatim until a unique end token <EOS> is also inserted.

4.2. Implementation

The environment we set up was a Ubuntu 20.04 system, and we made use of PyTorch [63] as the neural network engine for our implementation. These experiments were trained on NVIDIA GeForce RTX 3070Ti GPUs with 8.0 GB of device memory. The I3D network had been pre-trained on the Kinetics dataset used in the visual feature extraction stage. The input consisted of RGB features extracted at 25.0 fps and 64 optical flow features of size 224^2. The dimension of the output features was 1024. Additionally, audio features were extracted by VGGish pre-trained on AudioSet, where the pre-classification layer embedded 128 dimensions for each feature, and configured the batch size as 32. In addition, the learning rate was initialized to 10^{-4}, and Adam [64] was used as the optimizer.

For the multi-modal feature fusion module, we took features of different sizes and mapped them to an inner space with a 1024-dimensional vector. Then, we used two different sizes of features as 128 features for visual and 48 features for audio, and stacked $N = 2$ and $H = 4$ in multi-head attention. In Section 3.3, we configured the proposal header with $K_v = K_a = 10$, while for the kernel K, we used a different size. The size of the visual modality was determined after calculating the K mean. Next, the LSTM was connected to the proposed heads that received the hidden state and performed the decoding. These results were passed to *Softmax* to compute the probability of the next word determination. In practice, the word embedding size was 468 and the learning rate was 5×10^{-5}. The localization and target loss factor was 1.0. To maintain a balance between the two modalities, we set the size of the two hidden layers of the proposal header to 512, so the input size of the fully connected layer was also 512.

4.3. Results and Analysis

To demonstrate the performance of the proposed framework, we performed validation on the ActivityNet Captions dataset and compared it with various state-of-the-art methods. We provide ablation studies to validate the impacts of the individual modules in our framework on the experimental results. We also report the results of the qualitative analysis, which highlights the superiority of our proposed model.

4.3.1. Comparison to the State-of-the-Art

We compared the proposed model with various state-of-the-art methods on the dense video caption tasks, including: EEDVC [46], DCE [31], MFT [32], WLT [49], SDVC [33], EHVC [25], MDVC [55], BMT [54], PDVC [47], and EMVC [56]. The results of the comparison are shown in Table 2.

Among them, B@N represents the evaluation metric BLEU [65], which compares the degree of overlapping of N-grams between translated results of predicted and reference. It is widely used to evaluate the level of text expression in neural machine translation (NMT). Furthermore, METEOR [36] considers the recall rate and accuracy rate based on BLEU, and uses the F-Value as the final evaluation metric. Moreover, CIDEr [66] mainly calculates the similarity between predicted and reference sentences, whose principle meets the evaluation requirements in the field of image and video caption. Higher values of these evaluation metrics indicate better performance of the generated text description.

Table 2. Comparison of the performance of our proposed method with state-of-the-art methods on the ActivityNet Captions dataset. The **bold** fonts indicate the best results.

Models	B@1	B@2	B@3	B@4	METEOR	CIDEr
EEDVC [46]	9.96	4.81	2.91	1.44	6.91	9.25
DCE [31]	10.81	4.57	1.90	0.71	5.69	12.43
MFT [32]	13.31	6.13	2.84	1.24	7.08	21.00
WLT [49]	10.00	4.20	1.85	0.90	4.93	13.79
SDVC [33]	**17.92**	7.99	2.94	0.93	8.82	-
EHVC [25]	-	-	-	1.29	7.19	14.71
MDVC [55]	12.59	5.76	2.53	1.01	7.46	7.38
BMT [54]	13.75	7.21	3.84	1.88	8.44	11.35
PDVC [47]	-	-	-	**1.96**	8.08	28.59
EMVC [56]	14.65	7.10	3.23	1.39	9.64	13.29
Proposed	16.77	**8.15**	**4.03**	1.91	**10.24**	**32.82**

All experimental results are shown in Table 2, and the proposed work achieved better results than the others. Note that the SDVC model incorporates reinforcement learning in the training process, so the score of BLEU-1 was higher than our method. It is worth noting that the proposed model outperformed both BMT and EMVC in the scores of each item, and their approaches were also based on audiovisual feature fusion for intensive video caption generation. They differ from our model in that both models use the Bi-Transformer in both the encoder and decoder, and we use the Transformer framework only in the encoder; we instead use the LSTM as the decoder for generating descriptive text. In addition, some missing parts of the ActivityNet Captions dataset did not provide complete metrics for BMT and EMVC. Even so, this still proved that our method not only required less training time, but also performed competitively.

4.3.2. Ablation Study

We conducted a large number of comparative experiments to verify the impact of different components of the proposed model on the output results, including multi-modal feature extraction, a Transformer for multi-modal feature fusion, and an LSTM for caption generation.

Table 3 indicates the impact of multi-modal features on the generated captions. The results using fused audio and visual features always outperform the best under different evaluation metrics. In all three cases, the results using pure audio modality features were the weakest, which means that the visual modality may contain more information about the video content. However, the difference between visual and audio remained fixed, but the fusion effect was good, indicating that audio played a certain role as additional feature information.

Table 3. The impact of proposed multi-modal features on generated captions. The **bold** fonts indicate the best results.

Modality	B@1	B@2	B@3	B@4	METEOR	CIDEr
Visual-only	13.71	7.08	2.58	1.15	6.98	18.36
Audio-only	12.14	6.27	2.64	1.03	5.82	15.74
Proposed	**16.77**	**8.15**	**4.03**	**1.91**	**10.24**	**32.82**

We also conducted an ablation study to verify the effectiveness of using Transformer's multi-head attention to fully fuse multi-modal features in our model. We compared our method with a baseline method that used only multi-modal features for concatenation without attention. The results are shown in Table 4. We can see that our method outperforms the baseline method on all metrics, indicating that the proposed method can achieve a more complete and robust feature fusion by using Transformer. Moreover, we can observe that the proposed method can generate more comprehensive and accurate video captions than the baseline method, as it can capture more details and nuances from both visual and audio modalities.

Table 4. Comparing the impact of different fusion methods on proposed multi-modal features. The **bold** fonts indicate the best results.

Method	B@1	B@2	B@3	B@4	METEOR	CIDEr
Concatenate	14.84	5.19	3.61	1.66	7.53	25.47
Proposed	**16.77**	**8.15**	**4.03**	**1.91**	**10.24**	**32.82**

We tested the effect of using LSTM, GRU, and the Transformer as decoders on the abstract representation of the video. As shown in Table 5, we observe that most of the methods that used Transformer as a decoder to generate text descriptions achieved slightly higher scores than the LSTM we employed. However, the difference was not significant, and our method still outperformed most of the Transformer-based methods on some metrics. Moreover, our method has the advantage of faster training and inference speed than other methods, as it requires less memory and computation. The time to generate text descriptions for each video was correspondingly shorter, which is desirable for practical applications. Therefore, we conclude that our approach is still competitive and efficient for dense video caption.

Table 5. The effect of different decoder on dense video caption. The **bold** fonts indicate the best results.

Decoder	B@1	B@2	B@3	B@4	METEOR	CIDEr
Transformer	**18.14**	**8.29**	**4.12**	1.87	**10.31**	**33.46**
GRU	15.57	6.56	3.81	1.64	8.73	28.95
LSTM	16.77	8.15	4.03	**1.91**	10.24	32.82

4.3.3. Qualitative Analysis

Furthermore, we demonstrated text description generation using the proposed multi-modal feature fusion method on the ActivityNet Captions dataset. We also compared the video captions generated with visual or audio-only features as input, as shown in Figure 4.

As the results show in Figure 4, both our method and the method using only audio features captured the keyword "futsal" at the beginning of the video time [0:00–0:15]. In contrast, the method using only visual features missed this keyword and only expressed the conversation between two people. It is impossible to obtain the specific content of the chat without the audio features as a complement, so the generated captions are not acceptable. Furthermore, at video time [1:26–1:41], there was a gap between the audio and visual features provided in this solution, so the auxiliary function of the audio features was not highlighted. The entire video lasted more than 2 min, and the captions generated by our proposed model were more detailed and accurate than the reference captions provided by the ground truth.

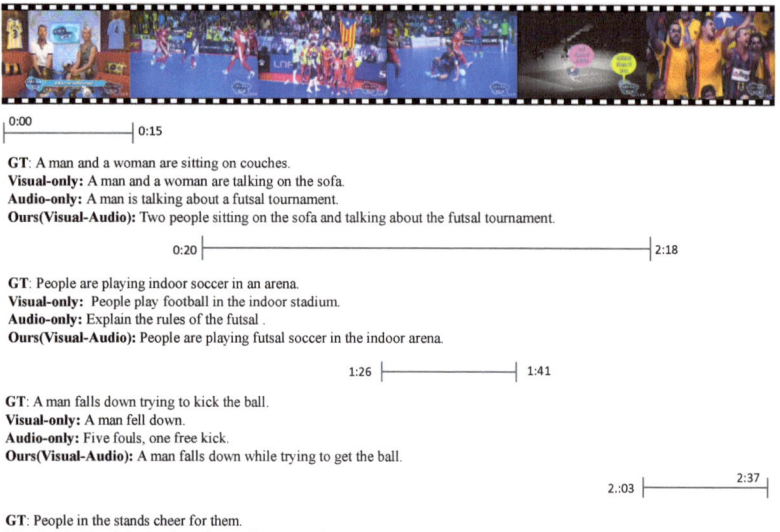

Figure 4. Results of a qualitative analysis of a video from the ActivityNet Caption validation dataset. The predicted results of the proposed model are compared to the visual-only model, the audio-only model, and the ground truth (GT) reference.

5. Conclusions

We propose a framework for enhancing the performance of dense video caption using a multi-modal feature fusion approach. The proposed framework employs I3D to extract visual features and VGGish to extract audio features from videos. The Transformer encodes the multi-modal features, and LSTM decodes them to generate descriptive text for the video. The overall framework is compact and efficient for training. To generate more accurate results, we fed the feature output from the encoder into the proposal head module. Moreover, to align the visual-audio features with different sequence lengths at each time step after the fusion, we use a Common Pool to predict and fuse each modality of every recurrent step. Furthermore, we use confidence scores to extract more consistent features for video content, which can improve the quality of the model-generated sentences.

Experiments show that the proposed multi-modal feature fusion model surpasses other approaches that use only visual or audio modality features. The model is also competitive with other dense video caption models. The text generated by the proposed model follows natural language rules, highlighting the importance of audio features for this task.

Our work still has shortcomings compared to the ground truth, and we cannot fully recognize the short-term killing parts. Future work could explore the addition of other modalities in video to enhance the expression of video content with more auxiliary information, increase the accuracy of computer understanding of videos, and improve the fine-grained caption generation.

Author Contributions: Conceptualization, X.H., W.K. and H.S.; methodology, X.H., W.K. and H.S.; software, W.K. and H.S.; validation, X.H., W.K. and H.S.; formal analysis, X.H., K.-H.C., W.K., and H.S. and W.W.; investigation, X.H.; resources, W.K. and H.S.; data curation, X.H., W.K. and H.S.; writing—original draft preparation, X.H., K.-H.C., W.K., H.S. and W.W.; writing—review and editing, X.H., K.-H.C., W.K., H.S. and W.W.; visualization, X.H., W.K. and H.S.; supervision, W.K. and H.S.; project administration, W.K. and H.S.; funding acquisition, K.-H.C., W.K. and H.S. All authors have read and agreed to the published version of the manuscript.

Funding: This work is partially supported by the National Key R&D Program of China (No. 2019YFB2101600), the National Natural Science Foundation of China (No. 61872025), the Macao Polytechnic University (RP/ESCA-03/2020, RP/FCA-06/2023), and the Open Fund of the State Key Laboratory of Software Development Environment (No. SKLSDE-2021ZX-03).

Institutional Review Board Statement: Not applicable.

Informed Consent Statement: Not applicable.

Data Availability Statement: Not applicable.

Acknowledgments: Thanks for the support from Macao Polytechnic University and HAWKEYE Group.

Conflicts of Interest: The authors declare no conflict of interest.

References

1. Jain, A.K.; Sahoo, S.R.; Kaubiyal, J. Online social networks security and privacy: Comprehensive review and analysis. *Complex Intell. Syst.* **2021**, *7*, 2157–2177. [CrossRef]
2. Wu, Y.; Sheng, H.; Zhang, Y.; Wang, S.; Xiong, Z.; Ke, W. Hybrid motion model for multiple object tracking in mobile devices. *IEEE Internet Things J.* **2022**, *10*, 4735–4748. [CrossRef]
3. Sheng, H.; Lv, K.; Liu, Y.; Ke, W.; Lyu, W.; Xiong, Z.; Li, W. Combining pose invariant and discriminative features for vehicle reidentification. *IEEE Internet Things J.* **2020**, *8*, 3189–3200. [CrossRef]
4. Shapiro, L.G. Computer vision: The last 50 years. *Int. J. Parallel Emerg. Distrib. Syst.* **2018**, *35*, 112–117. [CrossRef]
5. Wang, S.; Sheng, H.; Yang, D.; Zhang, Y.; Wu, Y.; Wang, S. Extendable multiple nodes recurrent tracking framework with RTU++. *IEEE Trans. Image Process.* **2022**, *31*, 5257–5271. [CrossRef]
6. Sheng, H.; Wang, S.; Zhang, Y.; Yu, D.; Cheng, X.; Lyu, W.; Xiong, Z. Near-online tracking with co-occurrence constraints in blockchain-based edge computing. *IEEE Internet Things J.* **2020**, *8*, 2193–2207. [CrossRef]
7. Zhang, W.; Ke, W.; Yang, D.; Sheng, H.; Xiong, Z. Light field super-resolution using complementary-view feature attention. *Comput. Vis. Media* **2023**. [CrossRef]
8. Chowdhary, K.R. Natural Language Processing. In *Fundamentals of Artificial Intelligence*; Springer: Delhi, India, 2020; pp. 603–649. [CrossRef]
9. Chan, K.H.; Im, S.K.; Pau, G. Applying and Optimizing NLP Model with CARU. In Proceedings of the 2022 8th International Conference on Advanced Computing and Communication Systems (ICACCS), Coimbatore, India, 25–26 March 2022. [CrossRef]
10. Ke, W.; Chan, K.H. A Multilayer CARU Framework to Obtain Probability Distribution for Paragraph-Based Sentiment Analysis. *Appl. Sci.* **2021**, *11*, 11344. [CrossRef]
11. Sheng, H.; Zheng, Y.; Ke, W.; Yu, D.; Cheng, X.; Lyu, W.; Xiong, Z. Mining hard samples globally and efficiently for person reidentification. *IEEE Internet Things J.* **2020**, *7*, 9611–9622. [CrossRef]
12. LeCun, Y.; Bengio, Y.; Hinton, G. Deep learning. *Nature* **2015**, *521*, 436–444. [CrossRef] [PubMed]
13. Sawarn, A.; Srivastava, S.; Gupta, M.; Srivastava, S. BeamAtt: Generating Medical Diagnosis from Chest X-rays Using Sampling-Based Intelligence. In *EAI/Springer Innovations in Communication and Computing*; Springer International Publishing: Berlin/Heidelberg, Germany, 2021; pp. 135–150. [CrossRef]
14. Pan, Y.; Wang, L.; Duan, S.; Gan, X.; Hong, L. Chinese image caption of Inceptionv4 and double-layer GRUs based on attention mechanism. *J. Phys. Conf. Ser.* **2021**, *1861*, 012044. [CrossRef]
15. Wang, S.; Sheng, H.; Zhang, Y.; Wu, Y.; Xiong, Z. A general recurrent tracking framework without real data. In Proceedings of the 2021 IEEE/CVF International Conference on Computer Vision (ICCV), Montreal, BC, Canada, 10–17 October 2021; pp. 13219–13228. [CrossRef]
16. Zhang, S.; Lin, Y.; Sheng, H. Residual networks for light field image super-resolution. In Proceedings of the 2019 IEEE/CVF Conference on Computer Vision and Pattern Recognition (CVPR), Long Beach, CA, USA, 15–20 June 2019; pp. 11046–11055. [CrossRef]
17. Jiao, Y.; Chen, S.; Jie, Z.; Chen, J.; Ma, L.; Jiang, Y.G. More: Multi-order relation mining for dense captioning in 3d scenes. In Proceedings of the Computer Vision—ECCV, Tel Aviv, Israel, 23–27 October 2022; Springer: Cham, Switzerland, 2022; pp. 528–545. [CrossRef]
18. Venugopalan, S.; Xu, H.; Donahue, J.; Rohrbach, M.; Mooney, R.; Saenko, K. Translating Videos to Natural Language Using Deep Recurrent Neural Networks. In Proceedings of the 2015 Conference of the North American Chapter of the Association for Computational Linguistics: Human Language Technologies, Denver, CO, USA, 31 May–5 June 2015; Association for Computational Linguistics: Toronto, ON, Canada, 2015. [CrossRef]
19. Huang, X.; Ke, W.; Sheng, H. Enhancing Efficiency and Quality of Image Caption Generation with CARU. In *Wireless Algorithms, Systems, and Applications*; Springer Nature: Cham, Switzerland, 2022; pp. 450–459. [CrossRef]
20. Aafaq, N.; Mian, A.S.; Akhtar, N.; Liu, W.; Shah, M. Dense video captioning with early linguistic information fusion. *IEEE Trans. Multimed.* **2022**. [CrossRef]

21. Venugopalan, S.; Rohrbach, M.; Donahue, J.; Mooney, R.; Darrell, T.; Saenko, K. Sequence to Sequence—Video to Text. In Proceedings of the 2015 IEEE International Conference on Computer Vision (ICCV), Santiago, Chile, 11–18 December 2015. [CrossRef]
22. Simonyan, K.; Zisserman, A. Very Deep Convolutional Networks for Large-Scale Image Recognition. *arXiv* **2014**, arXiv:1409.1556. [CrossRef]
23. Hochreiter, S.; Schmidhuber, J. Long Short-Term Memory. *Neural Comput.* **1997**, *9*, 1735–1780. [CrossRef] [PubMed]
24. Tran, D.; Bourdev, L.; Fergus, R.; Torresani, L.; Paluri, M. Learning Spatiotemporal Features with 3D Convolutional Networks. In Proceedings of the 2015 IEEE International Conference on Computer Vision (ICCV), Santiago, Chile, 11–18 December 2015. [CrossRef]
25. Wang, T.; Zheng, H.; Yu, M.; Tian, Q.; Hu, H. Event-Centric Hierarchical Representation for Dense Video Captioning. *IEEE Trans. Circuits Syst. Video Technol.* **2021**, *31*, 1890–1900. [CrossRef]
26. Hara, K.; Kataoka, H.; Satoh, Y. Can Spatiotemporal 3D CNNs Retrace the History of 2D CNNs and ImageNet? In Proceedings of the 2018 IEEE/CVF Conference on Computer Vision and Pattern Recognition, Salt Lake City, UT, USA, 18–22 June 2018. [CrossRef]
27. Zhang, Y.; Sheng, H.; Wu, Y.; Wang, S.; Lyu, W.; Ke, W.; Xiong, Z. Long-term tracking with deep tracklet association. *IEEE Trans. Image Process.* **2020**, *29*, 6694–6706. [CrossRef]
28. Wang, S.; Yang, D.; Wu, Y.; Liu, Y.; Sheng, H. Tracking Game: Self-adaptative Agent based Multi-object Tracking. In Proceedings of the Proceedings of the 30th ACM International Conference on Multimedia, Lisboa, Portugal, 10–14 October 2022; ACM: New York, NY, USA, 2022; pp. 1964–1972. [CrossRef]
29. Carreira, J.; Zisserman, A. Quo Vadis, Action Recognition? A New Model and the Kinetics Dataset. In Proceedings of the 2017 IEEE Conference on Computer Vision and Pattern Recognition (CVPR), Honolulu, HI, USA, 21–26 July 2017. [CrossRef]
30. Qiu, Z.; Yao, T.; Mei, T. Learning Spatio-Temporal Representation with Pseudo-3D Residual Networks. In Proceedings of the 2017 IEEE International Conference on Computer Vision (ICCV), Venice, Italy, 22–29 October 2017. [CrossRef]
31. Krishna, R.; Hata, K.; Ren, F.; Fei-Fei, L.; Niebles, J.C. Dense-Captioning Events in Videos. In Proceedings of the 2017 IEEE International Conference on Computer Vision (ICCV), Venice, Italy, 22–29 October 2017. [CrossRef]
32. Xiong, Y.; Dai, B.; Lin, D. Move Forward and Tell: A Progressive Generator of Video Descriptions. In Proceedings of the European Conference on Computer Vision (ECCV) 2018, Munich, Germany, 8–14 September 2018; Springer International Publishing: Berlin/Heidelberg, Germany, 2018; pp. 489–505. [CrossRef]
33. Mun, J.; Yang, L.; Ren, Z.; Xu, N.; Han, B. Streamlined Dense Video Captioning. In Proceedings of the 2019 IEEE/CVF Conference on Computer Vision and Pattern Recognition (CVPR), Long Beach, CA, USA, 15–20 June 2019. [CrossRef]
34. Yu, H.; Wang, J.; Huang, Z.; Yang, Y.; Xu, W. Video Paragraph Captioning Using Hierarchical Recurrent Neural Networks. In Proceedings of the 2016 IEEE Conference on Computer Vision and Pattern Recognition (CVPR), Las Vegas, NV, USA, 26 June–1 July 2016. [CrossRef]
35. Buch, S.; Escorcia, V.; Shen, C.; Ghanem, B.; Niebles, J.C. SST: Single-Stream Temporal Action Proposals. In Proceedings of the 2017 IEEE Conference on Computer Vision and Pattern Recognition (CVPR), Honolulu, HI, USA, 21–26 July 2017. [CrossRef]
36. Banerjee, S.; Lavie, A. METEOR: An Automatic Metric for MT Evaluation with Improved Correlation with Human Judgments. In Proceedings of the ACL Workshop on Intrinsic and Extrinsic Evaluation Measures for Machine Translation and/or Summarization, Ann Arbor, MI, USA, 29 June 2005; Association for Computational Linguistics: Toronto, ON, Canada, 2005; pp. 65–72.
37. Pan, Y.; Mei, T.; Yao, T.; Li, H.; Rui, Y. Jointly Modeling Embedding and Translation to Bridge Video and Language. In Proceedings of the 2016 IEEE Conference on Computer Vision and Pattern Recognition (CVPR), Las Vegas, NV, USA, 26 June–1 July 2016. [CrossRef]
38. Baraldi, L.; Grana, C.; Cucchiara, R. Hierarchical Boundary-Aware Neural Encoder for Video Captioning. In Proceedings of the 2017 IEEE Conference on Computer Vision and Pattern Recognition (CVPR), Honolulu, HI, USA, 21–26 July 2017. [CrossRef]
39. Yao, L.; Torabi, A.; Cho, K.; Ballas, N.; Pal, C.; Larochelle, H.; Courville, A. Video description generation incorporating spatio-temporal features and a soft-attention mechanism. *arXiv* **2015**, arXiv:1502.08029. [CrossRef]
40. Cherian, A.; Wang, J.; Hori, C.; Marks, T.K. Spatio-Temporal Ranked-Attention Networks for Video Captioning. In Proceedings of the 2020 IEEE Winter Conference on Applications of Computer Vision (WACV), Snowmass Village, CO, USA, 1–5 March 2020. [CrossRef]
41. Gabeur, V.; Sun, C.; Alahari, K.; Schmid, C. Multi-modal Transformer for Video Retrieval. In *Computer Vision—ECCV 2020*; Springer International Publishing: Berlin/Heidelberg, Germany, 2020; pp. 214–229. [CrossRef]
42. Yu, Z.; Han, N. Accelerated masked transformer for dense video captioning. *Neurocomputing* **2021**, *445*, 72–80. [CrossRef]
43. Lin, K.; Li, L.; Lin, C.C.; Ahmed, F.; Gan, Z.; Liu, Z.; Lu, Y.; Wang, L. SwinBERT: End-to-End Transformers with Sparse Attention for Video Captioning. *arXiv* **2021**, arXiv:2111.13196. [CrossRef]
44. Zhang, S.; Sheng, H.; Yang, D.; Zhang, J.; Xiong, Z. Micro-lens-based matching for scene recovery in lenslet cameras. *IEEE Trans. Image Process.* **2017**, *27*, 1060–1075. [CrossRef] [PubMed]
45. Zhong, R.; Zhang, Q.; Zuo, M. Enhanced visual multi-modal fusion framework for dense video captioning. *Res. Sq.* **2023**, *in press*. [CrossRef]

46. Zhou, L.; Zhou, Y.; Corso, J.J.; Socher, R.; Xiong, C. End-to-End Dense Video Captioning with Masked Transformer. In Proceedings of the 2018 IEEE/CVF Conference on Computer Vision and Pattern Recognition, Salt Lake City, UT, USA, 18–23 June 2018. [CrossRef]
47. Wang, T.; Zhang, R.; Lu, Z.; Zheng, F.; Cheng, R.; Luo, P. End-to-End Dense Video Captioning with Parallel Decoding. In Proceedings of the 2021 IEEE/CVF International Conference on Computer Vision (ICCV), Montreal, BC, Canada, 11–17 October 2021. [CrossRef]
48. Song, Y.; Chen, S.; Jin, Q. Towards diverse paragraph captioning for untrimmed videos. In Proceedings of the 2021 IEEE/CVF Conference on Computer Vision and Pattern Recognition, Nashville, TN, USA, 20–25 June 2021; pp. 11245–11254. [CrossRef]
49. Rahman, T.; Xu, B.; Sigal, L. Watch, Listen and Tell: Multi-Modal Weakly Supervised Dense Event Captioning. In Proceedings of the 2019 IEEE/CVF International Conference on Computer Vision (ICCV), Seoul, Republic of Korea, 27 October–2 November 2019. [CrossRef]
50. Jin, Q.; Chen, J.; Chen, S.; Xiong, Y.; Hauptmann, A. Describing videos using multi-modal fusion. In Proceedings of the 24th ACM International Conference on Multimedia, Amsterdam, The Netherlands, 15–19 October 2016; ACM: New York, NY, USA, 2016; pp. 1087–1091. [CrossRef]
51. Chen, S.; Jin, Q.; Chen, J.; Hauptmann, A.G. Generating Video Descriptions with Latent Topic Guidance. *IEEE Trans. Multimed.* **2019**, *21*, 2407–2418. [CrossRef]
52. Martinez, J.; Perez, H.; Escamilla, E.; Suzuki, M.M. Speaker recognition using Mel frequency Cepstral Coefficients (MFCC) and Vector quantization (VQ) techniques. In Proceedings of the CONIELECOMP 2012, 22nd International Conference on Electrical Communications and Computers, Cholula, Mexico, 27–29 February 2012. [CrossRef]
53. Hershey, S.; Chaudhuri, S.; Ellis, D.P.W.; Gemmeke, J.F.; Jansen, A.; Moore, R.C.; Plakal, M.; Platt, D.; Saurous, R.A.; Seybold, B.; et al. CNN architectures for large-scale audio classification. In Proceedings of the 2017 IEEE International Conference on Acoustics, Speech and Signal Processing (ICASSP), New Orleans, LA, USA, 5–9 March 2017. [CrossRef]
54. Iashin, V.; Rahtu, E. A better use of audio-visual cues: Dense video captioning with bi-modal transformer. *arXiv* **2020**, arXiv:2005.08271. [CrossRef]
55. Iashin, V.; Rahtu, E. Multi-modal dense video captioning. In Proceedings of the 2020 IEEE/CVF Conference on Computer Vision and Pattern Recognition Workshops (CVPRW), Seattle, WA, USA, 14–19 June 2020; pp. 958–959. [CrossRef]
56. Chang, Z.; Zhao, D.; Chen, H.; Li, J.; Liu, P. Event-centric multi-modal fusion method for dense video captioning. *Neural Netw.* **2022**, *146*, 120–129. [CrossRef] [PubMed]
57. Hao, W.; Zhang, Z.; Guan, H. Integrating both visual and audio cues for enhanced video caption. In Proceedings of the Thirty-Second AAAI Conference on Artificial Intelligence, New Orleans, LA, USA, 2–7 February 2018; Volume 32. [CrossRef]
58. Park, J.S.; Darrell, T.; Rohrbach, A. Identity-Aware Multi-sentence Video Description. In *Computer Vision—ECCV 2020*; Springer International Publishing: Berlin/Heidelberg, Germany, 2020; pp. 360–378. [CrossRef]
59. Carreira, J.; Noland, E.; Hillier, C.; Zisserman, A. A Short Note on the Kinetics-700 Human Action Dataset. *arXiv* **2019**, arXiv:1907.06987. [CrossRef]
60. Gemmeke, J.F.; Ellis, D.P.W.; Freedman, D.; Jansen, A.; Lawrence, W.; Moore, R.C.; Plakal, M.; Ritter, M. Audio Set: An ontology and human-labeled dataset for audio events. In Proceedings of the 2017 IEEE International Conference on Acoustics, Speech and Signal Processing (ICASSP), New Orleans, LA, USA, 5–9 March 2017. [CrossRef]
61. Chen, D.; Dolan, W. Collecting Highly Parallel Data for Paraphrase Evaluation. In Proceedings of the 49th Annual Meeting of the Association for Computational Linguistics: Human Language Technologies, Portland, OR, USA, 19–24 June 2011; pp. 190–200.
62. Zhou, L.; Xu, C.; Corso, J. Towards Automatic Learning of Procedures from Web Instructional Videos. In Proceedings of the Thirty-Second AAAI Conference on Artificial Intelligence, New Orleans, LA, USA, 2–7 February 2018; Volume 32. [CrossRef]
63. Paszke, A.; Gross, S.; Massa, F.; Lerer, A.; Bradbury, J.; Chanan, G.; Killeen, T.; Lin, Z.; Gimelshein, N.; Antiga, L.; et al. PyTorch: An Imperative Style, High-Performance Deep Learning Library. In *Advances in Neural Information Processing Systems 32*; Wallach, H., Larochelle, H., Beygelzimer, A., d'Alché Buc, F., Fox, E., Garnett, R., Eds.; Curran Associates, Inc.: Red Hook, NY, USA, 2019; pp. 8024–8035.
64. Kingma, D.P.; Ba, J. Adam: A Method for Stochastic Optimization. *arXiv* **2014**, arXiv:1412.6980. [CrossRef]
65. Papineni, K.; Roukos, S.; Ward, T.; Zhu, W.J. BLEU: A method for automatic evaluation of machine translation. In Proceedings of the 40th Annual Meeting on Association for Computational Linguistics—ACL'02, Philadelphia, PA, USA, 7–12 July 2002; Association for Computational Linguistics: Stroudsburg, PA, USA, 2002. [CrossRef]
66. Vedantam, R.; Zitnick, C.L.; Parikh, D. CIDEr: Consensus-based image description evaluation. In Proceedings of the 2015 IEEE Conference on Computer Vision and Pattern Recognition (CVPR), Boston, MA, USA, 7–12 June 2015. [CrossRef]

Disclaimer/Publisher's Note: The statements, opinions and data contained in all publications are solely those of the individual author(s) and contributor(s) and not of MDPI and/or the editor(s). MDPI and/or the editor(s) disclaim responsibility for any injury to people or property resulting from any ideas, methods, instructions or products referred to in the content.

Article

Chained Deep Learning Using Generalized Cross-Entropy for Multiple Annotators Classification

Jenniffer Carolina Triana-Martinez [1,†], Julian Gil-González [2,†], Jose A. Fernandez-Gallego [3], Andrés Marino Álvarez-Meza [1,*] and Cesar German Castellanos-Dominguez [1]

1. Signal Processing and Recognition Group, Universidad Nacional de Colombia, Manizales 170003, Colombia; jectrianama@unal.edu.co (J.C.T.-M.)
2. Department of Electronics and Computer Science, Pontificia Universidad Javeriana Cali, Cali 760031, Colombia
3. Programa de Ingeniería Electrónica, Facultad de Ingeniería, Universidad de Ibagué, Ibagué 730001, Colombia
* Correspondence: amalvarezme@unal.edu.co
† These authors contributed equally to this work.

Abstract: Supervised learning requires the accurate labeling of instances, usually provided by an expert. Crowdsourcing platforms offer a practical and cost-effective alternative for large datasets when individual annotation is impractical. In addition, these platforms gather labels from multiple labelers. Still, traditional multiple-annotator methods must account for the varying levels of expertise and the noise introduced by unreliable outputs, resulting in decreased performance. In addition, they assume a homogeneous behavior of the labelers across the input feature space, and independence constraints are imposed on outputs. We propose a Generalized Cross-Entropy-based framework using Chained Deep Learning (GCECDL) to code each annotator's non-stationary patterns regarding the input space while preserving the inter-dependencies among experts through a chained deep learning approach. Experimental results devoted to multiple-annotator classification tasks on several well-known datasets demonstrate that our GCECDL can achieve robust predictive properties, outperforming state-of-the-art algorithms by combining the power of deep learning with a noise-robust loss function to deal with noisy labels. Moreover, network self-regularization is achieved by estimating each labeler's reliability within the chained approach. Lastly, visual inspection and relevance analysis experiments are conducted to reveal the non-stationary coding of our method. In a nutshell, GCEDL weights reliable labelers as a function of each input sample and achieves suitable discrimination performance with preserved interpretability regarding each annotator's trustworthiness estimation.

Keywords: deep learning; multiple annotators; chained approach; generalized cross-entropy; classification

Citation: Triana-Martinez, J.C.; Gil-Gonzalez, J.; Fernandez-Gallego, J.A.; Álvarez-Meza, A.M.; Castellanos-Dominguez, C.G. Chained Deep Learning Using Generalized Cross-Entropy for Multiple Annotators Classification. *Sensors* **2023**, *23*, 3518. https://doi.org/10.3390/s23073518

Academic Editor: Anastasios Doulamis

Received: 24 January 2023
Revised: 22 February 2023
Accepted: 16 March 2023
Published: 28 March 2023

Copyright: © 2023 by the authors. Licensee MDPI, Basel, Switzerland. This article is an open access article distributed under the terms and conditions of the Creative Commons Attribution (CC BY) license (https://creativecommons.org/licenses/by/4.0/).

1. Introduction

Conventional Machine Learning (ML) and Deep Learning (DL) techniques utilize a prediction function that maps input data to output targets. In supervised tasks, output values (or "ground truth") are available for training, but in many real-world scenarios, these values may be unknown or too costly to obtain [1]. With the rise of DL-based approaches, there has been an increasing interest in their use as the primary tool in various classification and regression tasks [2]. However, a crucial factor that dramatically impacts the performance of DL models is the quantity and quality of labeled data used during training [3]. Concerning this, crowdsourcing is a widely recognized approach for obtaining labeled data cost-effectively and efficiently from multiple annotators. It involves the use of online platforms, such as Amazon Mechanical Turk (AMT), to recruit a large number of individuals (annotators) to label the data and provide their subjective interpretation of the unknown ground truth [4].

In the ML field, assigning labels to instances with the help of multiple annotators is a common practice. However, it presents a significant challenge when traditional supervised algorithms are applied because they rely on the assumption that the training labels provided by a single expert are reliable [5]. When multiple annotators with varying levels of expertise are employed, the reliability of the labels becomes uncertain, leading to decreased performance and inaccurate model predictions. Addressing this issue is crucial for developing practical ML and DL models that perform well on real-world datasets.

Traditional supervised learning algorithms cannot account for the varying levels of expertise and the noise introduced by unreliable labels, resulting in decreased performance. Further research is needed to develop methods that can effectively handle the problem of multi-annotator label aggregation and overcome this challenge. Moreover, several issues arise when using DL in a multiple annotator scenario. One of the main challenges is the variability of annotator effectiveness, which may depend on the sample instance presented. Even for a single task, annotators can provide inconsistent or incorrect outputs, leading to noisy labels [6]. These samples can negatively impact the model's performance by affecting the gradients and making it difficult to converge on a suitable solution [7,8].

Learning from Crowds (LFC) scenarios pose a significant challenge for ML models, and multiple approaches have been developed to tackle this issue. The most commonly used strategy is to adapt supervised learning algorithms and use majority voting for label aggregation. Yet, this approach has limitations since it assumes that annotators have the same level of reliability [9,10]. In addition, incorrect labels and outliers can influence the consensus, decreasing performance. Therefore, more advanced models, such as the Expectation-Maximization (EM) framework, have been considered to address these issues [11,12]. This approach simultaneously estimates true labels and annotator reliability, making more accurate predictions. Another strategy is to train a supervised learning algorithm while also modeling annotator behavior [13]. This approach yields better results than label aggregation and can be used to identify unreliable labelers and remove their outputs from the training set. On the other hand, recent work has shown that relaxing the independence assumption among annotators can lead to more accurate ground truth estimation [14,15]. Then, sophisticated models have been proposed, such as those that use regression tasks to model annotator behavior employing a multivariate Gaussian distribution [16,17]. Moreover, such techniques can help identify the relationships among experts and improve the overall accuracy of the predictions.

Furthermore, learning with noisy labels is a challenging problem in ML and DL. Recently, numerous methods have been proposed for learning noisy labels with DL-based approaches [18]. These methods can involve developing a robust architecture [19,20], enforcing a DL-model with robust regularization techniques [8] and identifying true labeled examples from noisy training data via multi-network [21,22] or multi-round learning [23,24]. In addition, sparse loss functions such as the Mean Absolute Error (MAE) are employed [25,26]. The MAE can help the model focus on correctly labeled examples; however, performance decreases when it is used on large and complex databases [26].

Here, we introduce a Chained Deep Learning (CDL) strategy to learn from multiple noisy annotators in classification tasks. Such an approach allows coding the non-stationary patterns of each annotator regarding the input space while preserving the inter-dependencies among experts. In addition, we combine the capabilities of CDL with a Generalized Cross-Entropy-based loss function aiming to build a model, termed the Generalized Cross-Entropy-based Framework using CDL (GCECDL), that is less prone to outlier annotations. Our proposal is similar to the works in [27,28] in that we use a deep-learning-based strategy to build a supervised learning model in the context of multiple annotators. Moreover, network self-regularization is accomplished by predicting each labeler's reliability within our chained scheme. On the other hand, the proposed research uses t-distributed Stochastic Neighbor Embedding- (t-SNE) [29] and Gradient-based Class Activation Maps [30] to interpret and validate the obtained results visually. Finally, experimental results related to multiple-annotator classification on several well-known datasets (synthetic and real-world scenarios) demonstrate that GCECDL outperforms state-of-the-art techniques.

The agenda for this paper is as follows: Section 2 summarizes the related work. Section 3 describes the methods. Sections 4 and 5 present the experiments and discuss the results. Finally, Section 6 outlines the conclusions and future work.

2. Literature Review

Several approaches have been developed to address LFC scenarios. In this light, we recognize two main groups: combining the annotations to estimate the gold standard or adapting supervised learning algorithms to that type of labels [31]. The primary method is called label aggregation. One of the most used techniques is known as Majority Voting (MV), which has been applied to different multi-labeler problems due to its simplicity [32]. In MV, the most frequent output among the experts is chosen as the final prediction. The latter is simple to implement and effective in some cases, but it also has limitations because MV relies on the assumption that annotators have the same level of reliability, which is challenging to fulfill in real-world scenarios. Additionally, the consensus can be heavily influenced by incorrect labels and outliers [10]. In this sense, EM methods have been proposed to handle imbalanced labeling to handle these issues [32,33]. The EM framework aims to estimate the true label and the annotator's reliability simultaneously, while methods to handle imbalanced labeling try to adjust for the differences in the annotator's expertise level. These more advanced models provide more robust solutions to the problem of multi-annotator label aggregation and can lead to better performance than MV.

An alternate approach to addressing multi-labeler tasks is to simultaneously train a supervised learning algorithm while also modeling the behavior of the annotators. It has been empirically demonstrated that this approach yields better results than label aggregation. Furthermore, features that train the learning algorithm can also be exploited to infer the ground truth [13]. One introductory study in this field is the EM-based framework presented by authors in [34], which estimates the sensitivity and specificity of annotators while also training a logistic regression classifier. This approach has served as a foundation for various algorithms that address multi-labeler tasks, including regression [35,36], binary classification [37,38], multi-class classification [4,39], and sequence labeling [40]. Likewise, some studies have adapted these ideas for DL techniques by incorporating an additional layer that contemplates multiple labelers [27,28].

In turn, the study in [15] represents an early exploration of the relationship between annotators' parameters and input features. The authors propose a method for binary classification with multiple labelers, where input data are grouped using a Gaussian Mixture Model (GMM). The algorithm posits that annotators have specific performance levels in terms of sensitivity and specificity. However, it does not incorporate information from multiple experts as input for GMM, which may result in labelers' parameter deviation. Likewise, the authors in [6] presented a binary classification algorithm that employs Bernoulli and Gaussian distributions to code the annotators' performance as a function of the input space. In addition, a linear relationship between the annotator's expertise and the input space is assumed, which can be problematic. For example, when assessing documents online, annotators may have varying levels of labeling accuracy. These differences could be due to their familiarity with specific topics related to the studied documents [41]. Moreover, in [36], a Gaussian Process (GP)-based regression procedure is proposed to incorporate multiple annotators. The annotators' parameters are estimated as a nonlinear function of the input space by using an additional GP. Nevertheless, since this approach is based on a classical formulation of GPs, its computational complexity is prohibited for large datasets [39]. Furthermore, relaxing the independence assumption among annotators has led to a more accurate estimation of the ground truth, as demonstrated in [14,15]. In [17], an unsupervised regression task is described where the labelers' behavior is modeled using a multivariate Gaussian distribution, with the covariance matrix encoding the annotators' interdependencies. Again, the authors in [38] proposed a binary classification method that utilizes a weighted combination of classifiers. The weights are estimated using a kernel alignment-based algorithm.

Of note, when multiple annotators are present, the training set may contain noisy labels, negatively impacting the model's generalization ability. Typical regularization approaches, such as dropout and batch normalization, only partially outweigh the overfitting drawback for DL [18]. An alternative is to use more robust loss functions to train DL models. Because of its fast convergence and generalization capability, most deep learning-based classifiers use Categorical Cross-Entropy (CE) as cost function. Nevertheless, MAE has been found to perform better when dealing with noisy labels [26]. However, the robustness of MAE can concurrently cause increased difficulty in training and lead to a performance drop. This limitation is particularly evident when using neural networks on complicated datasets. To combat this drawback, Zhang et al. [42] proposed the GCE, establishing a more general type of noise-robust loss function taking advantage of both MAE and CE, yielding good performance in the presence of noisy labels. Moreover, it can be readily applied to any existing neural network architecture.

Our proposal follows the lines of the work in [27,28] in that GCECDL uses a deep-based approach to build a supervised learning model in the context of multiple-annotator classification. Yet, while such approaches code the annotators' parameters as fixed points, we model them as functions to consider dependencies between the input features and the labelers' behavior. GCECDL is also similar to the works in [14,43]. Both approaches model the annotators' performance as a function of the input instances and consider the interdependencies among the labelers. Even so, unlike [14], where it is necessary to use as many classifiers as annotators, our approach only needs to train a single classifier from a DL representation, which is advantageous for a large number of labelers. Moreover, unlike [43], our loss function can deal with noisy labels and more difficult training scenarios by using a generalization between MAE and CE. Indeed, network self-regularization is accomplished by predicting each labeler's reliability.

3. Methods

3.1. Generalized Cross-Entropy

Let us consider a K-class classification problem from a given prediction function $f : \mathbb{R}^P \to [0,1]^K$, trained on the input–output set $\{X \in \mathbb{R}^{N \times P}, Y \in \{0,1\}^{N \times K}\}$ and holding P—dimensional input features in N row vectors $x_n \in \mathbb{R}^P$ corresponding to each ground truth label $y_n \in \{0,1\}^K, n \in \{1,2,\ldots,N\}$. The prediction function $f(\cdot)$ is commonly coupled by a softmax output to fulfill $1 - K$ one-hot labels. In turn, the well-known Mean Absolute Error (MAE) and Categorical Cross-Entropy (CE) losses, used typically for optimizing f, are defined as follows:

$$MAE(y, f(x)) = \|y - f(x)\|_1, \tag{1}$$

$$CE(y, f(x)) = \sum_{k=1}^{K} y_k \log(f_k(x)); \tag{2}$$

where $y_k \in y$, $f_k(x) \in f(x)$, and $\|\cdot\|_1$ stands for the l_1-norm. Of note, $\mathbf{1}^\top y = \mathbf{1}^\top f(x) = 1$, $\mathbf{1} \in \{1\}^K$ being an all-ones vector. In addition, the MAE loss can be rewritten for softmax outputs, yielding:

$$MAE(y, f(x)) = 2(1 - \mathbf{1}^\top (y \odot f(x))) \tag{3}$$

where $\odot$ stands for the Hadamard product.

On the one hand, CE is sensitive to label noise, being a nonsymmetric and unbounded loss function. On the other hand, MAE is noise-robust because of its symmetric property, that is [26]:

$$\sum_{k=1}^{K} MAE(y, f(x)|k) = \sum_{k=1}^{K} 2(1 - f_k(x)) = C, \quad \forall x \in \mathbb{R}^P, \forall f; \tag{4}$$

where $C \in \mathbb{R}^+$. The symmetric property of MAE for softmax-based outputs allows extending the L1-norm expression in Equation (1) to a vectorized form in Equation (3). Note that the symmetric property is only fulfilled for softmax-based representations. Therefore, the L1-norm favors sparse coding when computing the mismatch between target and prediction, favoring the filtering of noisy outputs, as commonly studied for L1-based filtering approaches [44].

Though MAE is symmetric and bounded, it also has some drawbacks when used as classification loss for deep learning networks trained on large datasets employing stochastic gradient-based techniques. Specifically, for a given network with parameter set θ, the MAE and CE gradients can be computed as:

$$\frac{\partial MAE(y, f(x;\theta)|k)}{\partial \theta} = -\nabla_\theta f_k(x;\theta), \tag{5}$$

$$\frac{\partial CE(y, f(x;\theta)|k)}{\partial \theta} = -\frac{1}{f_k(x;\theta)}\nabla_\theta f_k(x;\theta). \tag{6}$$

As seen in Equations (5) and (6), less congruent samples have greater weights in CE than predictions that agree more with ground truth labels; meanwhile, the MAE penalizes equally during gradient descent optimization. At first glance, MAE can deal with noisy labels; still, this can lead to longer and more difficult training scenarios, particularly for large databases.

Therefore, authors in [42] proposed a trade-off between MAE and CE using a Box–Cox transformation [45], yielding to the following Generalized Cross-Entropy (GCE) loss for training deep learning models:

$$GCE(y, f(x)) = 2\frac{1 - \left(\mathbf{1}^\top(y \odot f(x))\right)^q}{q}, \tag{7}$$

with $q \in (0, 1]$. Remarkably, the limiting case for $q \to 0$ in GCE is equivalent to the CE expression, and when $q = 1$, it equals the MAE loss. In addition, the GCE in Equation (7) holds the following gradient with regard to θ:

$$\frac{\partial GCE(y, f(x;\theta)|k)}{\partial \theta} = -f_k(x;\theta)^{q-1}\nabla_\theta f_k(x;\theta). \tag{8}$$

As depicted in Equation (8), the GCE's gradient weighs samples using the $f_k(x;\theta)^{q-1}$ factor, which could affect robustness against noisy labels depending on the hyperparameter value q. In summary, the larger the q value, the more noise robustness is attained. Therefore, a suitable q is required to find a trade-off between noisy robustness and better learning dynamics during network training.

Lastly, since tighter loss bounding would imply more robust noise tolerance, GCE can be extended to its truncated version as follows [42]:

$$TGCE(y, f(x); \tilde{\lambda}_x, \tilde{C}) = \tilde{\lambda}_x \frac{1 - \left(\mathbf{1}^\top(y \odot f(x))\right)^q}{q} + (1 - \tilde{\lambda}_x)\frac{1 - (\tilde{C})^q}{q}, \tag{9}$$

where $\tilde{\lambda}_x \in [0, 1]$ and $\tilde{C} \in (0, 1)$. Note that λ_x prunes samples regarding a noise tolerance ruled by $\tilde{C}$.

3.2. Chained Deep Learning Fundamentals

The seminal Chained Gaussian Processes (CGP) in [46] fixes a likelihood function with J parameters depending on the input–output set $\{X, Y\}$, as follows:

$$p(Y|X, \xi) = \prod_{n=1}^{N} p(y_n|\xi_1(x_n), \ldots, \xi_J(x_n)), \tag{10}$$

$\boldsymbol{\xi} = [\boldsymbol{\xi}_1 \ldots, \boldsymbol{\xi}_J]^\top \in \mathbb{R}^{NJ}$ being a parameter vector and $\boldsymbol{\xi}_j = [\xi_j(x_1) \ldots \xi_j(x_N)]^\top \in \mathbb{R}^N$. In addition, each $\xi_j(x) \in \mathcal{M}_j$ maps an input instance to the parameter space ($j \in \{1, 2, \ldots, J\}$). The likelihood in Equation (10) allows modeling the function parameters with J independent GPs (one GP prior per parameter).

Likewise, we can extend the concept of CGP to the field of DL. Hence, suppose a DNN with L layers, where the output layer contains J outputs (neurons). The DNN model can be represented by the following composite function $f(x) = [f_1(x), \ldots, f_J(x)]^\top \in \mathbb{R}^J$,

$$f(x) = (\varphi_L \circ \varphi_{L-1} \circ \cdots \circ \varphi_1)(x), \tag{11}$$

where $\circ$ stands for function composition. Accordingly, Chained Deep Learning (CDL) links each likelihood parameter $\xi_j(x)$ to one of the J outputs. Each CDL parameter can be estimated as: $\xi_j(x) = h_j(f_j(x))$, where $h_j : \mathbb{R} \to \mathcal{M}_j$ maps each $f_j(x)$ prediction to $\mathcal{M}_j$. Furthermore, each function $\varphi_l(\cdot)$, with $l \in \{1, 2, \ldots, L\}$, depends on a set of parameters $\boldsymbol{\phi} = [\boldsymbol{\phi}_1, \ldots, \boldsymbol{\phi}_L]^\top$, e.g., weights and biases, that can be optimized via gradient descent and back-propagation [47].

3.3. Generalized Cross-Entropy-Based Chained Deep Learning for Multiple Annotators

Nowadays, in several real-world classification problems, instead of the ground truth Y, multiple labels are provided by R experts with different levels of ability, e.g., Multiple Annotators (MA) [28]. For the sake of clarity, we assume that the r-th expert annotates $|\Omega_r| \leq N$ instances, $|\Omega_r|$ being the cardinality of the set Ω_r containing the indices of samples labeled by annotator r. Moreover, let Ψ_n be the index set gathering the annotators who labeled the n-th instance. Then, a multiple annotators dataset $\{X \in \mathbb{R}^{N \times P}, \tilde{Y} \in \{0, 1, \varnothing\}^{N \times K \times R}\}$, where $\tilde{y}_n^r \in \{0, 1, \varnothing\}^K$ is the $1 - K$ one-hot label of expert r for instance n, can be built to feed a CDL approach holding $J = R + K$ outputs. The former R outputs code each expert reliability $\lambda_n^r \in \{0, 1\}$, and the remaining K predictions approximate the ground truth y_n.

In this sense, given an input sample x, each annotator's reliability can be predicted by fixing a sigmoid activation to the first R neurons within layer L in Equation (11), as:

$$\hat{\lambda}_n^r = \frac{1}{1 + \exp(-f_r(x_n))}, \tag{12}$$

where $\hat{\lambda}_n^r \in [0, 1]$.

Moreover, a softmax function is set to the last K outputs in $\phi_L(\cdot)$ to predict the ground truth label, as follows:

$$\hat{y}_k = \frac{\exp(f_{R+k}(x))}{\sum_{i=R+1}^{J} \exp(f_i(x))}. \tag{13}$$

where $k = \{1, 2, \ldots, K\}$, $\hat{y}_k \in [0, 1]$, and $\sum_k \hat{y}_k = 1$.

Here, to circumvent noisy annotators while coding their non-stationary behavior along the input space, and to favor the CDL training ruled by the optimization of the parameter set $\boldsymbol{\phi}$, a TGCE-based loss as in Equation (9) is proposed for multiple-annotator classification, yielding:

$$\boldsymbol{\phi}^* = \arg\min_{\boldsymbol{\phi}} \sum_{n=1}^{N} \sum_{r \in \Psi_n} \left[\hat{\lambda}_n^r(\boldsymbol{\phi}) \sum_{k=1}^{K} \tilde{y}_{n,k}^r \left(\frac{1 - \hat{y}_{n,k}^q(\boldsymbol{\phi})}{q} \right) + (1 - \hat{\lambda}_n^r(\boldsymbol{\phi})) \left(\frac{1 - (\frac{1}{K})^q}{q} \right) \right] \tag{14}$$

where $\tilde{y}_{n,k}^r \in \tilde{y}_n^r$ and $q \in (0, 1]$.

As seen above, self-regularization is achieved through each expert's reliability estimation $\hat{\lambda}_n^r(\boldsymbol{\phi})$ in Equation (14), which prunes the TGCE loss ruled by q. Of note, when $\hat{\lambda}_n^r(\boldsymbol{\phi}) \to 1$, $\hat{y}_n(\boldsymbol{\phi}) \in [0, 1]^K$ holds the 1-K ground truth predictions as in Equation (13). As a consequence, only samples with $\hat{\lambda}_n^r(\boldsymbol{\phi}) \to 1$ are kept for updating the CDL parameters. In contrast, noisy or unreliable annotations ($\hat{\lambda}_n^r(\boldsymbol{\phi}) \to 0$) are avoided to update the network parameters. Therefore, our GCE-based CDL, termed GCECDL, allows coding the non-stationary patterns of each annotator regarding each input instance space while preserving the interdependencies among experts through a CDL approach. In summary, our approach benefits CDL and GCE by circumventing noisy experts with non-stationary patterns. Figure 1 summarizes the GCECDL sketch.

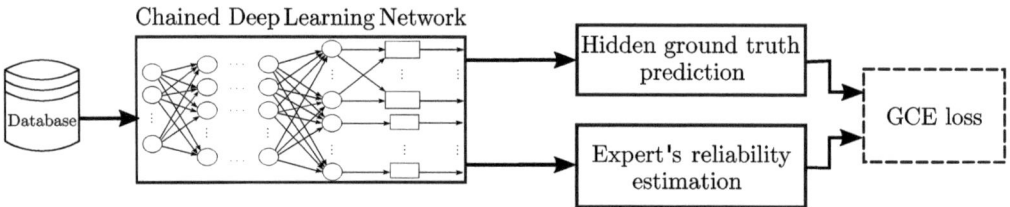

Figure 1. GCECDL sketch for multiple-annotator classification holding ground truth prediction and instance-based expert reliability estimation.

4. Experimental Set-Up

The following section comprehensively describes the tested datasets and the key experimental conditions utilized.

4.1. Tested Datasets

Our GCECDL approach, designed for multiple-annotator classification, is evaluated using synthetic and real-world datasets. The experiments aim to uncover the key insights and advantages of GCECDL for coding non-stationary and unreliable expert labels on complex datasets.

We generate synthetic data for a 1-dimensional, 3-class classification problem by randomly sampling 5000 points from a uniform distribution within the interval $[0,1]$ and using these points to construct the input feature matrix X. The true label $y_{n,k}$ for each sample is determined by taking the maximum value of $t_{n,k}$ for k in the set 1, 2, 3, where $t_{n,1} = \sin(2\pi x_n)$, $t_{n,2} = -\sin(2\pi x_n)$, and $t_{n,3} = -\sin(2\pi(x_n + 0.25)) + 0.5$. We also create a test set by extracting 2000 equally spaced samples from the same interval.

We then look for datasets where the input data come from real-world applications. Still, the labels from multiple annotators are obtained synthetically. The synthetic labeling is carried out to control the labeling process. In particular, six binary and multi-class classification task datasets are studied from the famous UCI repository (http://archive.ics.uci.edu/ml, accessed on 19 August 2022). The chosen datasets include: Occupancy Detection (Occupancy), Skin Segmentation (Skin), Tic-Tac-Toe Endgame (tic-tac-toe), Iris Plants (Iris), Wine (Wine), and Image Segmentation (Segmentation). Moreover, the Fashion-MNIST dataset [48], as well as the Balance and New Thyroid datasets, are also selected from the Keel-dataset Repository (https://sci2s.ugr.es/keel/category.php?cat=clas, accessed on 3 October 2022). In addition, the publicly available bearing data collected by the Case Western Reserve University (Western) are used. The aim is to build a system to diagnose an electric motor's status based on two accelerometers. The feature extraction is performed as in [49]. We also evaluate our proposed GCECDL classifier on two large image classification sets: MNIST of Handwritten Digits (MNIST) [50], an easily interpretable image database of labeled handwritten digits with 60,000 images for training and 100,000 for test sets, and the Cats vs. Dogs database, consisting of 25,000 images of dogs and cats [51], each class being represented by 1 and 0, respectively.

Finally, we include three real-world datasets provided with human annotations. First, the CIFAR-10H comprises over 500 k crowdsourced human categorization judgments obtained through AMT and includes ten categories: airplane, automobile, bird, cat, deer, dog, frog, horse, ship, and truck [52]. In our study, we applied a rigorous data-filtering process and discarded any samples that at least one annotator did not label. This filtering step resulted in 19.233 labeled samples, out of which 10.000 were used for testing. The second dataset is LabelMe, which aims to classify images into eight different classes: highway, inside city, tall building, street, forest, coast, mountain, and open country. It consists of 2688 images; each image was labeled by an average of 2547 workers, with a mean accuracy of 69.2%. We used the prepared dataset from [28], which performs a feature extraction stage based on a pre-trained VGG-16 deep neural network [53]. The third one is the Music genre database [54], comprising one thousand 30-second samples of songs categorized into classical, country, disco, hip-hop, jazz, rock, blues, reggae, pop, and metal. Each class contains 100 representative samples. A random selection of 700 samples was published on the AMT platform for workers to classify them from one to ten based on their genre. Feature extraction was performed following the method outlined in [55], resulting in an input space of 124 features. Table 1 summarizes the tested synthetic, semi-synthetic, and real-world datasets.

Table 1. Description of the number of features (P), instances (N), and classes (K) of tested datasets for multiple-annotator classification.

Name	Input Shape (P)	Number of Instances (N)	Number of Classes (K)
1D Synthetic	1	500	3
Ocupancy	7	20,560	2
Skin	4	245,057	2
Tic-Tac-Toe	9	958	2
Balance	4	625	3
Iris	4	150	3
New Thyroid	5	215	3
Wine	13	178	3
Fashion-Mnist	784	70,000	10
Segmentation	18	2310	7
Western	7	3413	4
Cat vs. Dog	$200 \times 200 \times 3$	25,000	2
Mnist	$28 \times 28 \times 1$	70,000	10
CIFAR-10H	$32 \times 32 \times 3$	19,233	10
LabelMe	$256 \times 256 \times 3$	2688	8
Music	124	1000	10

4.2. Provided and Simulated Annotations

To test our GCECDL classifier, we simulate annotator labels as corrupted versions of the hidden ground truth. Here, the simulations are performed by assuming: (i) dependencies among annotators and (ii) the labelers' performances are modeled as a function of the input features. In turn, the semiparametric latent factor model is used to build the labels, as follows [56]:

- Define Q deterministic functions $\hat{\mu}_q : \chi \to \mathbb{R}$ and their combination parameters $\hat{\omega}_{l_r,q} \in \mathbb{R}, \forall r \in R, n \in N$.
- Compute $\hat{f}_{l_r,n} = \sum_{q=1}^{Q} \hat{\omega}_{l_r,q} \hat{\mu}_q(\hat{x}_n)$, where $\hat{x}_n \in \mathbb{R}$ is the n-th component of $\hat{x} \in \mathbb{R}^N$, $\hat{x}$ being the 1-D representation of the input features in X by using t-SNE approach [29].
- Calculate $\lambda_r^n = \text{sigmoid}(\hat{f}_{l_r,n})$, where $\text{sigmoid}(\cdot) \in [0,1]$ is the sigmoid function.
- Finally, find the r-th label as $y_r^n = \begin{cases} y_n & \text{if } \lambda_r^n \geq 0.5 \\ \breve{y}_n & \text{if } \lambda_r^n < 0.5 \end{cases}$, where $\breve{y}_n$ is a flipped version of the actual label y_n.

4.3. Performance Measures, Method Comparison, and Training

As quantitative assessment concerning the classification performance, the overall Accuracy (ACC) and the Balanced Accuracy (BACC) are reported on the testing set, which can be written as follows:

$$ACC[\%] = 100 \frac{TP + TN}{TP + TN + FP + FN} \quad (15)$$

$$BACC[\%] = \frac{100}{2}\left(\frac{TP}{TP+FN} + \frac{TN}{TN+FP}\right) \quad (16)$$

where TP, FN, and FP represent the true positive, false negative, and false positive predictions, respectively, after comparing the actual and estimated labels y_n and $\hat{y}_n$ for a given input sample x_n.

In addition, we consider the Normalized Mutual Information (NMI) between the output and the target [57]. The NMI measures the amount of shared information between two variables and quantifies the strength of their relationship, yielding:

$$\text{NMI}[\%] = 100\frac{1}{N}\sum_{n=1}^{N}\frac{2I(y_n,\hat{y}_n)}{H(y_n)+H(\hat{y}_n)}, \quad (17)$$

where $I(\cdot,\cdot)$ stands for mutual information and $H(\cdot)$ for marginal entropy. Furthermore, we estimate the Area Under the ROC Curve (AUC) that can be computed by varying the decision boundary concerning the sensitivity (Sen) and specificity (Spe) measures, as follows [47]:

$$AUC[\%] = 100\frac{Sen+Spe}{2} \quad (18)$$

For concrete testing, we use a cross-validation scheme with 10 repetitions holding 70% of the samples for training and the remaining 30% for testing (except for the Mnist, F-Mnist, CIFAR-10H, and music dataset, where training and testing sets are clearly defined).

Moreover, Table 2 displays the tested state-of-the-art algorithms for comparison purposes. The abbreviations are fixed as follows: Regularized Chained Deep Neural Network Classifier for Multiple Annotators (RCDNN) [43], Deep Learning Majority Voting (DL-MV), and Deep Learning from Crowds (DL-CW(MW)) [28]. Our Python codes are publicly available for DL-CL(MW), DL-MV, RCDNN, and GCECDL at https://github.com/Jectrianama/GCCE_TEST (accessed on 19 December 2022). Regarding DL-CL(MW), we use the codes at http://www.fprodrigues.com/ (accessed on 19 December 2022). Of note, DL-GOLD is a deep learning model trained with the true label, which is used only to provide an upper bound.

In turn, to better grasp the behavior of our GCECDL classifier over every dataset, we implemented a grid-search scheme to fix the q value within the grid $[0.001, 0.1, 0.2, 0.3, 0.4, 0.5, 0.75]$.

Table 2. A brief overview of the state-of-the-art methods tested.

Algorithm	Description
DL-GOLD	A DL classification model using the real labels (upper bound).
DL-MV	A DL classification model using the MV of the labels as the ground truth.
RCDNN [43]	A regularized chained deep neural network which predicts the ground truth and annotators' performance from input space samples.
DL-CL(MW) [28]	A crowd Layer for DL, where annotators' parameters are constant across the input space.

The proposed GCECDL architecture for multiple annotators comprises (i) a fully connected network for tabular data (see Figure 2) and (ii) a convolutional network for image data (see Figure 3). For all provided layers, elastic-net-based weight regularizers are used. As usual, the optimization problem is solved using a back-propagation algorithm. Moreover, to favor scalability, we utilize a mini-batch-based gradient descent approach with automatic differentiation (the Adam-based optimizer is fixed). In addition, we employed callbacks during the training process to monitor the model's performance. Specifically, we used an EarlyStopping callback to stop the training process if the validation performance did not improve for a specified number of epochs and a LearningRateScheduler callback, allowing the model to converge more quickly by avoiding becoming stuck in a suboptimal solution. These callbacks allowed us to optimize the performance of our neural network and sidestep overfitting. Finally, we selected the best performance between models with or without callbacks for each database. All experiments were conducted in Python 3.8, with the Tensorflow 2.4.1 API, on a Google Colaboratory environment.

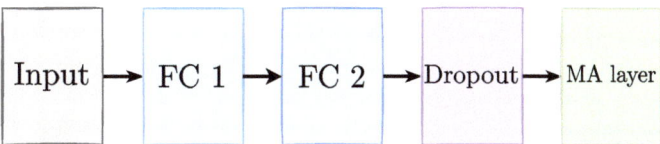

Figure 2. GCECDL-based fully connected architecture for tabular databases. *FC* stands for a fully connected (dense) layer. *FC1* and *FC2* use a selu activation. A dropout layer (Dropout) is included to avoid overfitting. The MA layer contains two fully connected layers that output the hidden ground truth label and the annotator's reliability, fixing a softmax and a sigmoid activation, respectively.

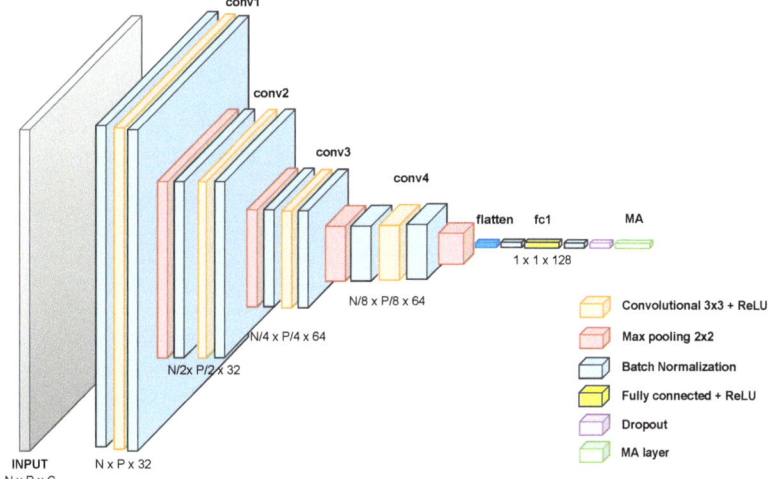

Figure 3. GCECDL-based convolutional architecture for image databases. Four convolutional layers with 3×3 patches, 2×2 max pooling, and ReLU activations are included. The MA layer outputs the hidden ground truth label and the annotator's reliability fixing a softmax and a sigmoid activation, respectively.

5. Results and Discussion

5.1. Reliability Estimation and Visual Inspection Results

We first perform a controlled experiment to test the GCECDL's capability when dealing with binary and multiclass classification. We use the one-dimensional synthetic dataset described in Section 4. In addition, five labelers (R = 5) are simulated with different levels of expertise. To simulate the error variances, we define $Q = 3$ $\hat{\mu}_q(\cdot)$ functions (see Section 4.2), yielding:

$$\hat{\mu}_1(x) = 4.5\cos(2\pi x + 1.5\pi) - 3\sin(4.3\pi x + 0.3\pi) \tag{19}$$

$$\hat{\mu}_2(x) = 4.5\cos(1.5\pi x + 0.5\pi) + 5\sin(3\pi x + 1.5\pi) \tag{20}$$

$$\hat{\mu}_3(x) = 1, \tag{21}$$

where $x \in [0, 1]$. In addition, the combination weights are gathered within the following combination matrix $\hat{\mathbf{W}} \in \mathbb{R}^{Q \times R}$:

$$\hat{\mathbf{W}} = \begin{bmatrix} 0.4 & 0.7 & -0.5 & 0.0 & -0.7 \\ 0.4 & -1.0 & -0.1 & -0.8 & 1.0 \\ 3.1 & -1.8 & -0.6 & -1.2 & 1.0 \end{bmatrix} \quad (22)$$

holding elements $\hat{\omega}_{l_r,q}$, which are used to combine functions $\hat{\mu}_1$, $\hat{\mu}_2$, and $\hat{\mu}_3$.

For visual inspection purposes, Figure 4 shows the predictive label's probability (PLP) and the AUC, for all studied approaches regarding the one-dimensional synthetic database.

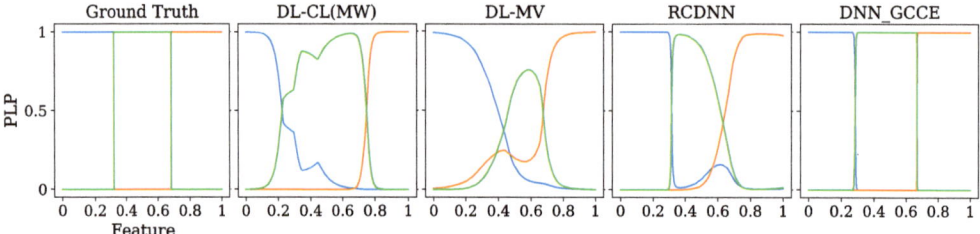

Figure 4. Synthetic results—1D dataset. The predictive label's probability (PLP) is shown, comparing the prediction of our GCECDL (AUC = 0.99) against: DL-CL(MW) (AUC = 0.79), DL-MV (AUC = 0.9), and RCDNN (AUC = 0.99). Label classes are represented as blue, green, and orange.

As seen in Figure 4, DL-CL(MW) and RCDNN have a different shape than the ground truth. Additionally, DL-MV has the worst accuracy for two out of the three classes. Upon further analysis of the results of our GCECDL method, we note that its predictive accuracy is quite close to the absolute ground truth, which is the theoretical upper limit. Thus, GCECDL offers a more suitable representation of the labelers' behavior compared to its competitors. This is because GCECDL takes into account both the annotators' dependencies and the relationship between the input features and the annotators' performance.

To support the previous statement, Figure 5 illustrates the per-annotator reliability estimated by our model and the simulated accuracy of each annotator. As can be seen, our model provides an excellent representation for annotators one and five and an acceptable representation for annotators two, three, and four. This is a direct result of modeling the labelers' parameters as functions of the input features. This outcome demonstrates how our approach effectively identifies the areas where a specific labeler aligns with the regions of higher accuracy of the simulated annotators.

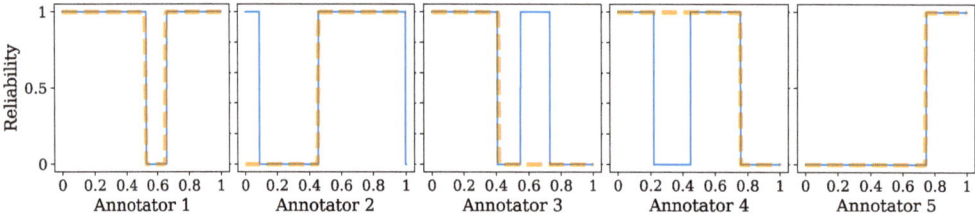

Figure 5. GCECDL-based annotators' performance (reliability) estimation for the synthetic 1D experiment. Orange dashed line depicts (from left to right) the simulated accuracy for each annotator based on Equations (19)–(22). The blue line shows (from left to right) the estimated annotator's reliability (λ_r).

In this next step, we conduct two crucial experiments utilizing the MNIST dataset, as outlined in Section 4. These experiments include an examination of explainable multiple-annotator classification and a t-SNE-based 2D visualization of the data. To achieve this, we employed the Gradient-weighted Class Activation Mapping (Grad-CAM++) approach to extract normalized class activation mapping from the image data, as described in [30]. We then plotted heatmap images related to the FC 2 layer, which represents the high-level visual features that are extracted from the characteristics. These feature maps are then projected onto a two-dimensional space using the t-SNE algorithm. The visual analysis of these results shows that the same color image samples cluster together in the 2D, low-dimensional space according to their class, while preserving the spatial relationships from the input space.

Figure 6 presents a visualization of the gold standard and simulated annotators ($R = 3$ for illustrative purposes) plotted over the resulting 2D features projection for the training set. It can be observed that the features extracted possess a high degree of separability and discriminative ability, as every class (0–9) is represented by a distinct cluster. For illustration purposes, a few images are depicted over each corresponding projection. The last two rows show a selection of simulated labels and their different scores (annotator reliability). We can see the different levels of expertise obtained from the confusion matrix. The first annotator, whose accuracy score with respect to the ground truth labels is 97%, is depicted over the projection. We can observe how most samples have a correct version of the ground truth. However, it tends to fail more for classes 0, 2, 3, and 8. This behavior becomes more pronounced for the last two annotators, whose accuracy drops to 41% and 11%, respectively. The mismatch between the labels and the ground truth is more evident in the top figure. We expected to compare this with the estimated reliability obtained by our model.

Then, Figure 7 illustrates the hidden ground truth prediction and reliability estimation generated by our GCECDL approach. As shown, GCECDL demonstrates a high level of suitability for the MNIST digit classification problem by achieving an ACC score of 0.99, which highlights its generalization capability, even in cases where the ground truth is unknown. This is because the proposed model takes into account both the relationship between the input space and the annotator's behavior, as well as the dependencies among their labels, which improves the quality of expert codification, as described in [4,38]. To provide further insights, we also generated visual explanations for a subset of the samples in the test dataset. To achieve this, Grad-CAM++ was applied to a given image and class K to determine the regions of the image that are most competitive for classification. As seen in the CAM, the important regions highlighted in red reveal that our model can effectively exploit the most relevant features to correctly identify the image's class. The second row of the figure depicts some visual explanations on the 2D projections. Notably, class 4 can be confused by the model with a seven or a nine in a few samples.

In the last row of Figure 7, we can infer that our method effectively identifies the zones where the labelers have the best accuracy. This is not unexpected as the annotators' accuracy (simulated) is compared with their reliability (estimated); therefore, the clusters where a specific labeler obtains the highest accuracy should align with the clusters where the estimated reliability is closest to 1 (yellow). To further support this statement, we depict the estimated probability function through a kernel density estimation (KDE) plot, to show the reliability estimated per annotator. For example, regarding annotator 1 (blue), as most of its estimates are reliable, the KDE increases when the reliability is 1. However, for annotator two (orange), its peak KDE value is slightly lower when the reliability is one. Similarly, annotators 3 (green) and 4 (red) exhibit an inverse behavior, as their performance is more doubtful.

In addition, it is important to note that our proposed GCECDL encodes the interdependence between annotators. By comparing the simulated annotators and the labelers' performance in Figures 6 and 7, it is clear that our proposed model closely follows the performance pattern of the labelers' capabilities. Therefore, when a suitable labeler is presented, the model provides a high estimation for the same labeler in the zones where the labelers have the best accuracy. Conversely, when malicious labelers are present, the model reflects this in poor reliability estimation. This highlights how the ability to label and reliability estimation are closely related. Furthermore, it is worth noting that annotators with high uncertainty tend to have CAMs with more energy, which supports the aforementioned statement empirically.

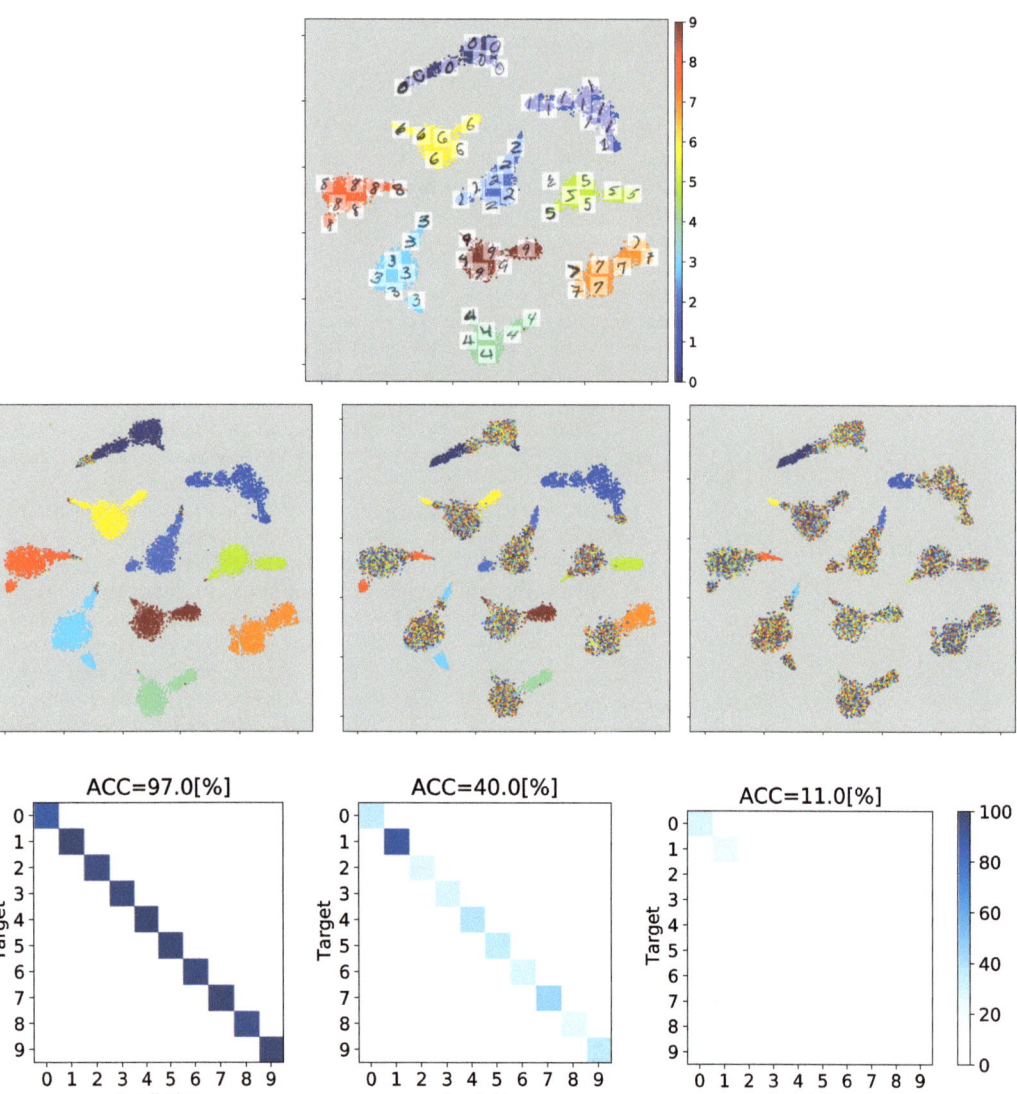

Figure 6. Simulated annotators' for the MNIST database. A 2D t-SNE-based projection of the Mnist training set from the FC1 layer feature maps is shown (some exemplary images are depicted over their 2D projections, where numbers from zero to nine stand for each MNIST class). The ground truth labels are used as a color. The second row depicted the 2D projection of each annotator's labels in color. The third row presents the confusion matrix of each expert and their achieved accuracy.

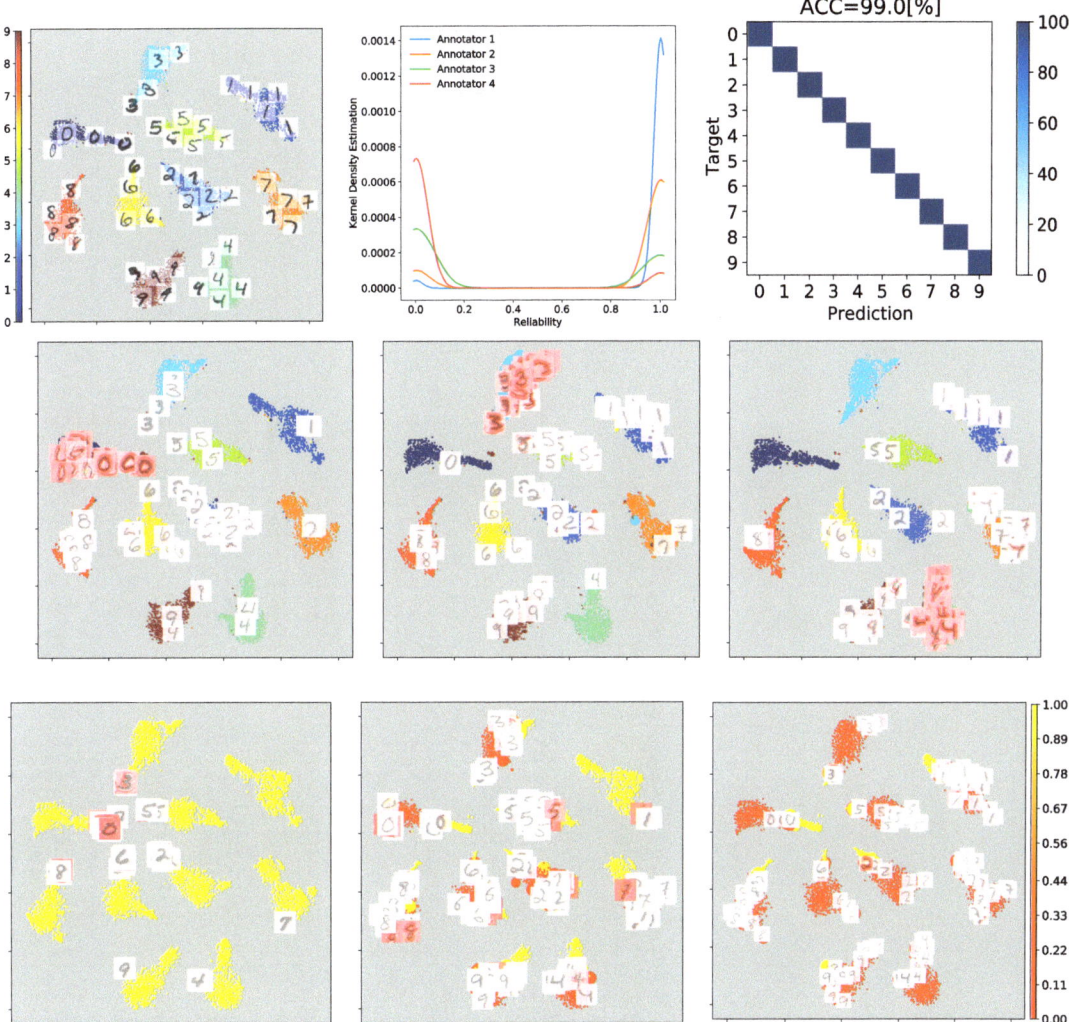

Figure 7. Visual inspection results for the MNIST database. A 2D t-SNE-based projection of the Mnist test set from the FC1 layer feature maps is shown (some exemplary images are depicted over their 2D projections, where numbers from zero to nine stand for each MNIST class). The ground truth labels are shown in color. The second row depicted the 2D projection of the ground truth prediction in color with some explanation maps for visualization purposes. The third row presents the 2D projection of each annotator's reliability estimation in color.

5.2. Method Comparison Results for Multiple-Annotator Classification

Table 3 presents the results of our experiments on real-world datasets. The non-parametric Friedman test results for every quantitative assessment measure establish their statistical significance. The null hypothesis is rejected, indicating that all algorithms attain different performances as described in [58]. Additionally, the significance threshold is fixed at $p < 0.05$. The DL-GOLD standard is not included in the test to compare state-of-the-art approaches exclusively. It is worth noting that most classification schemes exhibit a highly satisfactory performance for the datasets with simulated annotators, as reflected in the scores.

Table 3. Method comparison results—multiple-annotator classification. Bold: the highest score for each measure excluding the upper bound (target) classifier DL-GOLD. For our GCECDL, the fixed q-value is presented in parentheses. Friedman test: we obtain a Chi-square of 27.08, 23.48, 28.03, and 7.92 (p-value = 5.6×10^{-6}, 3.56×10^{-6}, 3.20×10^{-5}, and 0.047) for ACC, BACC, NMI, and AUC measures, respectively.

Database	Measure	DL-GOLD	DL-CL(MW)	DL-MV	RCDNN	GCECDL
Ocupancy	ACC [%]	97.59 ± 0.11	90.89 ± 0.92	90.31 ± 0.80	96.61 ± 0.42	**97.50 ± 0.06** (0.1)
	BACC [%]	95.92 ± 0.09	74.30 ± 3.04	75.50 ± 0.99	93.22 ± 1.27	**95.86 ± 0.10** (0.1)
	NMI [%]	83.91 ± 0.35	57.98 ± 2.16	52.32 ± 3.56	77.93 ± 2.51	**83.64 ± 0.31** (0.1)
	AUC [%]	97.96 ± 0.05	87.17 ± 1.52	87.75 ± 0.49	99.03 ± 0.14	**97.92 ± 0.05** (0.1)
Skin	ACC [%]	99.74 ± 0.13	56.48 ± 0.36	80.55 ± 3.23	89.45 ± 2.06	**96.06 ± 0.59** (0.1)
	BACC [%]	99.62 ± 0.15	45.07 ± 0.46	79.80 ± 1.79	64.27 ± 10.66	**90.60 ± 4.31** (0.1)
	NMI [%]	96.86 ± 1.15	18.31 ± 0.21	13.56 ± 1.06	38.99 ± 8.25	**71.14 ± 5.22** (0.1)
	AUC [%]	99.81 ± 0.08	91.80 ± 1.03	80.45 ± 6.22	82.13 ± 5.33	**99.30 ± 0.10** (0.1)
Tic-Tac-Toe	ACC [%]	98.47 ± 0.17	80.00 ± 3.73	78.26 ± 1.85	90.49 ± 17.16	**97.28 ± 1.17** (0.2)
	BACC [%]	96.00 ± 0.27	66.24 ± 5.98	56.09 ± 2.14	78.74 ± 38.93	**93.10 ± 2.78** (0.2)
	NMI [%]	88.57 ± 1.52	36.88 ± 6.07	23.90 ± 2.05	71.42 ± 24.79	**83.52 ± 6.13** (0.2)
	AUC [%]	98.05 ± 0.14	96.48 ± 1.00	85.37 ± 1.21	91.71 ± 21.93	**96.55 ± 1.39** (0.2)
Balance	ACC [%]	91.91 ± 2.50	85.27 ± 1.93	73.72 ± 1.51	**91.33 ± 1.14**	90.21 ± 0.83 (0.1)
	BACC [%]	60.00 ± 5.68	74.9 ± 4.09	43.39 ± 0.8	46.93 ± 1.00	**75.27 ± 7.33** (0.1)
	NMI [%]	73.34 ± 3.32	61.98 ± 3.58	59.43 ± 2.04	66.52 ± 3.30	**68.00 ± 2.19** (0.1)
	AUC [%]	83.66 ± 4.91	**84.97 ± 4.08**	70.66 ± 3.35	64.51 ± 17.36	81.11 ± 8.25 (0.1)
Iris	ACC [%]	97.34 ± 0.94	95.11 ± 2.59	95.33 ± 1.59	96.89 ± 1.15	**97.34 ± 0.94** (0.1)
	BACC [%]	96.47 ± 1.24	91.67 ± 4.54	91.56 ± 1.84	94.24 ± 2.03	**96.06 ± 2.11** (0.1)
	NMI [%]	91.69 ± 1.91	86.82 ± 5.59	87.61 ± 2.86	89.68 ± 3.74	**91.15 ± 3.07** (0.1)
	AUC [%]	100.00 ± 0.00	100.00 ± 0.00	100.00 ± 0.00	99.83 ± 0.00	**100.00 ± 0.00** (0.1)
New Thyroid	ACC [%]	96.15 ± 0.77	95.69 ± 1.15	**97.69 ± 1.03**	95.08 ± 0.62	96.00 ± 1.84 (0.01)
	BACC [%]	92.5 ± 1.94	92.61 ± 2.8	**93.88 ± 2.65**	88.41 ± 2.22	90.52 ± 4.36 (0.01)
	NMI [%]	82.98 ± 3.44	82.38 ± 3.86	**90.44 ± 3.61**	80.40 ± 2.42	83.61 ± 6.85 (0.01)
	AUC [%]	96.19 ± 0.96	95.46 ± 1.78	**97.80 ± 0.6**	94.72 ± 1.67	95.19 ± 2.23 (0.01)
Wine	ACC [%]	97.59 ± 0.85	95.74 ± 1.67	73.15 ± 3.73	94.26 ± 2.10	**96.48 ± 2.38** (0.01)
	BACC [%]	96.59 ± 1.08	94.17 ± 2.52	64.44 ± 5.42	91.99 ± 2.59	**96.88 ± 2.92** (0.01)
	NMI [%]	91.50 ± 3.07	85.54 ± 4.03	44.18 ± 4.39	82.91 ± 5.97	**89.87 ± 5.37** (0.01)
	AUC [%]	98.88 ± 0.37	99.88 ± 0.37	87.76 ± 2.88	97.13 ± 1.26	**100.0 ± 0.00** (0.01)
F-Mnist	Acc [%]	86.45 ± 0.25	79.54 ± 9.16	85.83 ± 0.35	86.98 ± 0.71	**88.26 ± 0.26** (0.01)
	BACC [%]	84.92 ± 0.27	77.28 ± 10.11	84.23 ± 0.38	85.50 ± 0.77	**86.94 ± 0.28** (0.01)
	NMI [%]	77.16 ± 0.21	72.92 ± 3.84	76.46 ± 0.38	78.75 ± 0.48	**79.89 ± 0.21** (0.01)
	AUC [%]	88.75 ± 1.61	85.61 ± 13.95	90.42 ± 1.46	89.33 ± 2.93	**89.11 ± 2.27** (0.01)
Segmentation	ACC [%]	95.17 ± 0.25	94.65 ± 0.95	91.11 ± 0.94	91.85 ± 0.67	**94.88 ± 0.8** (0.01)
	BACC [%]	94.49 ± 0.59	93.81 ± 1.09	89.79 ± 1.09	90.69 ± 0.77	**94.15 ± 0.95** (0.01)
	NMI [%]	91.07 ± 0.77	**90.38 ± 1.18**	84.46 ± 1.65	86.27 ± 0.70	89.92 ± 1.25 (0.01)
	AUC [%]	99.95 ± 0.13	99.76 ± 0.20	98.71 ± 0.44	98.94 ± 0.33	**99.79 ± 0.22** (0.01)
Western	ACC [%]	99.25 ± 0.38	95.94 ± 4.39	95.11 ± 1.15	97.89 ± 0.40	**98.31 ± 0.61** (0.01)
	BACC [%]	99.22 ± 0.42	94.37 ± 5.54	93.47 ± 1.45	97.25 ± 0.57	**98.00 ± 0.97** (0.01)
	NMI [%]	98.61 ± 0.70	96.80 ± 1.70	94.33 ± 0.97	97.16 ± 0.30	**97.35 ± 0.60** (0.01)
	AUC [%]	100.00 ± 0.00	100.00 ± 0.00	98.97 ± 0.85	99.99 ± 0.01	**100.00 ± 0.00** (0.01)
Cats vs. Dogs	ACC [%]	85.00 ± 0.44	68.97 ± 0.84	58.69 ± 1.02	70.84 ± 4.71	**74.15 ± 0.58** (0.1)
	BACC [%]	70.19 ± 0.22	37.44 ± 0.35	16.50 ± 0.51	43.51 ± 2.08	**48.24 ± 0.54** (0.1)
	NMI [%]	40.66 ± 0.41	11.03 ± 0.25	2.50 ± 0.12	16.02 ± 1.04	**20.79 ± 0.33** (0.1)
	AUC [%]	93.17 ± 0.22	68.94 ± 0.80	58.63 ± 0.98	70.98 ± 4.57	**82.99 ± 0.30** (0.1)
Mnist	ACC [%]	99.32 ± 0.06	87.99 ± 2.73	92.88 ± 0.54	99.09 ± 0.05	**99.11 ± 0.08** (0.01)
	BACC [%]	99.22 ± 0.06	86.49 ± 3.11	91.97 ± 0.59	98.98 ± 0.06	**99.02 ± 0.09** (0.01)
	NMI [%]	97.95 ± 0.15	82.82 ± 2.29	83.43 ± 1.02	97.28 ± 0.11	**97.39 ± 0.21** (0.01)
	AUC [%]	99.81 ± 0.08	99.82 ± 0.02	97.88 ± 0.33	99.71 ± 0.06	**99.68 ± 0.08** (0.01)
CIFAR-10H	ACC [%]	71.72 ± 1.12	60.80 ± 1.59	68.24 ± 1.05	**69.53 ± 0.63**	69.24 ± 0.67 (0.01)
	BACC [%]	68.46 ± 0.28	56.56 ± 1.72	64.69 ± 1.18	**66.18 ± 0.72**	65.88 ± 0.76 (0.01)
	NMI [%]	64.2 ± 0.23	43.09 ± 1.21	49.55 ± 1.4	**51.44 ± 0.81**	50.95 ± 0.78 (0.01)
	AUC [%]	96.08 ± 0.11	89.42 ± 0.7	94.81 ± 0.35	**95.17 ± 0.16**	95.01 ± 0.19 (0.01)
LabelMe	ACC [%]	90.91 ± 0.44	83.11 ± 0.96	76.94 ± 1.15	**89.09 ± 0.41**	88.97 ± 0.55 (0.01)
	BACC [%]	90.03 ± 0.42	81.85 ± 0.86	75.08 ± 1.25	**88.01 ± 0.42**	87.97 ± 0.57 (0.01)
	NMI [%]	81.36 ± 0.79	73.92 ± 0.80	68.36 ± 0.54	**77.82 ± 0.55**	78.00 ± 0.70 (0.01)
	AUC [%]	99.37 ± 0.06	96.34 ± 0.88	97.65 ± 0.13	**99.14 ± 0.06**	99.11 ± 0.08 (0.01)
Music	ACC [%]	76.4 ± 0.88	57.13 ± 2.80	62.8 ± 1.06	**65.80 ± 2.83**	65.63 ± 2.41 (0.01)
	BACC [%]	74.32 ± 0.91	51.47 ± 2.91	57.67 ± 1.27	**61.96 ± 3.32**	61.83 ± 2.84 (0.01)
	NMI [%]	65.86 ± 1.16	57.25 ± 1.55	53.87 ± 1.49	**57.02 ± 2.63**	57.36 ± 1.40 (0.01)
	AUC [%]	96.34 ± 0.13	88.62 ± 3.08	93.58 ± 0.21	**94.07 ± 0.94**	94.13 ± 0.50 (0.01)
Average Ranking	ACC [%]	-- ± --	3.26 ± --	3.40 ± --	2.00 ± --	1.33 ± --
	BACC [%]	-- ± --	3.00 ± --	3.40 ± --	2.33 ± --	1.26 ± --
	NMI [%]	-- ± --	3.14 ± --	3.46 ± --	2.21 ± --	1.20 ± --
	AUC [%]	-- ± --	2.67 ± --	3.13 ± --	2.47 ± --	1.73 ± --

Our approach outperforms the selected state-of-the-art methods across most evaluation measures. For example, our proposal achieves the highest accuracy on 12 of 15 datasets, as shown in Table 3. Similarly, GCECDL outperforms tested strategies regarding BACC, NMI, and AUC for most datasets. This outcome is due to the fact that GCECDL properly codes annotators' reliability by considering the correlations among their decision, even for noisy outputs. Remarkably, RCDNN reaches the second highest average performance. Indeed, it obtains the highest performance on CIFAR-10H, LabelMe, and Music. In general, RCDNN is similar to GCECDL, but it is significantly affected when the annotators' performance is below 60%. Furthermore, it struggles with databases with a high class imbalance ratio. This behavior highlights that RCDNN is more susceptible to noisy or imbalanced scenarios.

Moreover, we observe that high NMI values for some datasets suggest that the labels provided by the annotators are consistent and reliable. Our GCECDL method effectively captured these dependencies. For example, in the MNIST dataset, a high NMI value of 97.39% is attained, indicating that the labels provided by the different annotators and the hidden ground truth prediction are highly consistent. In contrast, in the Cats vs. Dogs dataset, we observe a lower NMI value of 20.79%, pointing to more significant variability in the annotations and highlighting the challenges of dealing with noisy or inconsistent annotations.

Finally, the DL-CL(MW) approach outperforms the DL-MV scheme, which can be observed in the average ranking. Furthermore, it is worth noting that DL-CL(MW) includes the introduction of the CrowdLayer, which allows for training neural networks directly from multiple labels without encoding the annotators' behavior. On the other hand, DL-MV presents the lowest performance among the studied methods. It can be explained by the fact that DL-MV is the most naive approach, and most annotators were simulated with a low level of expertise, negatively impacting the outcome of the majority voting strategy.

6. Conclusions

This article introduced a Generalized Cross-Entropy-based Chained Deep Learning model, termed GCECDL, to deal with multiple-annotator scenarios. Our method follows the ideas of [43,46], where each parameter is modeled in a multi-labeler likelihood by using the outputs of a deep neural network. Nonetheless, unlike [43]—where a CCE-based loss was used—we also introduced a noise-robust loss function based on GCE [42] as a tradeoff between MAE and CCE. Thus, GCECDL codes the non-stationary patterns of each annotator regarding the input space. We tested our approach for classification tasks using fully synthetic and real-world databases from well-known repositories, including structured data and images. According to the results, our GCECDL can achieve robust predictive properties for the used datasets defeating the selected state-of-the-art models. We attribute this behavior to the coupled MAE and CE within GCE, exploiting the symmetry property of MAE for softmax-based outputs and the L1-norm and cross-entropy tradeoff in weighting noisy annotations as a function of the input space. In addition, our chained architecture yields a self-regularization strategy within a DL framework that favors proper labeler's reliability estimation and ground truth prediction. On the other hand, we created visual explanations using GradCam++ [30] to identify the most influential regions that demonstrate the model's ability to correctly predict the hidden ground truth and assess the reliability of the annotators. Furthermore, using t-SNE [29] to project the extracted features onto a two-dimensional space allowed us to retain the spatial relationships from the original input space.

As future work, extending the GCECDL for regression tasks is a promising research area, as demonstrated by the model introduced in [35]. Our next step is to experiment with various activation functions and deeper convolutional and recurrent architectures to tackle complex tasks such as computer vision, natural language processing, and graph-based modeling [59]. Additionally, we plan to develop a model for identifying non-stationary patterns from multidomain input spaces holding noisy targets in an agricultural context, using multispectral imagery, climatic data, and infield data, instead of relying on annotators [60]. Furthermore, we plan to test more Explainable AI methods to provide deeper insight into our model performance [61–63], e.g., Layer-wise Relevance Propagation that captures both negative and positive relevance. Finally, actionable and explainable AI extensions based on our GCECDL would be an exciting research line [64].

Author Contributions: Conceptualization, J.C.T.-M., J.G.-G. and A.M.Á.-M.; data curation, J.C.T.-M.; methodology, J.C.T.-M., J.G.-G. and A.M.Á.-M.; project administration, C.G.C.-D.; supervision, C.G.C.-D., J.A.F.-G. and A.M.Á.-M.; resources, J.C.T.-M., J.G.-G. and J.A.F.-G. All authors have read and agreed to the published version of the manuscript.

Funding: This research was funded (APC) by the project "Herramienta de apoyo a la predicción de los efectos de anestésicos locales vía neuroaxial epidural a partir de termografía por infrarrojo" (Code 111984468021-Minciencias). Also, G. Castellanos thanks to the project "Desarrollo de una herramienta de visión por computador para el análisis de plantas orientado al fortalecimiento de la seguridad alimentaria" (HERMES 54339-Universidad Nacional de Colombia and Universidad de Caldas). J. Triana thanks to the program "Beca de Excelencia Doctoral del Bicentenario-2019-Minciencias" and the project "Rice remote monitoring: climate change resilience and agronomical management practices for regional adaptation—RiceClimaRemote" (Flanders Research Institute for agriculture, fisheries and food ILVO-Government of Flanders-Belgium, Universidad de Ibagué, and Agrosavia).

Institutional Review Board Statement: Not applicable.

Informed Consent Statement: Not applicable.

Data Availability Statement: Publicly available datasets were analyzed in this study. These data can be found at the UCI repository (http://archive.ics.uci.edu/ml, (accessed on 19 August 2022)), Keel-dataset repository (https://sci2s.ugr.es/keel/category.php?cat=clas, (accessed on 3 October 2022)), and Rodrigues repository (http://www.fprodrigues.com/, (accessed on 20 Sepetember 2022)). Fashion-Mnist is freely available at dataset repository (https://github.com/zalandoresearch/fashion-mnist, (accessed on 27 August 2022)).

Conflicts of Interest: The authors declare no conflict of interest.

References

1. Zhang, J.; Sheng, V.S.; Wu, J. Crowdsourced label aggregation using bilayer collaborative clustering. *IEEE Trans. Neural Netw. Learn. Syst.* **2019**, *30*, 3172–3185. [CrossRef]
2. Parvat, A.; Chavan, J.; Kadam, S.; Dev, S.; Pathak, V. A survey of deep-learning frameworks. In Proceedings of the 2017 International Conference on Inventive Systems and Control (ICISC), Coimbatore, India, 19–20 January 2017; pp. 1–7.
3. Liu, Y.; Zhang, W.; Yu, Y. Truth inference with a deep clustering-based aggregation model. *IEEE Access* **2020**, *8*, 16662–16675.
4. Gil-Gonzalez, J.; Orozco-Gutierrez, A.; Alvarez-Meza, A. Learning from multiple inconsistent and dependent annotators to support classification tasks. *Neurocomputing* **2021**, *423*, 236–247. [CrossRef]
5. Sung, H.E.; Chen, C.K.; Xiao, H.; Lin, S.D. A Classification Model for Diverse and Noisy Labelers. In *Proceedings of the Pacific-Asia Conference on Knowledge Discovery and Data Mining*; Springer: Berlin/Heidelberg, Germany, 2017; pp. 58–69.
6. Yan, Y.; Rosales, R.; Fung, G.; Subramanian, R.; Dy, J. Learning from multiple annotators with varying expertise. *Mach. Learn.* **2014**, *95*, 291–327. [CrossRef]
7. Xu, G.; Ding, W.; Tang, J.; Yang, S.; Huang, G.Y.; Liu, Z. Learning effective embeddings from crowdsourced labels: An educational case study. In Proceedings of the 2019 IEEE 35th International Conference on Data Engineering (ICDE), Macao, China, 8–11 April 2019; pp. 1922–1927.
8. Tanno, R.; Saeedi, A.; Sankaranarayanan, S.; Alexander, D.C.; Silberman, N. Learning from noisy labels by regularized estimation of annotator confusion. In Proceedings of the IEEE/CVF Conference on Computer Vision and Pattern Recognition, Long Beach, CA, USA, 15–20 June 2019; pp. 11244–11253.
9. Davani, A.M.; Díaz, M.; Prabhakaran, V. Dealing with disagreements: Looking beyond the majority vote in subjective annotations. *Trans. Assoc. Comput. Linguist.* **2022**, *10*, 92–110. [CrossRef]
10. Kara, Y.E.; Genc, G.; Aran, O.; Akarun, L. Modeling annotator behaviors for crowd labeling. *Neurocomputing* **2015**, *160*, 141–156. [CrossRef]
11. Cao, P.; Xu, Y.; Kong, Y.; Wang, Y. Max-mig: An information theoretic approach for joint learning from crowds. *arXiv* **2019**, arXiv:1905.13436.
12. Chen, Z.; Wang, H.; Sun, H.; Chen, P.; Han, T.; Liu, X.; Yang, J. Structured Probabilistic End-to-End Learning from Crowds. In Proceedings of the IJCAI, Yokohama, Japan, 7–21 January 2021; pp. 1512–1518.

13. Ruiz, P.; Morales-Álvarez, P.; Molina, R.; Katsaggelos, A.K. Learning from crowds with variational Gaussian processes. *Pattern Recognit.* **2019**, *88*, 298–311. [CrossRef]
14. G. Rodrigo, E.; Aledo, J.A.; Gámez, J.A. Machine learning from crowds: A systematic review of its applications. *Wiley Interdiscip. Rev. Data Min. Knowl. Discov.* **2019**, *9*, e1288.
15. Zhang, P.; Obradovic, Z. Learning from inconsistent and unreliable annotators by a gaussian mixture model and bayesian information criterion. In Proceedings of the Joint European Conference on Machine Learning and Knowledge Discovery in Databases, Athens, Greece, 5–9 September 2011; pp. 553–568.
16. Zhang, J. Knowledge learning with crowdsourcing: A brief review and systematic perspective. *IEEE/CAA J. Autom. Sin.* **2022**, *9*, 749–762. [CrossRef]
17. Zhu, T.; Pimentel, M.A.; Clifford, G.D.; Clifton, D.A. Unsupervised Bayesian inference to fuse biosignal sensory estimates for personalizing care. *IEEE J. Biomed. Health Inform.* **2018**, *23*, 47–58. [CrossRef] [PubMed]
18. Song, H.; Kim, M.; Park, D.; Shin, Y.; Lee, J.G. Learning from noisy labels with deep neural networks: A survey. *IEEE Trans. Neural Netw. Learn. Syst.* **2022**. [CrossRef] [PubMed]
19. Cheng, L.; Zhou, X.; Zhao, L.; Li, D.; Shang, H.; Zheng, Y.; Pan, P.; Xu, Y. Weakly supervised learning with side information for noisy labeled images. In *Proceedings of the Computer Vision–ECCV 2020: 16th European Conference, Glasgow, UK, 23–28 August 2020, Proceedings, Part XXX 16*; Springer: Berlin/Heidelberg, Germany, 2020; pp. 306–321.
20. Lee, K.; Yun, S.; Lee, K.; Lee, H.; Li, B.; Shin, J. Robust inference via generative classifiers for handling noisy labels. In Proceedings of the International Conference on Machine Learning, PMLR, Long Beach, CA, USA, 9–15 June 2019; pp. 3763–3772.
21. Chen, P.; Liao, B.B.; Chen, G.; Zhang, S. Understanding and utilizing deep neural networks trained with noisy labels. In Proceedings of the International Conference on Machine Learning, PMLR, Long Beach, CA, USA, 9–15 June 2019; pp. 1062–1070.
22. Yu, X.; Han, B.; Yao, J.; Niu, G.; Tsang, I.; Sugiyama, M. How does disagreement help generalization against label corruption? In Proceedings of the International Conference on Machine Learning, PMLR, Long Beach, CA, USA, 9–15 June 2019; pp. 7164–7173.
23. Lyu, X.; Wang, J.; Zeng, T.; Li, X.; Chen, J.; Wang, X.; Xu, Z. TSS-Net: Two-stage with sample selection and semi-supervised net for deep learning with noisy labels. In Proceedings of the Third International Conference on Intelligent Computing and Human-Computer Interaction (ICHCI 2022), SPIE, Guangzhou, China, 12–14 August 2022; Volume 12509, pp. 575–584.
24. Shen, Y.; Sanghavi, S. Learning with bad training data via iterative trimmed loss minimization. In Proceedings of the International Conference on Machine Learning, PMLR, Long Beach, CA, USA, 9–15 June 2019; pp. 5739–5748.
25. Ghosh, A.; Manwani, N.; Sastry, P. Making risk minimization tolerant to label noise. *Neurocomputing* **2015**, *160*, 93–107. [CrossRef]
26. Ghosh, A.; Kumar, H.; Sastry, P.S. Robust loss functions under label noise for deep neural networks. In Proceedings of the AAAI Conference on Artificial Intelligence, San Francisco, CA, USA, 4–9 February 2017; Volume 31.
27. Albarqouni, S.; Baur, C.; Achilles, F.; Belagiannis, V.; Demirci, S.; Navab, N. Aggnet: Deep learning from crowds for mitosis detection in breast cancer histology images. *IEEE Trans. Med. Imaging* **2016**, *35*, 1313–1321. [CrossRef]
28. Rodrigues, F.; Pereira, F. Deep learning from crowds. In Proceedings of the AAAI Conference on Artificial Intelligence, New Orleans, LA, USA, 2–7 February 2018; Volume 32.
29. Van der Maaten, L.; Hinton, G. Visualizing data using t-SNE. *J. Mach. Learn. Res.* **2008**, *9*, 2579–2605.
30. Chattopadhay, A.; Sarkar, A.; Howlader, P.; Balasubramanian, V.N. Grad-cam++: Generalized gradient-based visual explanations for deep convolutional networks. In Proceedings of the 2018 IEEE Winter Conference on Applications of Computer Vision (WACV), Lake Tahoe, NV, USA, 12–15 March 2018; pp. 839–847.
31. Rizos, G.; Schuller, B.W. Average jane, where art thou?–recent avenues in efficient machine learning under subjectivity uncertainty. In Proceedings of the International Conference on Information Processing and Management of Uncertainty in Knowledge-Based Systems, Lisbon, Portugal, 15–19 June 2020; pp. 42–55.
32. Zhang, J.; Wu, X.; Sheng, V.S. Imbalanced multiple noisy labeling. *IEEE Trans. Knowl. Data Eng.* **2014**, *27*, 489–503. [CrossRef]
33. Dawid, A.P.; Skene, A.M. Maximum likelihood estimation of observer error-rates using the EM algorithm. *J. R. Stat. Soc. Ser. C (Appl. Stat.)* **1979**, *28*, 20–28. [CrossRef]
34. Raykar, V.C.; Yu, S.; Zhao, L.H.; Valadez, G.H.; Florin, C.; Bogoni, L.; Moy, L. Learning from crowds. *J. Mach. Learn. Res.* **2010**, *11*, 1297–1322.
35. Groot, P.; Birlutiu, A.; Heskes, T. Learning from multiple annotators with Gaussian processes. In Proceedings of the International Conference on Artificial Neural Networks, Espoo, Finland, 14–17 June 2011; pp. 159–164.
36. Xiao, H.; Xiao, H.; Eckert, C. Learning from multiple observers with unknown expertise. In Proceedings of the Pacific-Asia Conference on Knowledge Discovery and Data Mining, Gold Coast, Australia, 14–17 April 2013; pp. 595–606.
37. Morales-Alvarez, P.; Ruiz, P.; Coughlin, S.; Molina, R.; Katsaggelos, A.K. Scalable variational Gaussian processes for crowdsourcing: Glitch detection in LIGO. *IEEE Trans. Pattern Anal. Mach. Intell.* **2020**, *44*, 1534–1551. [CrossRef]
38. Gil-Gonzalez, J.; Alvarez-Meza, A.; Orozco-Gutierrez, A. Learning from multiple annotators using kernel alignment. *Pattern Recognit. Lett.* **2018**, *116*, 150–156. [CrossRef]
39. Morales-Álvarez, P.; Ruiz, P.; Santos-Rodríguez, R.; Molina, R.; Katsaggelos, A.K. Scalable and efficient learning from crowds with Gaussian processes. *Inf. Fusion* **2019**, *52*, 110–127. [CrossRef]
40. Rodrigues, F.; Pereira, F.; Ribeiro, B. Sequence labeling with multiple annotators. *Mach. Learn.* **2014**, *95*, 165–181. [CrossRef]
41. Wang, X.; Bi, J. Bi-convex optimization to learn classifiers from multiple biomedical annotations. *IEEE/ACM Trans. Comput. Biol. Bioinform.* **2016**, *14*, 564–575. [CrossRef]

42. Zhang, Z.; Sabuncu, M. Generalized cross entropy loss for training deep neural networks with noisy labels. *Adv. Neural Inf. Process. Syst.* **2018**, *31*, 1–11.
43. Gil-González, J.; Valencia-Duque, A.; Álvarez-Meza, A.; Orozco-Gutiérrez, Á.; García-Moreno, A. Regularized chained deep neural network classifier for multiple annotators. *Appl. Sci.* **2021**, *11*, 5409. [CrossRef]
44. Zhao, X.; Li, X.; Bi, D.; Wang, H.; Xie, Y.; Alhudhaif, A.; Alenezi, F. L1-norm constraint kernel adaptive filtering framework for precise and robust indoor localization under the internet of things. *Inf. Sci.* **2022**, *587*, 206–225. [CrossRef]
45. Box, G.E.; Cox, D.R. An analysis of transformations. *J. R. Stat. Soc. Ser. B (Methodol.)* **1964**, *26*, 211–243. [CrossRef]
46. Saul, A.; Hensman, J.; Vehtari, A.; Lawrence, N. Chained Gaussian processes. In Proceedings of the Artificial Intelligence and Statistics, Cadiz, Spain, 9–11 May 2016; pp. 1431–1440.
47. Géron, A. *Hands-On Machine Learning with Scikit-Learn, Keras, and TensorFlow: Concepts, Tools, and Techniques to Build Intelligent Systems*; O'Reilly Media: Sebastopol, CA, USA, 2019.
48. Xiao, H.; Rasul, K.; Vollgraf, R. Fashion-mnist: A novel image dataset for benchmarking machine learning algorithms. *arXiv* **2017**, arXiv:1708.07747.
49. Hernández-Muriel, J.A.; Bermeo-Ulloa, J.B.; Holguin-Londoño, M.; Álvarez-Meza, A.M.; Orozco-Gutiérrez, Á.A. Bearing health monitoring using relief-F-based feature relevance analysis and HMM. *Appl. Sci.* **2020**, *10*, 5170. [CrossRef]
50. LeCun, Y.; Boser, B.; Denker, J.; Henderson, D.; Howard, R.; Hubbard, W.; Jackel, L. Handwritten digit recognition with a back-propagation network. *Adv. Neural Inf. Process. Syst.* **1989**, *2*, 396–404.
51. Dogs vs. Cats—Kaggle.com. Available online: https://www.kaggle.com/c/dogs-vs-cats (accessed on 6 January 2023).
52. Peterson, J.C.; Battleday, R.M.; Griffiths, T.L.; Russakovsky, O. Human uncertainty makes classification more robust. In Proceedings of the IEEE/CVF International Conference on Computer Vision, Seoul, Republic of Korea, 27–28 October 2019; pp. 9617–9626.
53. Simonyan, K.; Zisserman, A. Very deep convolutional networks for large-scale image recognition. *arXiv* **2014**, arXiv:1409.1556.
54. Tzanetakis, G.; Cook, P. Musical genre classification of audio signals. *IEEE Trans. Speech Audio Process.* **2002**, *10*, 293–302. [CrossRef]
55. Rodrigues, F.; Pereira, F.; Ribeiro, B. Learning from multiple annotators: Distinguishing good from random labelers. *Pattern Recognit. Lett.* **2013**, *34*, 1428–1436. [CrossRef]
56. Gil-Gonzalez, J.; Giraldo, J.J.; Alvarez-Meza, A.; Orozco-Gutierrez, A.; Alvarez, M. Correlated Chained Gaussian Processes for Datasets with Multiple Annotators. *IEEE Trans. Neural Netw. Learn. Syst.* **2021**. [CrossRef]
57. MacKay, D.J.; Mac Kay, D.J. *Information Theory, Inference and Learning Algorithms*; Cambridge University Press: Cambridge, UK, 2003.
58. Demšar, J. Statistical comparisons of classifiers over multiple data sets. *J. Mach. Learn. Res.* **2006**, *7*, 1–30.
59. Li, Z.L.; Zhang, G.W.; Yu, J.; Xu, L.Y. Dynamic Graph Structure Learning for Multivariate Time Series Forecasting. *Pattern Recognit.* **2023**, *138*, 109423. [CrossRef]
60. Leroux, L.; Castets, M.; Baron, C.; Escorihuela, M.J.; Bégué, A.; Seen, D.L. Maize yield estimation in West Africa from crop process-induced combinations of multi-domain remote sensing indices. *Eur. J. Agron.* **2019**, *108*, 11–26. [CrossRef]
61. Montavon, G.; Binder, A.; Lapuschkin, S.; Samek, W.; Müller, K.R. Layer-wise relevance propagation: An overview. In *Explainable AI: Interpreting, Explaining and Visualizing Deep Learning*; Springer Nature: Cham, Switzerland, 2019; pp. 193–209.
62. Holzinger, A.; Saranti, A.; Molnar, C.; Biecek, P.; Samek, W. Explainable AI methods-a brief overview. In *Proceedings of the xxAI-Beyond Explainable AI: International Workshop, Held in Conjunction with ICML 2020, Vienna, Austria, 18 July 2020, Revised and Extended Papers*; Springer: Berlin/Heidelberg, Germany, 2022; pp. 13–38.
63. Bennetot, A.; Donadello, I.; Qadi, A.E.; Dragoni, M.; Frossard, T.; Wagner, B.; Saranti, A.; Tulli, S.; Trocan, M.; Chatila, R.; et al. A practical tutorial on explainable ai techniques. *arXiv* **2021**, arXiv:2111.14260.
64. Saranti, A.; Hudec, M.; Mináriková, E.; Takáč, Z.; Großschedl, U.; Koch, C.; Pfeifer, B.; Angerschmid, A.; Holzinger, A. Actionable Explainable AI (AxAI): A Practical Example with Aggregation Functions for Adaptive Classification and Textual Explanations for Interpretable Machine Learning. *Mach. Learn. Knowl. Extr.* **2022**, *4*, 924–953. [CrossRef]

Disclaimer/Publisher's Note: The statements, opinions and data contained in all publications are solely those of the individual author(s) and contributor(s) and not of MDPI and/or the editor(s). MDPI and/or the editor(s) disclaim responsibility for any injury to people or property resulting from any ideas, methods, instructions or products referred to in the content.

Article

Deep Learning for Combating Misinformation in Multicategorical Text Contents

Rafał Kozik [1,*], Wojciech Mazurczyk [2], Krzysztof Cabaj [2], Aleksandra Pawlicka [3], Marek Pawlicki [1] and Michał Choraś [1]

1. Faculty of Telecommunications, Computer Science and Electrical Engineering, Bydgoszcz University of Science and Technology, 85-796 Bydgoszcz, Poland
2. Institute of Computer Science, Division of Software Engineering and Computer Architecture, Warsaw University of Technology, 00-661 Warsaw, Poland
3. Faculty of Applied Linguistics, University of Warsaw, 00-927 Warsaw, Poland
* Correspondence: rafal.kozik@pbs.edu.pl

Abstract: Currently, one can observe the evolution of social media networks. In particular, humans are faced with the fact that, often, the opinion of an expert is as important and significant as the opinion of a non-expert. It is possible to observe changes and processes in traditional media that reduce the role of a conventional 'editorial office', placing gradual emphasis on the remote work of journalists and forcing increasingly frequent use of online sources rather than actual reporting work. As a result, social media has become an element of state security, as disinformation and fake news produced by malicious actors can manipulate readers, creating unnecessary debate on topics organically irrelevant to society. This causes a cascading effect, fear of citizens, and eventually threats to the state's security. Advanced data sensors and deep machine learning methods have great potential to enable the creation of effective tools for combating the fake news problem. However, these solutions often need better model generalization in the real world due to data deficits. In this paper, we propose an innovative solution involving a committee of classifiers in order to tackle the fake news detection challenge. In that regard, we introduce a diverse set of base models, each independently trained on sub-corpora with unique characteristics. In particular, we use multi-label text category classification, which helps formulate an ensemble. The experiments were conducted on six different benchmark datasets. The results are promising and open the field for further research.

Keywords: deep learning; fake news; ensemble of classifiers; text classification; misinformation

1. Introduction

Today, the level of expertise, education, or experience is no longer an obstacle to spreading opinions and becoming a content creator. Through the Internet, content creators are able to remain fully anonymous if they so choose. Nevertheless, this situation has laid the groundwork for the fake news trend to surface. Though the disinformation problem is as old as modern civilization, the evolution of digital media has dramatically transformed the way deception is spread. Eventually, it became a highly influential weapon.

With the evolution of machine learning and, in particular, the development of deep learning NLP-based methods, an array of new tools has been proposed to combat fake news problems [1–3].

However, most of the analyzed solutions propose a monolithic approach that focuses on fake news detection as a single-task learning problem in which the entire ML model is trained mainly from scratch [4–6]. Moreover, only some approaches consider domain segmentation before fake news detection, as, typically, the existing methods mix different types of models for feature extraction and other classification solutions.

Therefore, in this paper, the authors put forward a more scalable solution where a committee of classifiers is composed to address the problem of fake news detection. In more detail, the authors propose an alternative, novel approach in which the construction of a diversified pool of base models is learned independently on a sub-corpus of texts with unique characteristics (resulting directly from multi-label classification). This improves the detection efficiency of fake news in textual form.

The paper is structured as follows. First, the related work is discussed in Section 2. Next, in Section 3, the authors introduce the framework for the proposed deep learning solution. The experimental evaluation results are included in Section 4. Finally, the paper closes with final remarks, at the same time outlining the perspectives on future work in Section 5.

2. Related Work

At the beginning of this millennium, it has started to become evident that the diversification of sources providing online news, as well as the utilization of social network sites to filter and consume them, may lead to the risk of containing its users within inevitable "bubbles" where only information that goes along with their intuition is presented to them [7]. Moreover, online opinion leaders have begun to appear and increasingly influence online communities, often using fake news [8].

The rapid expansion of fake news has put more pressure on legitimate news sources to ensure reliable information and create tools to verify the presented facts. In this vein, various projects started to appear. Firstly, fact-checkers, i.e., specialized websites, have been widely adopted by consumers. This includes, for instance, FactCheck.org in the USA, Maldita.es in Spain, Demagog.org.pl in Poland, and FactCheck Initiative in Japan. These websites allow for, e.g., the presentation of multimedia content metadata so anyone can verify its originality. However, there are better solutions than this, as metadata can be easily modified, and there is no information on what has been changed in such content. Secondly, there are solutions like Trustproject (https://thetrustproject.org/), where over 120 news organizations, including well-known media companies such as the BBC, South China Morning Post, and Bay Area News Group, are working towards greater transparency and accountability in the global news industry. Trustproject provides a protocol encompassing eight indicators of trust. For example, one of them concerns the capability of the news addressee to assess the journalist's level of expertise. Although news consumers have welcomed solutions like Trustproject, it must be noted that their procedures still need to be automated owing to the lack of suitable tools.

Apart from the above, many technological developments have also been used to try to detect fake news. These solutions have already been analyzed in several surveys on this topic, e.g., in [9–11].

One such solution is digital watermarking, which may be used to overcome issues such as copyright protection, content authentication, detection of tampering, and so on [12]. For example, one of the watermarking applications is fingerprinting, which allows one to apply a unique fingerprint, which may then be utilized to identify the recipient of the content; this is true for each individual copy of the disseminated content. Owing to that, it becomes possible to deter illegal redistribution as it makes it possible for the content owner to find out where the redistributed copy came from [13,14]. However, it has been proved that the application of digital watermarking techniques is of great value in the domain of fake news identification and tracing; this line of research is currently pursued within the DISSIMILAR project [15].

Many works have been devoted to tackling the problem of fake news detection in various types of content (text, multimedia, etc.) using AI-based solutions.

At present, the research community is casting a discerning eye on one particular form of fake news: deepfakes. In these fake videos, it is practically impossible to distinguish whether they are forged or not. This brings an urgent need to create automated detection methods. The existing literature on this subject discusses several attempts at countering deepfakes [16] or fake news in images [17]. Conventionally established methodologies for identifying counterfeit videos center on discerning the subtle features present within these manipulated recordings. For instance, the technique introduced in [18] relies on identifying eye blinking, as this physiological signal is typically incorrectly mapped in synthesized fake videos. Conversely, the approach presented in [19] offers a visualization of the CNN layers and filters, demonstrating that the eyes and mouth are instrumental in identifying faces tampered with by prevailing deepfake software utilities [20,21]. A recent paper presents another solution, i.e., DeepTag, an end-to-end deep watermarking framework incorporating a GAN simulator that can employ common distortions to facial images [22]. Consequently, one can extract the watermark to discern the original, untouched facial image.

It is also worth noting that to accelerate advancements in the detection of fake media, the DFDC (DeepFake Detection Challenge) dataset [23] was created and released by several major industry players like Amazon Web Services (AWS), Facebook, Microsoft, and the Partnership on AI's Media Integrity Steering Committee and academics. The main aim of this challenge is to boost research to design and develop innovative new solutions to expedite deepfakes and manipulated media detection.

Additionally, significant research has been devoted to detecting fake news in textual information. In [24], for instance, the authors introduce an innovative model for automated fake news credibility assessment. Leveraging a combination of overt and latent attributes derived from text, they establish a deep diffusive network model that concurrently learns representations for news pieces, their authors, and topics.

In [4], Khan et al. investigate the performance of various classical machine learning techniques and neural network models for textual fake news detection using three different datasets. They concluded that the best-performing solution was Naive Bayes with n-gram (bigram TF-IDF) features.

Next, in [5], the authors simultaneously analyze the correlations of publisher bias, news stance, and relevant user engagements. Then, based on these observations, they introduce a novel fake news detection framework. The experimental results obtained on two comprehensive real-world fake news datasets prove the effectiveness of the proposed approach.

Another fake news detection approach is described in [6]. The authors use propagation features to detect fake news on Twitter. They note that real news is significantly larger than fake news and is typically spread by users who have been active on Twitter for a long time and have more followers and fewer followings. Then, using these features, they train a Random Forest classifier, which results in 87% accuracy. Finally, they also use Geometric Deep Learning solutions and create a graph neural network that directly learns from the propagation graphs and results in 73.3% accuracy.

Huh et al. [25] unveil a learning algorithm adept at identifying visual alterations in images. Remarkably, it is trained solely on a substantial collection of authentic photographs and capitalizes on the inherent photo EXIF metadata as the guiding beacon for model training. This allows for evaluating if an image is produced by a single imaging pipeline. The experimental analysis revealed that the proposed approach achieves state-of-the-art performance on the chosen image forensics benchmark datasets.

In [1], the authors present a deep hierarchical co-attention network. This network is designed to learn feature representations for fake news detection and to identify relevant sentences or comments. The experimental evaluation conducted on real-world datasets showed that the proposed framework is effective. Another attention-based approach was proposed in [26].

Then, in [27], another approach to supervised learning-based fake news detection is presented. Apart from analysing the usability of features proposed in the literature, the authors point to a new set of features and evaluate the prediction performance of existing solutions and features for performing automatic detection. The obtained experimental results prove that the prediction performance of the utilized features in combination with state-of-the-art classifiers has suitable discriminative power for detecting fake news.

In [28], Perez-Rosas et al. discuss the automatic identification of textual fake news. Apart from proposing a new detection approach capable of achieving accuracy ca. 76%, the authors also introduce two datasets that can help the research community in benchmarking fake news detection methods. Additionally, the paper also includes the results of a comparative analysis of the automatic and manual identification of fake news.

A unique strategy for fake news detection was introduced in [28]. This method merges the publishing and sharing patterns of both publishers and users to enhance the accuracy of detecting fake news and predicting its credibility. When tested on three actual datasets, the method showcased its capability by achieving a detection accuracy surpassing 91%.

The following solution to fake news detection proposed in [29] emphasizes considering relational features like sentiment, named entities, and facts extracted from structured and unstructured data. Based on the presented experimental evaluation, the obtained results are generally better for all tested classifiers.

O'Brien et al. [2] try to overcome a lack of transparency in the decision-making process (black-box problem) of deep neural networks by showing that the emergent representations of such networks can pinpoint subtle differences between fake and real news. This makes convolutional neural networks a powerful tool for detecting fake news. The authors' evaluation also shows the generalization capabilities of such solutions in detecting fake news in novel subjects based solely on language patterns.

The UMLARD model was proposed in [30]. It focuses on rumor detection. In particular, the authors tackled the problem of learning representations from different user perspectives. They utilized a fusion mechanism to enhance prediction. The authors reported performance improvements with respect to various baseline methods.

In [31], the researchers introduce a Deep Normalized Attention-based mechanism designed to enhance the extraction of dual emotion features. The proposal also uses Adaptive Genetic Weight Update-Random Forest (AGWu-RF), which is utilized for the classification. Similarly, in [32] the authors propose a novel rumor detection model called graph contrastive learning with feature augmentation (FAGCL). This technique allows the author to inject noise into the feature space, which enables the model to learn contrastively by constructing asymmetric structures.

A fast fake news detection model was proposed in [33]. This approach is particularly focused on cyber–physical social services. The authors mostly rely on Chinese texts. Moreover, they argue that fake news items are generally short texts and can be effectively accompanied by relevant keywords. In order to extract feature vectors for the analyzed texts, a convolution neural network (CNN) is facilitated.

In the quest to combat misinformation, traditional methods often rely on intricate features or credibility networks, requiring expert knowledge and extensive engineering. Recent advances in deep learning have offered promising approaches, but they struggle with over-reliance on content features and neglect the individual user's impact in spreading rumors. Addressing these challenges, the proposed UMLARD model (User-aspect Multi-view Learning with Attention for Rumor Detection) effectively captures different user perspectives, combines them using a distinctive fusion mechanism, and integrates these learned features with content features for improved rumor detection. Experiments on real-world datasets demonstrate that UMLARD surpasses existing methods, providing both enhanced performance and interpretability.

Finally, in [34], the authors develop a deep learning-based system for concept extraction and relation identification. The solution is based on the BERT ensemble using a majority voting strategy.

The analysis of the state-of-the-art reveals various limitations of the existing methods, which the authors intend to address in this work. These can be summarized as follows:

- Observation: The majority of the analyzed solutions propose a monolithic approach that focuses on fake news detection as a single-task learning problem (STL), where the entire ML model is trained mainly from scratch.
Proposed solution: In this paper, the authors propose a more scalable solution where a committee of classifiers is composed to address the fake news detection problem.
- Observation: Very few approaches consider domain segmentation prior to fake news detection. Instead, the authors frequently mix different types of models for feature

extraction (e.g., RNN, CNN) and different classification methods (e.g., DNN, SVM, RF, etc.).

Proposed solution: In this approach, a diversified pool of base models is considered that were trained independently on a sub-corpus of texts with unique characteristics.

3. Proposed Solution

In this section, the authors outline the architecture of the approach proposed in this paper. Then, they characterize how data harvesting and open media crawling are typically performed. Finally, the utilized document representation and classification are described.

3.1. Architecture

The architecture of the proposed solution is presented in Figure 1. The textual data were collected from open data sources (e.g., news providers) utilizing data harvesters (described in Section 3.2). Next, the collected data were normalized regarding the format so that all documents were stored in a similar and coherent way (e.g., title, document, body, author, and date when the article was published). It must be emphasized here that the feature vectors (the document encoding step) are extracted only from the document body. This way, one can avoid judging the document by its author or origins. The extracted features rely on the BERT language model described in Section 3.4. The novelty of the architecture is the fact that the document source is predicted, which drives the votes obtained by several (domain-related) content classification models.

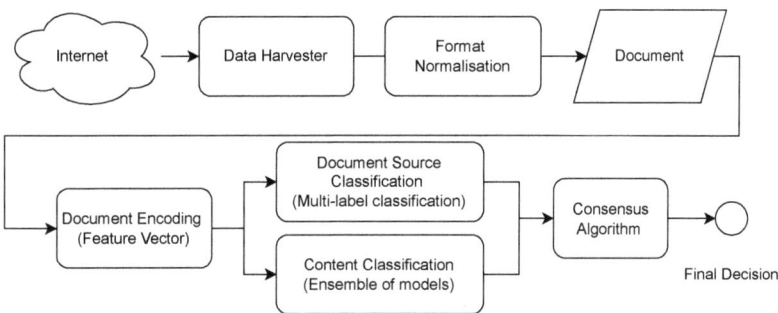

Figure 1. Information flow used by the proposed approach.

3.2. Data Harvesting and Open Media Crawling

In fake news-related research, an important question is always associated with data sources. One solution that could be utilized for this purpose is media crawling, often called harvesting. In this technique, specialized software, which behaves as a browser with custom logic, automatically visits a predefined list of resources on the Internet and downloads certain content (e.g., files) for further analysis. These solutions have been used during various kinds of research for years. The first papers describing such systems in the IEEExplore database are from the late 1990s, for example, [35,36].

Access to data gathered by web crawlers can be achieved by various means. A short survey presented in [37] defines three methods of how a researcher can obtain data via web harvesters. The first one utilizes dedicated extensions to the web browser, which record some valuable information to JSON, XLS, CSV, or HTML files. Currently, such a solution can be used within Firefox and Google Chrome browsers. The second solution is related to dedicated Open Source, commercial, or is even delivered as SaaS products. The provided software harvests all previously defined data sources. The third method utilizes some auxiliary tools and libraries that the researcher can use to develop a custom harvester specialized in the nature of the conducted research. This solution requires the highest effort from the three ways mentioned above. However, it also gives the best results. Further on in this section, the latter kind of solution is discussed.

Figure 2 presents the architecture of a custom harvester which was developed during the DISSIMILAR project. The system consists of two standard parts: (i) the front-end and (ii) the back-end.

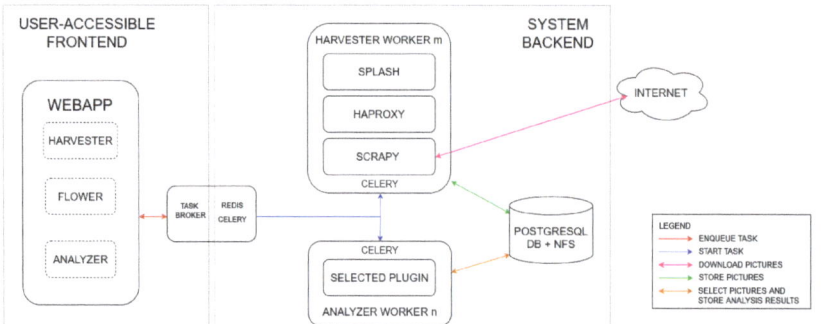

Figure 2. Architecture of the harvester, which consists of front-end (for administration) and back-end (for data harvesting) parts. The core of the system utilizes Celery task distribution framework, which controls data crawling process.

While the front-end is designed for system administration, the back-end focuses on data crawling. The connection to web servers is performed via Scrapy library (https://scrapy.org/, accessed on 21 October 2022). Since modern web pages, in most cases, are dynamically created using JavaScript, Splash framework (https://splash.readthedocs.io/en/stable/, accessed on 21 October 2022) is used to execute scripts and provide the final version of the generated HTML code.

In order to have control over the harvesting process, we developed the processing pipeline on top of the Celery task distribution subsystem. This allows for controlling dedicated workers delegated to harvesting and analyzing processes. Moreover, such an architecture enables the distribution of web scraping on multiple machines.

3.3. BERT-Based Document Representation and Classification

As part of the research, a collection of benchmark sets (commonly used by various researchers) was compiled. This allowed us to create an integrated language corpus that eventually mixed various categories of content (e.g., politics, health, news, etc.). Because the datasets differ in terms of various characteristics (e.g., text length), there is a separate problem related to the extraction of features from documents.

To address this, a unified and common methodology for all datasets' feature extraction was adopted. In particular, the same BERT language model (see Figure 3) was adapted to establish word, sentence, and document representation. In that regard, each input sentence is first transformed by a tokenizer, which breaks the sequence of words into individual tokens and then adds unique [CLS] (Classification) and [SEP] (Separator) tokens at the beginning and the end, respectively. Moreover, each token is replaced with the corresponding unique identifiers, which are established from the so-called embedding table. To simplify the text processing and classification pipeline, the tokenizer is configured to either truncate or pad sentences to a maximum of 512 tokens.

In this paper, the authors adapted the pre-trained DistilBERT model, which is a 'lightweight' variant of BERT. While training the entire model (DistilBERT + classification head), the BERT layers are left frozen during this process, and only the weights in the classification head are updated. In order to decrease the feature vector dimension, average pooling over the tokens returned by the BERT model was applied.

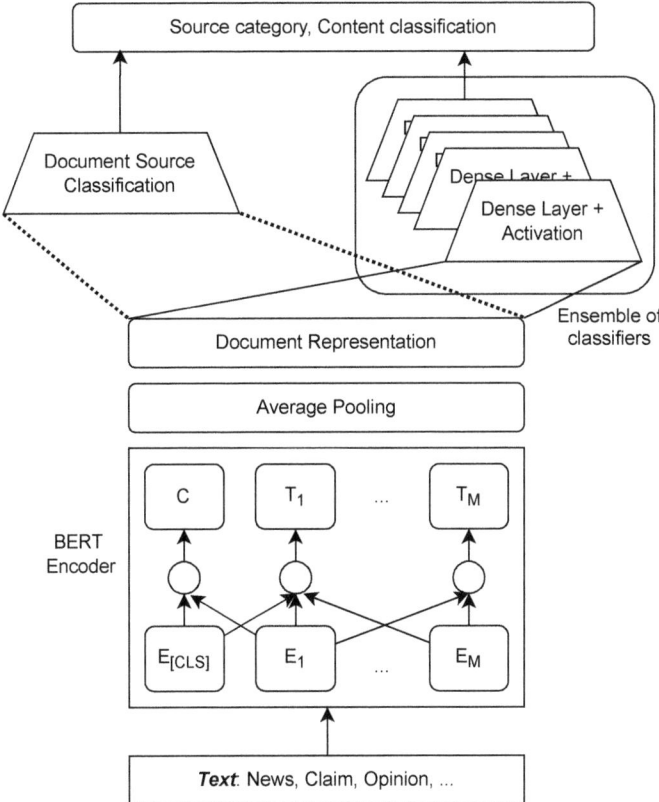

Figure 3. Document representation and classification model.

3.4. Document Source Classification

As is visible in Figure 3, on top of the document representation layer are two classification pathways. The right pathway is actually a committee of classifiers. In this case, each model in the ensemble is specialized in a different type of document. The one on the left is responsible for document source classification. In this paper, we used six different datasets, as detailed in Section 4.2, and therefore, each dataset is considered as a different source. The role of the document source classifier is to build consensus in the community in the above-mentioned processing pathway. More precisely, the classifier outputs the weights which are assigned to the responses of the base models. For example, if we have a document entirely devoted to COVID, (most likely) the document source classifier will delegate the prediction (will associate the highest weight) to the responses provided by the base model, which has competencies in this domain. Moreover, we used a diverse set of base models that we independently trained on sub-corpora with unique characteristics. Therefore, (from our perspective) multi-label text category classification is the entity that formulates the abovementioned ensemble.

4. Results

First, this section describes the data and methodology used to conduct the experimental evaluation. Next, the results for a single dataset are presented. Then, the outcomes of the text source classification accuracy are outlined. Finally, the authors focus on the results from the ensemble of BERT models.

4.1. Methodology and Tools Used for Experiments

In order to compare different approaches and configurations on different datasets, the 5x2-fold cross-validation method was employed. In this method, conventional two-fold cross-validation is utilized, with outcomes being averaged. Additionally, the standard deviation from the mean is computed to highlight both the variability and the importance of the distinctions. The 5x2-fold cross-validation (CV) method's nested structure helps achieve a more robust model. The outer loop helps assess how well the model generalizes across different subsets of the data, and the inner loop (two-fold CV) can help reduce the variance in estimates compared to a single-fold validation.

To assess the performance, we used common metrics for comparing classification effectiveness, namely the following:

- Precision = $\frac{TP}{TP+FP}$.
- Recall (Sensitivity) = $\frac{TP}{TP+FN}$.
- Specificity = $\frac{TN}{TN+FP}$.
- Balanced Accuracy = $\frac{Sensitivity + Specificity}{2}$.
- F1-score = $2 * \frac{Precision * Recall}{Precision + Recall}$.
- G-mean = $\sqrt{Sensitivity * Specificity}$.

In the above equations, TP, FP, FN, and TN indicate True Positive, False Positive, False Negative, and True Negative, respectively. TP (True Positive) is the number of fake news cases correctly classified as such. FN (False Negative) is the number of fake news cases incorrectly classified as real news. FP (False Positive) is the number of real news cases that are incorrectly identified as Fake, and TN (True Negative) is the number of real news cases correctly classified as such.

4.2. Data Used for Experiments

Six different datasets related to the fake news detection problem were used in all experiments. The details of the datasets, as well as the numbers they will be referred to as in this paper, are presented in Table 1.

Table 1. Summary of datasets related to news authenticity.

No.	Dataset Name	Size	Content Type	Description	Reference
1	COVID-19 Fake News	10,200 items (~9700 fake, ~500 real)	News (COVID-19)	Fake and legitimate news items related to COVID-19.	[38]
2	MM-Covid	11,000 items (~4000 fake, ~7000 real)	News (Multilingual)	Fake and legitimate news items with social context.	[39]
3	Q-Prop	50,000 articles	News	Articles labeled as "legitimate" or "propaganda".	[40]
4	ISOT	44,898 documents (23,481 fake, 21,417 real)	News	Trustworthy and fake documents with additional metadata.	[41]
5	GRAFN	13,000 posts	News (Global Politics)	Text and metadata from 244 websites.	[42]
6	PubHealth	Not specified	Health (Public /Biomedical)	Documents with labels and explanations related to public health topics.	[43]

4.3. Single Dataset Performance

In this subsection, for the classification, we report the accuracy obtained with a BERT-based classifier (as defined in Section 3) on the considered datasets. The results are reported in Table 2. In all cases, the authors repeated five times two-fold cross-validation (5x2 CV) on a single dataset. The presented outcomes are shown as averages along with the standard deviation from the mean value. In the case of the *F1*, *Precision*, and *Recall* metrics, a weighted average calculated on both classes (fake news and real news) is reported. Here, it can be noticed that for all datasets, the BERT model achieves relatively high values of performance metrics (e.g., Balanced Accuracy exceeds 75%).

Table 2. Performance of BERT-based classifier trained and evaluated on the same datasets (results reported using 5x2-fold cross-validation methodology).

Dataset	Bal. Acc.	F1	G-Mean	Precision	Recall
Covid-FN	87.6 ± 0.8	91.9 ± 0.4	87.0 ± 0.9	92.2 ± 0.3	92.2 ± 0.3
MMCovid	78.8 ± 0.5	80.2 ± 0.5	78.6 ± 0.5	80.2 ± 0.5	80.2 ± 0.5
QProp	77.3 ± 1.3	96.7 ± 0.2	74.1 ± 1.8	96.7 ± 0.2	97.0 ± 0.2
ISOT	76.7 ± 0.4	92.6 ± 0.1	73.8 ± 0.6	92.5 ± 0.1	93.1 ± 0.1
GRAFN	99.3 ± 0.0	99.2 ± 0.0	99.3 ± 0.0	99.2 ± 0.0	99.2 ± 0.0
PubHealth	79.9 ± 0.2	88.5 ± 0.1	78.4 ± 0.2	88.4 ± 0.1	88.9 ± 0.1

In such cases, the results can be promising and considered optimistic. However, the truth is that the model needs to be better generalized to other datasets. It is quite vivid when observing Figure 4. The diagram depicts the base model's performance pre-trained on one dataset and evaluated on the other. As a pre-trained model, one understands the best model amongst all validation folds. The bright diagonal in Figure 4 indicates that the classifier performs well on the original dataset. One can also notice some brighter areas, which may suggest that some datasets share similar contents (e.g., ISOT and GRAFN, which are highly populated with political news).

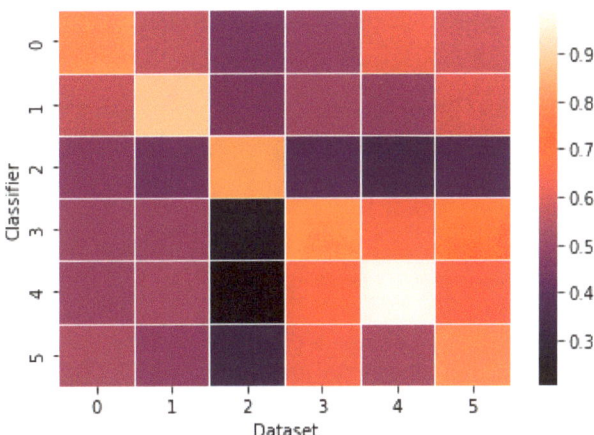

Figure 4. Balanced Accuracy achieved with BERT classifier (pre-trained on a single dataset and evaluated on another: 0—CovidFN, 1—MMCovid, 2—QProp, 3—ISOT, 4—GRAFN, 5—PubHealth).

4.4. Text Source Classification Accuracy

In Table 3, various metrics for multi-label source classification are presented. As in the previous cases, the 5x2 cross-validation was employed to compare and report the classification performance. The goal of the multi-label classification model is to correctly recognize the dataset (the source of the analyzed text). The worst results were obtained for QProp, while for the other datasets, performance metrics achieved relatively high values (e.g., F1 scores close to 90%). In Table 3, the Accuracy, Balanced Accuracy, and the G-mean are reported only as an average because these involve all labels when calculating their values (e.g., Balanced Accuracy requires an average of recall scores per class to be calculated).

Table 3. Performance of the proposed text source classification model (dataset name is considered here as a label returned by the model).

	F1	Precision	Recall
CovidFN	97.5 ± 0.1	97.2 ± 0.2	97.8 ± 0.2
MMCovid	88.7 ± 0.3	89.6 ± 0.7	87.8 ± 0.3
QProp	14.2 ± 0.9	45.6 ± 3.4	8.4 ± 0.6
ISOT	92.7 ± 0.1	91.8 ± 0.2	93.7 ± 0.1
GRAFN	92.1 ± 0.1	91.4 ± 0.1	92.8 ± 0.2
PubHealth	87.0 ± 0.2	87.9 ± 0.4	86.1 ± 0.4
Average	91.0 ± 0.1	90.9 ± 0.1	91.5 ± 0.1
Accuracy	91.5 ± 0.1		
Balanced Accuracy	77.8 ± 0.1		

4.5. Ensemble of BERT Models

Finally, in this subsection, the results obtained for the ensemble of BERT models are showcased. More precisely, the ensemble is constructed of a diversified pool of base models which learned independently from a sub-corpus of texts. The unique characteristics result from multi-label classification, which was presented in the previous section.

In Table 4, the authors compare the proposed approach to other techniques known from the literature, namely the following:

- Batch MTL—a single classification model is trained from scratch on the entire dataset.
- Majority Voting—an ensemble of the BERT model is constructed, and the consensus is established using a hard voting approach (e.g., [34]).
- Weighted Voting—an ensemble of BERT models combined with a soft voting approach (e.g., [44]).

Table 4. Classification performance of BERT model ensembles integrated using text source classifier.

Approach	Balanced Accuracy	Avg. F1
Proposed Ensemble Method	82.7 ± 0.1	84.8 ± 0.1
Batch MTL	77.7 ± 0.1	84.0 ± 0.1
Majority Voting BERT Ensemble	65.7 ± 0.1	67.3 ± 0.1
Weighted BERT Ensemble	54.4 ± 0.1	33.5 ± 0.2

Based on the results presented in Table 4, it is visible that the method proposed in this paper outperformed other state-of-the-art solutions. This demonstrates the potential of the proposed approach. Concerning the Balanced Accuracy, the introduced technique turned out to be better by almost 30% than the Weighted BERT Ensemble and 5% better than Batch MTL. Regarding the average F1 metric, the improvement over the Weighted BERT Ensemble is even more remarkable, i.e., about 50%, but at the same time comparable to Batch MTL (0.8% increase).

5. Conclusions

Advanced deep learning techniques offer potent tools for addressing the issue of false news. However, these systems often struggle with poor model generalization in real-world scenarios due to data limitations. Unfortunately, many existing solutions adopt a monolithic approach, treating fake news detection as a single-task learning problem and utilizing a variety of models and classification methods without considering domain segmentation.

To overcome these limitations, our paper proposes an innovative solution involving a committee of classifiers in order to tackle the fake news detection challenge. In that regard, we introduce a diverse set of base models, each independently trained on sub-corpora with unique characteristics. In order to evaluate different approaches and configurations on the considered datasets, we employed the 5x2-fold cross-validation (CV) method. This nested CV structure enhances model robustness by assessing generalization across different data subsets and reducing variance in estimates.

The experimental results (on six different benchmark datasets) show that the suggested approach is promising, opening further research areas. In particular, the proposed approach outperforms the Weighted BERT Ensemble by nearly 30% in Balanced Accuracy and exhibits a 50% improvement in the F1 metric. It also sustains competitiveness with Batch MTL, with a slight 0.8% increase in the F1 score.

The promising results and performance indicate the future direction and value of using ensembles of classifiers in misinformation detection systems.

Author Contributions: Conceptualization, R.K. and M.C.; Methodology, W.M., K.C., A.P. and M.C.; Formal analysis, M.C.; Investigation, M.P. and M.C.; Resources, M.C. All authors have read and agreed to the published version of the manuscript.

Funding: This research was funded by the National Center for Research and Development within the INFOSTRATEG program, number of application for funding: INFOSTRATEG-I/0019/2021-00, as well as within EIG CONCERT-Japan call to the project Detection of fake newS on SocIal MedIa pLAtfoRms "DISSIMILAR" through the grant EIG CONCERT-JAPAN/05/2021.

Institutional Review Board Statement: Not applicable.

Informed Consent Statement: Not applicable.

Data Availability Statement: No new data were created or analyzed in this study.

Conflicts of Interest: The authors declare no conflict of interest.

References

1. Shu, K.; Cui, L.; Wang, S.; Lee, D.; Liu, H. DEFEND: Explainable Fake News Detection. In Proceedings of the 25th ACM SIGKDD International Conference on Knowledge Discovery and Data Mining, KDD '19, Anchorage, AK, USA, 4–8 August 2019; Association for Computing Machinery: New York, NY, USA, 2019; pp. 395–405. [CrossRef]
2. O'Brien, N.; Latessa, S.; Evangelopoulos, G.; Boix, X. The Language of Fake News: Opening the Black-Box of Deep Learning Based Detectors. In Proceedings of the Workshop on "AI for Social Good", NIPS 2018, Montreal, QC, Canada, 2–8 December 2018.
3. Goldani, M.H.; Safabakhsh, R.; Momtazi, S. Convolutional neural network with margin loss for fake news detection. *Inf. Process. Manag.* **2021**, *58*, 102418. [CrossRef]
4. Khan, J.Y.; Khondaker, M.T.I.; Afroz, S.; Uddin, G.; Iqbal, A. A benchmark study of machine learning models for online fake news detection. *Mach. Learn. Appl.* **2021**, *4*, 100032. [CrossRef]
5. Shu, K.; Wang, S.; Liu, H. Exploiting Tri-Relationship for Fake News Detection. *arXiv* **2017**, arXiv:1712.07709.
6. Meyers, M.; Weiss, G.; Spanakis, G. Fake News Detection on Twitter Using Propagation Structures. In Proceedings of the Disinformation in Open Online Media, Leiden, The Netherlands, 26–27 October 2020; van Duijn, M., Preuss, M., Spaiser, V., Takes, F., Verberne, S., Eds.; Springer International Publishing: Cham, Switzerland, 2020; pp. 138–158.
7. Sunstein, C.R. *Echo Chambers: Bush V. Gore, Impeachment, and Beyond*; Princeton Digital Books+: Princeton, NJ, USA, 2001.
8. Guo, L.; Rohde, J.A.; Wu, H.D. Who is responsible for Twitter's echo chamber problem? Evidence from 2016 U.S. election networks. *Inf. Commun. Soc.* **2020**, *23*, 234–251. [CrossRef]
9. Zhou, X.; Zafarani, R. Fake News Detection: An Interdisciplinary Research. In Proceedings of the Companion Proceedings of The 2019 World Wide Web Conference, WWW '19, San Francisco, CA, USA, 13–17 May 2019; Association for Computing Machinery: New York, NY, USA, 2019; p. 1292. [CrossRef]
10. Mridha, M.F.; Keya, A.J.; Hamid, M.A.; Monowar, M.M.; Rahman, M.S. A Comprehensive Review on Fake News Detection With Deep Learning. *IEEE Access* **2021**, *9*, 156151–156170. [CrossRef]
11. Parikh, S.B.; Atrey, P.K. Media-Rich Fake News Detection: A Survey. In Proceedings of the 2018 IEEE Conference on Multimedia Information Processing and Retrieval (MIPR), Miami, FL, USA, 10–18 April 2018; pp. 436–441. [CrossRef]
12. Megías, D. Data Hiding: New Opportunities for Security and Privacy? In Proceedings of the European Interdisciplinary Cybersecurity Conference (EICC 2020), Rennes, France, 18 November 2020; pp. 1–6. [CrossRef]
13. Blakley, G.R.; Meadows, C.; Purdy, G.B. Fingerprinting Long Forgiving Messages. In *Advances in Cryptology—CRYPTO '85 Proceedings*; Williams, H.C., Ed.; Springer: Berlin/Heidelberg, Germany, 1986; pp. 180–189.
14. Boneh, D.; Shaw, J. Collusion-Secure Fingerprinting for Digital Data. *IEEE Trans. Inf. Theory* **1998**, *44*, 1897–1905. [CrossRef]

15. Megías, D.; Kuribayashi, M.; Rosales, A.; Cabaj, K.; Mazurczyk, W. Architecture of a fake news detection system combining digital watermarking, signal processing, and machine learning. *J. Wirel. Mob. Netw. Ubiquitous Comput. Dependable Appl.* **2022**, *13*, 33–55. [CrossRef]
16. Korus, P.; Memon, N. Content Authentication for Neural Imaging Pipelines: End-To-End Optimization of Photo Provenance in Complex Distribution Channels. In Proceedings of the 2019 IEEE/CVF Conference on Computer Vision and Pattern Recognition (CVPR), Long Beach, CA, USA, 15–20 June 2019; pp. 8613–8621. [CrossRef]
17. Jing, J.; Wu, H.; Sun, J.; Fang, X.; Zhang, H. Multimodal fake news detection via progressive fusion networks. *Inf. Process. Manag.* **2023**, *60*, 103120. [CrossRef]
18. Li, Y.; Chang, M.C.; Lyu, S. In Ictu Oculi: Exposing AI Created Fake Videos by Detecting Eye Blinking. In Proceedings of the 2018 IEEE International Workshop on Information Forensics and Security (WIFS), Hong Kong, China, 11–13 December 2018; IEEE: Piscataway, NJ, USA, 2018. [CrossRef]
19. Afchar, D.; Nozick, V.; Yamagishi, J.; Echizen, I. MesoNet: A Compact Facial Video Forgery Detection Network. In Proceedings of the 2018 IEEE International Workshop on Information Forensics and Security (WIFS), Hong Kong, China, 11–13 December 2018; IEEE: Piscataway, NJ, USA, 2018. [CrossRef]
20. Korshunov, P.; Marcel, S. DeepFakes: A New Threat to Face Recognition? Assessment and Detection. *arXiv* **2018**, arXiv:1812.08685.
21. Tolosana, R.; Rodriguez, R.V.; Fierrez, J.; Morales, A.; Garcia, J.O. Deepfakes and beyond: A Survey of face manipulation and fake detection. *Inf. Fusion* **2020**, *64*, 131–148. [CrossRef]
22. Wang, R.; Juefei-Xu, F.; Luo, M.; Liu, Y.; Wang, L. FakeTagger: Robust Safeguards against DeepFake Dissemination via Provenance Tracking. In Proceedings of the 29th ACM International Conference on Multimedia, Virtual Event, 20–24 October 2020. [CrossRef]
23. Dolhansky, B.; Howes, R.; Pflaum, B.; Baram, N.; Ferrer, C.C. The Deepfake Detection Challenge (DFDC) Preview Dataset. *arXiv* **2019**, arXiv:1910.08854.
24. Zhang, J.; Dong, B.; Yu, P.S. FakeDetector: Effective Fake News Detection with Deep Diffusive Neural Network. In Proceedings of the 2020 IEEE 36th International Conference on Data Engineering (ICDE), Dallas, TX, USA, 20–24 April 2020; IEEE Computer Society: Los Alamitos, CA, USA, 2020; pp. 1826–1829. [CrossRef]
25. Huh, M.; Liu, A.; Owens, A.; Efros, A.A. Fighting Fake News: Image Splice Detection via Learned Self-Consistency. In Proceedings of the Computer Vision–ECCV 2018, Munich, Germany, 8–14 September 2018; Ferrari, V., Hebert, M., Sminchisescu, C., Weiss, Y., Eds.; Springer International Publishing: Cham, Switzerland, 2018; pp. 106–124.
26. Bazmi, P.; Asadpour, M.; Shakery, A. Multi-view co-attention network for fake news detection by modeling topic-specific user and news source credibility. *Inf. Process. Manag.* **2023**, *60*, 103146. [CrossRef]
27. Reis, J.C.S.; Correia, A.; Murai, F.; Veloso, A.; Benevenuto, F. Supervised Learning for Fake News Detection. *IEEE Intell. Syst.* **2019**, *34*, 76–81. [CrossRef]
28. Pérez-Rosas, V.; Kleinberg, B.; Lefevre, A.; Mihalcea, R. Automatic Detection of Fake News. In Proceedings of the 27th International Conference on Computational Linguistics, Santa Fe, NM, USA, 20–26 August 2018; Association for Computational Linguistics: Toronto, ON, Canada, 2018; pp. 3391–3401.
29. Brașoveanu, A.M.P.; Andonie, R. Semantic Fake News Detection: A Machine Learning Perspective. In Proceedings of the Advances in Computational Intelligence, Gran Canaria, Spain, 12–14 June 2019; Rojas, I.; Joya, G.; Catala, A., Eds.; Springer International Publishing: Cham, Switzerland, 2019; pp. 656–667.
30. Chen, X.; Zhou, F.; Trajcevski, G.; Bonsangue, M. Multi-view learning with distinguishable feature fusion for rumor detection. *Knowl. Based Syst.* **2022**, *240*, 108085. [CrossRef]
31. Luvembe, A.; Li, W.; Li, S.; Liu, F.; Xu, G. Dual emotion based fake news detection: A deep attention-weight update approach. *Inf. Process. Manag.* **2023**, *60*, 103354. [CrossRef]
32. Li, S.; Li, W.; Luvembe, A.M.; Tong, W. Graph Contrastive Learning With Feature Augmentation for Rumor Detection. *IEEE Trans. Comput. Soc. Syst.* **2023**, 1–10. [CrossRef]
33. Zhang, Q.; Guo, Z.; Zhu, Y.; Vijayakumar, P.; Castiglione, A.; Gupta, B.B. A Deep Learning-based Fast Fake News Detection Model for Cyber-Physical Social Services. *Pattern Recognit. Lett.* **2023**, *168*, 31–38. [CrossRef]
34. Yang, X.; Zhang, H.; He, X.; Bian, J.; Wu, Y. Extracting Family History of Patients from Clinical Narratives: Using Deep Learning Models (Preprint). *JMIR Med. Inform.* **2020**, *8*, e22982. [CrossRef]
35. Bogonikolos, N.; Fragoudis, D.; Likothanassis, S. "ARCHIMIDES": An intelligent agent for adaptive-personalized navigation within a WEB server. In Proceedings of the 32nd Annual Hawaii International Conference on Systems Sciences, Maui, HI, USA, 5–8 January 1999; p. 9. [CrossRef]
36. Bradford, C.; Marshall, I. A bandwidth friendly search engine. In Proceedings of the IEEE International Conference on Multimedia Computing and Systems, Florence, Italy, 7–11 June 1999; Volume 2, pp. 720–724. [CrossRef]
37. Diouf, R.; Sarr, E.N.; Sall, O.; Birregah, B.; Bousso, M.; Mbaye, S.N. Web Scraping: State-of-the-Art and Areas of Application. In Proceedings of the 2019 IEEE International Conference on Big Data (Big Data), Los Angeles, CA, USA, 9–12 December 2019; pp. 6040–6042. [CrossRef]
38. Banik, S. COVID Fake News Dataset [Data set]. Zenodo. 2020. Available online: https://zenodo.org/records/4282522 (accessed on 16 October 2023).
39. Li, Y.; Jiang, B.; Shu, K.; Liu, H. MM-COVID: A Multilingual and Multimodal Data Repository for Combating COVID-19 Disinformation. *arXiv* **2020**, arXiv:2011.04088.

40. Barrón-Cedeño, A.; Da San Martino, G.; Jaradat, I.; Nakov, P. Proppy: Organizing the news based on their propagandistic content. *Inf. Process. Manag.* **2019**, *56*, 1849–1864. [CrossRef]
41. Ahmed, H.; Traore, I.; Saad, S. Detecting opinion spams and fake news using text classification. *Secur. Priv.* **2018**, *1*, e9, . [CrossRef]
42. Risdal, M. Getting Real about Fake News. Kaggle. 2016. Available online: https://www.kaggle.com/mrisdal/fake-news (accessed on 16 October 2023).
43. Kotonya, N.; Toni, F. Explainable Automated Fact-Checking for Public Health Claims. In Proceedings of the 2020 Conference on Empirical Methods in Natural Language Processing (EMNLP), Online, 16–20 November 2020; Association for Computational Linguistics: Toronto, ON, Canada, 2020; pp. 7740–7754.
44. Mahabub, A. A robust technique of fake news detection using Ensemble Voting Classifier and comparison with other classifiers. *SN Appl. Sci.* **2020**, *2*, 525. [CrossRef]

Disclaimer/Publisher's Note: The statements, opinions and data contained in all publications are solely those of the individual author(s) and contributor(s) and not of MDPI and/or the editor(s). MDPI and/or the editor(s) disclaim responsibility for any injury to people or property resulting from any ideas, methods, instructions or products referred to in the content.

Article

Convolutional Neural Network Approach Based on Multimodal Biometric System with Fusion of Face and Finger Vein Features

Yang Wang †, Dekai Shi † and Weibin Zhou *

School of Electronic Information and Automation, Tianjin University of Science and Technology, Tianjin 300453, China
* Correspondence: zhouweibin@tust.edu.cn
† These authors contributed equally to this work.

Abstract: In today's information age, how to accurately identify a person's identity and protect information security has become a hot topic of people from all walks of life. At present, a more convenient and secure solution to identity identification is undoubtedly biometric identification, but a single biometric identification cannot support increasingly complex and diversified authentication scenarios. Using multimodal biometric technology can improve the accuracy and safety of identification. This paper proposes a biometric method based on finger vein and face bimodal feature layer fusion, which uses a convolutional neural network (CNN), and the fusion occurs in the feature layer. The self-attention mechanism is used to obtain the weights of the two biometrics, and combined with the RESNET residual structure, the self-attention weight feature is cascaded with the bimodal fusion feature channel Concat. To prove the high efficiency of bimodal feature layer fusion, AlexNet and VGG-19 network models were selected in the experimental part for extracting finger vein and face image features as inputs to the feature fusion module. The extensive experiments show that the recognition accuracy of both models exceeds 98.4%, demonstrating the high efficiency of the bimodal feature fusion.

Keywords: CNN; dual-channel biometric identification system; biometric fusion; face recognition; finger vein recognition; identification system

Citation: Wang, Y.; Shi, D.; Zhou, W. Convolutional Neural Network Approach Based on Multimodal Biometric System with Fusion of Face and Finger Vein Features. *Sensors* **2022**, *22*, 6039. https://doi.org/10.3390/s22166039

Academic Editors: Erik Blasch and Yufeng Zheng

Received: 15 June 2022
Accepted: 10 August 2022
Published: 12 August 2022

Publisher's Note: MDPI stays neutral with regard to jurisdictional claims in published maps and institutional affiliations.

Copyright: © 2022 by the authors. Licensee MDPI, Basel, Switzerland. This article is an open access article distributed under the terms and conditions of the Creative Commons Attribution (CC BY) license (https://creativecommons.org/licenses/by/4.0/).

1. Introduction

In recent years, with the increasing level of science and technology, people direct progressively more attention to information security [1]. Traditional authentication methods, such as using account numbers and passwords are easily replaced by impersonation once they are stolen. Biometrics is an emerging security technology and one of the most promising technologies in this century. Compared with traditional identification, biometrics has many advantages, such as the technology will not be lost, stolen or copied. Biometrics mainly refers to a technology that authenticates identity through measurable physical or behavioral biometrics [2]. Biological features are divided into two categories: Physical features and behavioral features [3]. Physical features, such as fingerprints, retinas, and faces, are mostly inborn, and behavioral features, such as gait and keystrokes, are mostly acquired due to habit [4].

Primarily, a biometric identification system involves four processes: Image acquisition, image processing feature extraction, feature value comparison, and individual identification. There are two biometric systems, single-modal and multimodal. A single-modal biometric system refers to the use of a single biometric feature to identify users. Although the current biometric technologies, such as face recognition [1], fingerprint recognition [5], and iris recognition [6] are relatively mature, with the outbreak of COVID-19, single biometric technologies, such as fingerprint recognition have suffered serious challenges. Face recognition hidden behind masks fails, and fingerprint recognition with protective gloves cannot be unlocked. Single biometric technologies represented by fingerprints and faces are increasingly difficult in application.

With the upgrading and development of biometric identification technology, biometric identification is constantly changing from single to multimodal. Through careful design and fusion algorithm, it can realize the combination of face, fingerprint, fingerprint vein, iris, voice print, and other biometrics, which can lead to complementary information and further improvement of recognition accuracy. Multimodal biometrics refer to the integration or fusion of multiple human biometrics, using the respective advantages of biometrics and combining different feature fusion algorithms to make the identification process more accurate and secure [7]. Compared with single biometric identification technology, multimodal biometric identification technology has the advantages of high identification accuracy, higher security, and wider application range.

Although the current single-modal biometric technology has been relatively mature, this single-mode biometric recognition technology is not only affected by the external environment in practical applications, but also affected by the limitations of single-mode biometrics itself, which greatly limits its application scenarios and reduces the accuracy of identity recognition. For example, in fingerprint recognition, fingerprint damage and finger wetness will affect the correct rate of collection and recognition. In face recognition, the growth of age and wearing masks will affect its recognition accuracy. To overcome these problems, we propose a bimodal recognition method that fuses face and finger veins. Compared with the three-modal, four-mode, and other modes, the finger vein and face bimodal not only reduce the complexity of time, algorithm, design, and production, but also complement the advantages of in vivo biometrics (finger veins) and in vitro biometrics (faces). It breaks the application limitations of single biometrics and improves the security and accuracy of identity information.

In this study, we chose face and finger veins as the targets for the following reasons: Faces are chosen since facial features are the most pronounced, and more data are used, making them more precise. Face recognition does not require any contact, it can be identified by air. Finger veins are chosen since they are internal physiological features that are difficult to forge and have a high level of safety [2]. It has the advantages of enabling non-contact measurement, good hygiene, and easy acceptance for users. In addition, the use of each person's finger vein is unique and cannot be forged, especially for places with high safety requirements.

Traditional biometric systems primarily have four components: Pre-processing, feature extraction, matching, and decision-making phases [8]. Feature extraction methods can affect the system significantly. In view of the excellent performance of convolutional neural networks in image recognition and image feature extraction tasks, this study aims to deeply study the application of convolutional neural network algorithms in finger vein and face biometric recognition. In this paper, a bimodal biometric system based on a deep learning model of finger veins and face images is proposed. First, the finger vein and face images were acquired, and the region-of-interest (ROI) was intercepted on the finger vein images. For the problem of the lack of data in the finger veins, the image enhancement method was adopted, and then the finger vein and face images were input into the dual-channel convolutional neural network to extract the features. Feature fusion is performed before the fully connected layer, and weights are assigned according to the confidence level of each feature, and finally user identification is performed.

The rest of the paper is organized as follows: Section 2 provides a brief description of the relevant research. Section 3 proposes the method used in this article. Section 4 discusses and analyzes the results. Finally, Section 5 concludes the paper and discusses potential future work.

2. Related Work

Multimodal biometric technology refers to the fusion of more than two kinds of single biometric features as a new feature for identification [9]. According to the location of the fusion, it can be divided into image layer fusion, feature layer fusion, matching layer fusion, and decision layer fusion [10].

Haghiatt et al. [11] proposed discriminative correlation analysis (DCA), which performs an effective feature fusion by maximizing the pairwise correlations across the two feature sets and, at the same time, eliminating the between-class correlations and restricting the correlations to be within the classes. Shaheed et al. [12] considered the issues of convolutional neural networks (CNNs). The authors presented a pre-trained CNN network named Xception model based on depth-wise separable CNNs with residual connection, which is considered as a more effective, less complex neural network to extract robust features. Their proposed method for the SDUMLA database achieved an accuracy of 99% with an F1-score of 98%. While on THU-FVFDT2, the proposed method obtained an accuracy of 90% with an F1-score of 88%. Qi et al. [13] proposed a biometric method based on three biometric patterns of face, iris, and palm print, and introduced a strategy of fusion of multiple features. Cherrat et al. [14] proposed a hybrid system based on a multi-biometric fingerprint, finger vein, and face recognition system that combines a convolutional neural network (CNN), Softmax, and a random forest (RF) classifier, using K-means and density-based spatial clustering of application with Noise (DBSCAN) algorithms that employ image preprocessing to separate foreground and background areas. Experiments have shown that multimodality can provide accurate and efficient matching, which can be significantly better than the single-modal representation accuracy.

Few studies focus on identifying users through behavioral biometrics, and in the process of research, feature recognition and feature extraction are more difficult since user behavior is non-repeatable and variable. Abinaya et al. [15] used two different unmoded biometric keystrokes (typography) and acoustics (speech) to identify the use of pre-trained deep learning models, using weighted linear feature-level fusion, and trained by CNN classifier simulation. The results show that the system has good accuracy. Ding et al. [16] proposed a deep learning structure consisting of a neural network (CNN) and a three-layer stacked autoencoder, and then concatenated the extracted features to form a high-dimensional feature vector. This method achieves a recognition rate of up to 99.0% on the Labeled Faces in the Wild (LFW) database. Chawla et al. [17] investigated the finger vein recognition problem. They employed one-shot learning model namely the Triplet loss network model and evaluated its performance. The extensive set of experiments that they have conducted yield classification and correct identification accuracies in ranges upwards of 95% and equal error rates less than 4%.

In recent years, research on finger vein recognition has been particularly active. For example, Alay et al. [18] combined three CNN models of iris, finger vein, and face to fuse at the feature and fractional levels, respectively, using the VGG-16 model, where the accuracy of feature-level fusion was 100% and the accuracy of different score-level fusion methods was 100%. Ammour et al. [8] proposed a new feature extraction technique for a multimodal biometric system using face–iris traits. The iris feature extraction is carried out using an efficient multi-resolution 2D Log-Gabor filter to capture textural information in different scales and orientations. On the other hand, the facial features are computed using the powerful method of singular spectrum analysis (SSA) in conjunction with the wavelet transform. Kim et al. [19] studied a biometric system capable of recognizing finger veins and finger shapes, and constructed a CNN biometric system, using fusion methods, such as weighted sum and Bayesian rules to fuse two features on the fractional level.

Yang et al. [20] proposed a research method for the fusion of finger veins and fingerprints, developed a feature-level fusion strategy for three fusion options, and combined the details of finger vein fingerprint image features. Soleymani et al. [21] proposed a fusion method for the fingerprints of the human face and the iris. Multiple features are extracted from each modal-specific CNN in multiple different convolutional layers for joint feature fusion, optimization, and classification. Waisy et al. [22] proposed a deep learning method based on the left and right iris images of people, using a hierarchical fusion method to fuse the results, which is called IrisConvNet. The recognition rate of the database used is

100% and the recognition time is less than 1 s. Ren et al. [23] proposed a new finger vein image encryption scheme, which applies Rivest–Shamir–Adleman encryption technology to finger vein image encryption. In addition, a complete cancelable finger vein recognition system with template protection is proposed to ensure the security of the user's vein template while maintaining the recognition performance. Jomaa et al. [4] aimed at the serious vulnerability of fingerprint system in presentation attack (PA), they proposed a novel architecture based on end-to-end deep learning to fuse between neuroprinting and ECG, improving PA detection in fingerprint biometrics and using EfficientNet to generate fingerprint features. Experimental results show that the architecture produces better average classification accuracy than the single fingerprint. Tyagi et al. [24] proposed highly accurate and robust multimodal biometric identification as well as recognition systems based on fusion of face and finger vein modalities. The feature extractions for both face and finger vein are carried out by exploiting deep convolutional neural networks.

In view of the selected biological features that are easy to be affected by external factors and different fusion locations, this study selected face and finger vein, complementing the advantages of internal biological features (finger vein) and external biological features (face), and integrating in the feature layer with the most rich and effective information. Not only does it make up for the shortcomings of single biometric recognition and improve the accuracy of identity recognition, but also compared with the three-mode, four-mode, and other modes, the finger vein and face bimodal reduces the time, algorithm, design, and production complexity.

3. Method

In this study, a two-channel CNN feature fusion framework is proposed as shown in Figure 1, and the fusion occurs at the feature layer. The framework is mainly divided into three parts, namely the feature extraction part, the feature fusion part, and the classification recognition part. In the feature extraction, the collected finger vein pictures and face pictures are mainly pre-processed, such as the interception of the finger vein area of interest and data enhancement, and then the biometrics are convoluted into the neural network model, and the biometrics of the images are extracted through multi-layer convolution and pooling layers. In the feature fusion module, fusion Conv is first performed to reduce dimensions and then pass through the Softmax layer, which obtains the self-attention weights and multiplies the features obtained by feature extraction, and then channel Concat cascades the two features together to obtain the fusion features of finger vein and face (Fv + Face_feature). To prevent the loss of some feature information during feature fusion, the finger vein features and face features after feature extraction are obtained, as well as feature cascading after the feature three-channel Concat. The classification recognition part is mainly classified in the fully connected layer.

3.1. Network

Mode convolutional neural network is a feedforward neural network, which is the mainstream technology in image processing today [24]. Using CNN for processing pictures can not only effectively reduce the dimension of large data volume to small data amount, but also can effectively retain the characteristics of pictures, which complies with the principle of image processing. Similar to other neural networks, CNN networks also contain several parts: Input layer, hidden layer, and output layer. The convolutional layer [25] is the core layer of building a convolutional neural network, the convolutional layer is composed of multiple convolutional units, and the parameters of each convolutional unit are optimized by backpropagation algorithms. Convolutional operations are mainly to extract the features of the image, the first few layers are mainly to extract the low-level features of the image, such as color and other information, with the increase in the convolutional layer, the multi-layer network can extract more complex image features. Linear rectification mainly refers to the ReLU function of the activation function operation, which can realize the non-linear mapping function of the network, increase the expression ability of the network, and have a good effect on feature extraction. The input of the pooling layer is multiple feature mapping and pooling input operation. After convolution, the dimensional features of the image still divide the feature matrix into several single blocks to take its maximum or average, which can play a role in dimensionality reduction and reduce the calculation of the network and avoid overfitting. Overfitting is also known as over-fitting. Due to over-fitting of the training samples, the ability to accept samples other than the training samples is poor, and the model cannot have good generalization ability. The input of the pooling layer is multiple feature mapping and pooling the input. After convolution, the dimensional features of the image are still many, and the feature matrix is divided into several single blocks to take their maximum or average values, which can play a role in reducing dimensionality, reducing network computation, and avoiding overfitting. As the last layer of the CNN model, the fully connected layer combines all local features and the feature matrix of each channel into vector representations, and calculates the score of each final class, as shown in Figure 2.

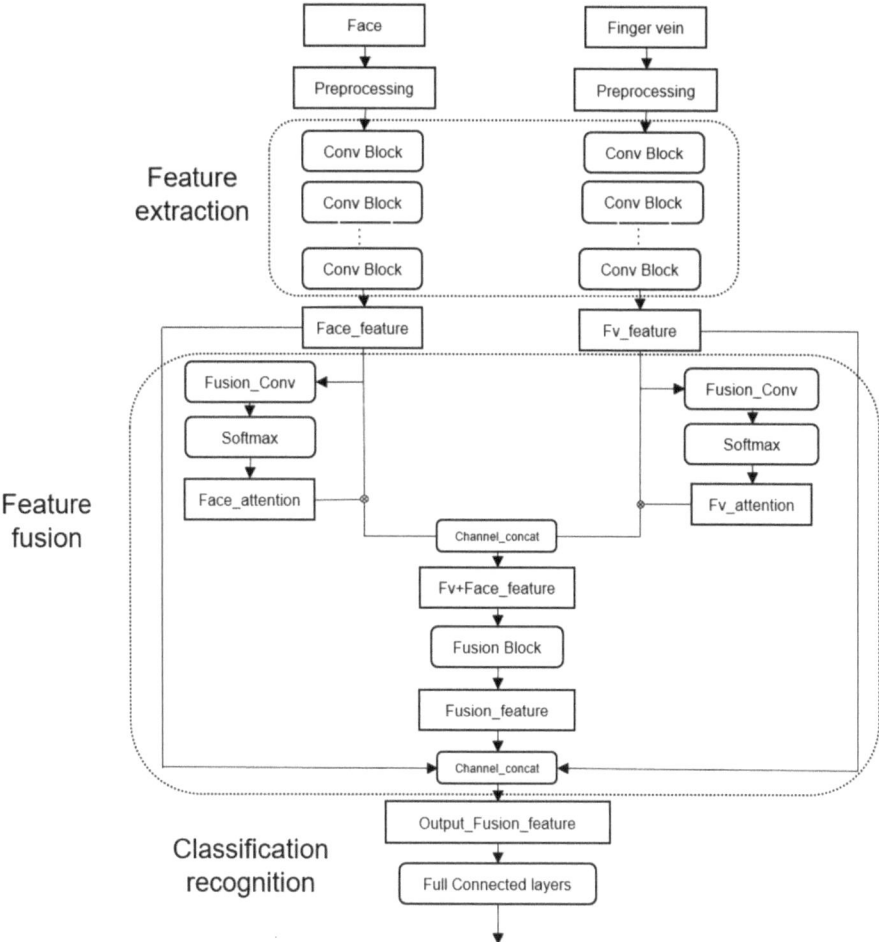

Figure 1. Bimodal feature fusion framework.

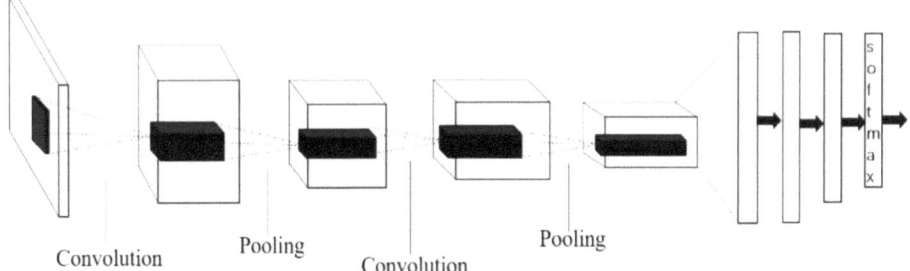

Figure 2. Schematic diagram of convolutional neural network structure.

To demonstrate the high efficiency of bimodal feature fusion, the feature extraction networks of two representative AlexNet [25] and VGG-19 [26] networks are selected as the benchmark [27,28], discarding their fully connected layers and using only the convolutional and pooling layers before the fully connected layers.

AlexNet: The network, designed by 2012 ImageNet competition winner Hinton and his student Alex Krizhevsky, consists of five convolutional layers (conv) and three fully connected layers (fc) as shown in Figure 3. The activation function uses ReLU, and the entire network has more than 62 million trainable parameters. The input is a 224 × 224 × 3 image, and the output is a 1000-dimensional vector corresponding to the probability of each classification. The first, second, and fifth convolution layers add a pooling layer (MaxPool) with a kernel of 3 × 3 and a stride of 2, which can improve the accuracy of the pooling layer, as shown in Figure 3.

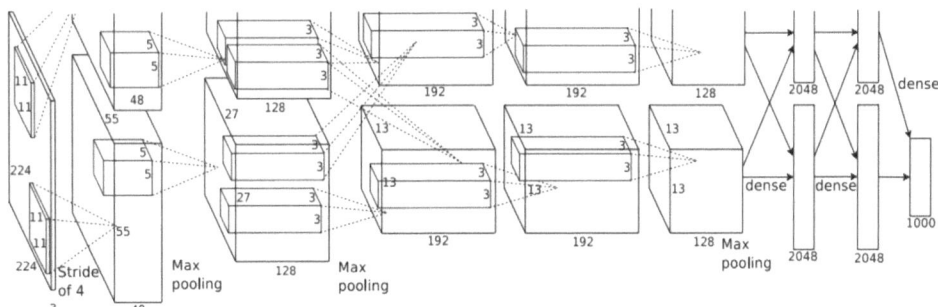

Figure 3. Schematic diagram of the AlexNet model architecture.

VGG-19: The network has 19 layers with 16 convolutional layers and 3 fully connected layers as shown in Figure 4. The convolution kernels are all 3 × 3 in size, which reduces the network parameters by repeatedly stacking the 3 × 3 convolution kernels rather than a large convolution kernel. Compared with the AlexNet network, VGG-19 is able to extract deeper features of images. Therefore, in the feature extraction module, the first 16 convolutional layers of VGG-19 are used to extract finger vein face image features.

3.2. Data Preprocessing

Face images use the face public dataset CASIA-WebFace [29], which performs pre-processing, such as resize and normalization in the network. Since the final experimental data worked well, no other complex pre-processing operations were performed.

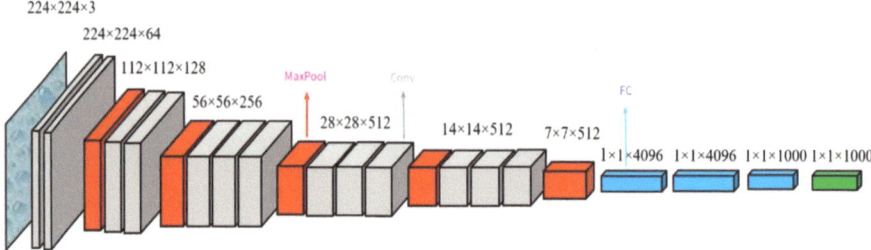

Figure 4. Schematic diagram of the VGGNet model architecture.

For finger veins, since the SDUMLA-FV dataset [30] does not provide the region-of-interest (ROI) of the finger vein, an interception of the ROI region of the dataset is required to remove excessive background information which is likely to be useless. First, the Prewitt edge detection operator is used to detect the upper and lower edges in the vertical direction of the original diagram of the finger vein, and for the phenomenon of pseudo-edges, the pseudo-edges are removed by setting the connection domain threshold. Use least squares linear regression to fit the central axis of the finger, and correct the image rotation according to the angle between the fitted line and the horizontal line. Fit the inner tangent of the upper and lower edges of the finger. According to the brightness change trend in the horizontal direction of the image, select the knuckle (brightness peak). Finally, the venous ROI area of the finger is intercepted.

To obtain clear finger vein lines, contrast-limited adaptive histogram equalization (CLAHE) is also required for the captured region-of-interest (ROI) images, and a Gabor filter is added after the CLAHE image enhancement to remove noise. The ROI original image, after CLAHE image enhancement and Gabor filtering, can get clear vein lines compared with the original image. The finger vein dataset provided only 6 images of veins per finger, and to prevent overfitting during CNN model training, we amplified the data for each type of finger vein, including random translation, rotation, cropping, brightness adjustment, and contrast adjustment of the image, expanding the original 6 images per class to 36 pictures. The FV-USM dataset [31] provides ROI images, thus only image enhancement and augmentation are required for this dataset, as shown in Figure 5.

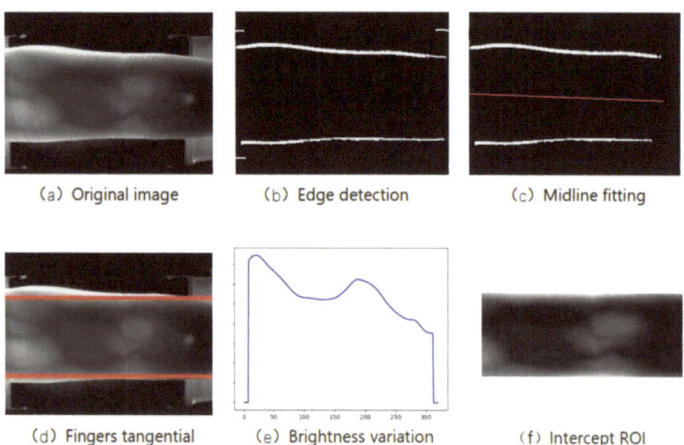

Figure 5. SDUMLA-FV dataset image preprocessing.

3.3. Feature Layer Fusion

In the feature layer fusion method, the face image feature (Face_feature) and the finger vein image feature (Fv_feature) are taken as the input to the self-attention mechanism to obtain their respective attention weights. The vein and face features are combined with their respective attention weights, and they are cascaded to form cascade features (Fv + Face_feature). The cascade features (Fv + Face_feature) are taken as the input of the fusion module to convolute the cascade features to extract the deeper features and obtain the fusion features (Fusion feature). To prevent the loss of some feature information during feature fusion and ensure the maximization of effective feature information, the fusion features are fused again with the single finger venous features and the single face features into new features and output in Figure 6.

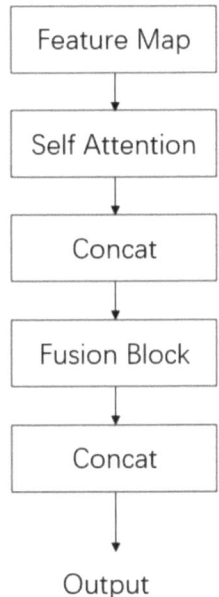

Figure 6. Schematic diagram of the feature fusion module framework.

The self-attention mechanism proposed in this article consists of three convolutional layers and one Softmax layer, and the structure of the convolutional layer is shown in Table 1. The convolutional kernel size of the first convolution is 1×1, padding is 0, and the convolutional kernel size of the second and third convolutions is 3×3, and padding is 1. After the convolutional layer, the ReLU activation function is used, and the ReLU activation function can not only quickly converge the network, but also solve the problem of gradient disappearance. Table 1 shows the convolutional structure in the self-attention mechanism network, and Table 2 is the convolutional structure in the Fusion Block fusion module proposed in this paper.

Table 1. Convolutional architecture in self-attention networks.

Layer Name	In Channel	Out Channels	Convolutional Kernel Size	Padding	Stride
Conv1	a	a/2	1	0	1
Conv2	a/2	a/4	3	1	1
Conv3	a/4	1	3	1	1

Table 2. Convolutional layer structure in fusion block.

Layer Name	In Channel	Out Channels	Convolutional Kernel Size	Padding	Stride
Conv1	$2 \times a$	a	1	0	1
Conv2	a	a/2	3	1	1
Conv3	a/2	a	3	1	1

The ReLU function formula is as follows:

$$ReLU(x) = \max(0, x) \tag{1}$$

The input of the whole dual-channel convolutional neural network is two images (one finger vein image and one face image), and new features are formed after the feature extraction module and feature fusion module. New features were used as input to the classification identification module.

4. Experimental Results and Analysis

4.1. Experimental Environment and Data Distribution

The language used in this experiment is Python3.8, the software environment is Ubuntu18.04 (64-bit), and the framework used for deep learning is Pytorch 1.7.1, CUDA11.0, cuDNN8.0, using two NVIDIA GeForce RTX 3080Ti GPUs.

To demonstrate the effectiveness of the method studied in this paper, we used three publicly available datasets, CASIA-WebFace [29], Finger Vein USM (FV-USM) [31], and SDUMLA-FV [30], to test the proposed bimodal feature fusion algorithm and compare it with single-modal biometric identification. The SDUMLA-FV dataset was created by Shandong University and captured a total of 3816 images. A total of 636 classes contained 106 finger vein images of the left and right index fingers, 6 middle fingers, and ring fingers. Each image is stored in "BMP" format (320 × 240) pixel size. FV-USM is a Malaysian Polytechnic finger vein dataset with 2952 maps. A total of 492 types of fingers contains 6 middle finger images and 123 images of the left and right hands, and the advantage of using this dataset is that it provides already intercepted region-of-interest (ROI) images. The CASIA-WebFace dataset is one of the most widely available datasets for applied face recognition, which collects face images on the network with a total of 10,575 classes and 494,414 images.

This experiment used two finger vein public datasets, FV-USM, SDUMLA-FV, and one face dataset, CASIA-WebFace. In single-mode experiments, each sample was amplified to 36 pictures using the data amplification method, and 636 categories were randomly selected for the face dataset. Additionally, each dataset divides the training, validation, and testing. Sixty percent of patterns are assigned to the training set, 10% to the validation set, and 30% to the test set. To increase the generalization capacity of the network, n × n is used to match each finger vein image belonging to the same class with each face image. For the dataset of 636 classes, the training set total is 636 × 21 × 21 = 280,476 images, validation set total is 636 × 3 × 3 = 5724 images, and test set total is 636 × 12 × 12 = 91,584 images. Then, the total amounts were divided into training sets, validation sets, and test sets.

4.2. Testing and Analysis of Biometric Systems

4.2.1. Performance Evaluation

This article mainly provides a bimodal fusion method. To evaluate the structural performance of this model, we measure the performance of the model by the following indicators.

Confusion Matrix: In the field of image recognition, it is mainly used to compare the relationship between the classification results and the actual predicted values. Table 3 shows the confusion matrix, positive represents positive samples, and negative represents negative samples. Each column represents the predicted value, and each row represents the actual value.

Table 3. Confusion matrix.

Actual Value \ Predicted Value	Positive	Negative
Positive	True Positive (TP)	False Negative (FN)
Negative	False Positive (FP)	True Negative (TN)

Accuracy rate: In all the samples, the proportion of the correct number of samples in the total number of samples is predicted; the mathematical expression is as follows:

$$accuracy\_rate = \frac{TP+TN}{TP+TN+FP+FN} \quad (2)$$

Precision: In all the samples with positive class prediction result, the proportion of the samples is actually positive class and the prediction result is also positive class number; the mathematical expression is as follows:

$$precision = \frac{TP}{TP+FN} \quad (3)$$

Recall rate: In all the samples that are actually positive, the proportion of the number of samples is predicted as positive; the mathematical expression is as follows:

$$recall = \frac{TP}{TP+FN} \quad (4)$$

ROC curve: The abscissa of ROC curve is negative positive rate (false positive rate, FPR). The ordinate is the true class rate (true positive rate, TPR).

$$FPR = \frac{FP}{FP+TN} \quad (5)$$

$$TPR = \frac{TP}{TP+FN} \quad (6)$$

The area under the ROC curve represents AUC (area under curve), and its value can directly reflect the performance of the classification model. The value range is between 0.1 and 1. The closer the value is to 1, the better the performance of the classification model.

P-R curve: The P-R curve is constructed with precision as the ordinate and recall as the abscissa, and the area enclosed by the coordinate axis indicates the average accuracy (average precision, AP). The value of AP reflects the performance of the classification model. The closer the AP value is to 1, then the classification model has good performance.

4.2.2. Experimental Results and Analysis of Single-Mode Identification

In the single-mode biometric experiment, three datasets, SDUMLA-FV, FV-USM, and CASIA-WebFace, were used to experiment on the CNN network framework (AlexNet, VGG-19). As can be seen from Table 4, in the two network models, the identification accuracy of single-modal experiment can reach 87.57% and the lowest is 45.61%. Compared with

Yuancheng's face recognition experiment [15], the accuracy of single-modal identification is not high and the number of parameters is large.

Table 4. Results of a single-mode experiment.

Model	Dataset Parameter Quantity	Test Set Accuracy		
		SDUMLA-FV	USM-FV	CASIA-WebFace
AlexNet	16,630,440	0.7020	0.4561	0.5395
VGG-19	143,667,240	0.8757	0.6734	0.5575

4.2.3. Experimental Results and Analysis of Multimodal Recognition

In Table 5, the accuracy of the recognition of face and finger veins in the feature layer fused under the two CNN network models is shown. Compared with the single-mode experimental results, the accuracy rate of AlexNet as a feature extraction network was above 99.3%. In the experiment using VGG-19 as the feature extraction network, the feature fusion experimental recognition accuracy of the SDUMLA-FV dataset and the CASIA-WebFace face dataset reached 99.98%, and the feature fusion experimental recognition accuracy of the USM-FV dataset and the CASIA-WebFace face dataset reached 98.42%. Through the experimental data, we can prove that the proposed method can greatly improve the accuracy and effectively shorten its model parameters.

Table 5. Results of multimodal experiments.

Model	Dataset Parameter Quantity	Test Set Accuracy	
		SDUMLA-FV + CASIA-WebFace	USM-FV + CASIA-WebFace
AlexNet-Fusion	9,858,994	0.9990	0.9935
VGG-19-Fusion	45,229,938	0.9998	0.9842

The following figure is the plotted ROC curve. There are two methods of ROC curve drawing, which are micro average and macro average. Micro average refers to calculating the prediction accuracy of each sample model, establishing a global confusion matrix, and then calculating the corresponding indicators globally to draw the micro average ROC curve of the dataset, regardless of the category of each test sample in the dataset. Macro-averaging refers to the separation of each class of the dataset, then calculates the accuracy of each class, establishing the confusion matrix, and finally averaging the average of the ROC curve and the AUC values.

Figure 7 shows the ROC curves of the FV-USM dataset and the SDUMLA-FV dataset combined with the CASIA-WebFace face dataset to use the bimodal feature fusion method on the AlexNet feature extraction network. As you can see from the figure, the AUC of the micro average and the macro average is both 1.

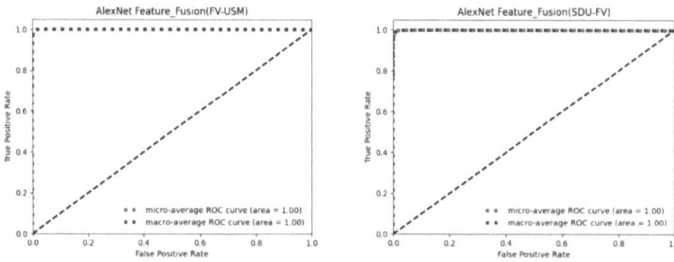

Figure 7. Fusion of identified ROC curves on AlexNet feature extraction network.

Figure 8 shows the ROC curves of experimental results for different datasets using the bimodal feature fusion method on the VGG-19 feature extraction network. As you can see from the figure, the AUC at both the micro average and the macro average is 1.

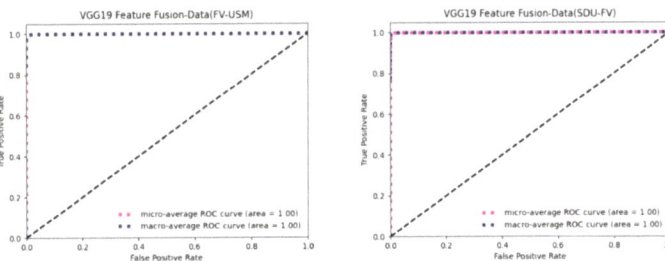

Figure 8. Fusion of identified ROC curves on the VGG-19 feature extraction network.

Table 6 shows a bimodal dataset consisting of two different finger vein datasets and a face dataset, and the area AUC of the ROC curve and the curve surrounded by the coordinate axis is summarized on three different feature extraction networks. As you can see from the summary table, the AUC is all 1.

Table 6. AUC summary of different datasets under different feature extraction networks.

Feature Extraction Network	AUC	
Dataset	AlexNet	VGG-19
FV-USM + CASIA-WebFace	1	1
SDUMLA-FV + CASIA-WebFace	1	1

In summary, according to Yasen's experimental analysis [25], the closer the ROC curve is to the upper left corner, the better the performance. Experimental data show that the bimodal feature fusion method proposed in this paper has good classification performance in different bimodal datasets and different feature extraction networks.

To evaluate the proposed bimodal feature fusion recognition method more objectively and accurately, we plotted the P-R curves for each experimental result. Figure 9 shows the P-R curve results of different datasets using the bimodal feature fusion method on the AlexNet feature extraction network. Among them, the bimodal feature fusion experiment of the FV-USM dataset has an area AP of 0.88 surrounded by coordinate values, and the area AP of the SDUMLA-FV dataset surrounded by coordinate axes is 0.9.

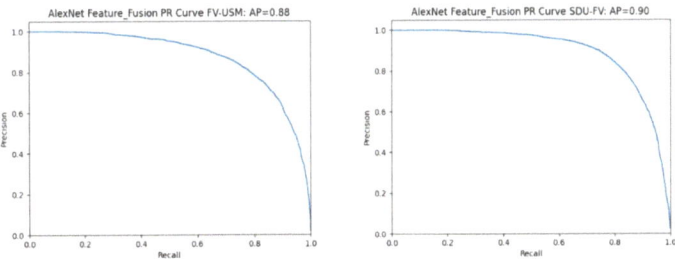

Figure 9. Fusion of identified P-R curves on the AlexNet feature extraction network.

Figure 10 shows the P-R curve results of different datasets using the bimodal feature fusion method on the VGG-19 feature extraction network. As can be seen from the figure, the area AP enclosed by the P-R curve and the coordinate values is 0.92 and 0.8, respectively.

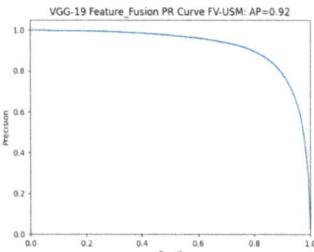
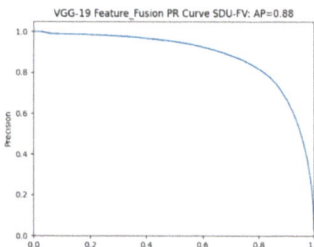

Figure 10. Fusion of identified P-R curves on the VGG-19 feature extraction network.

Table 7 shows a summary of the values of two different bimodal datasets on the P-R curve as well as the coordinate axis surrounded by the P-R curve and coordinate axis on two different feature extraction networks.

Table 7. Summary of AP values of different datasets under different feature extraction networks.

Feature Extraction Network	AP	
Dataset	AlexNet	VGG-19
FV-USM + CASIA-WebFace	0.88	0.92
SDUMLA-FV + CASIA-WebFace	0.90	0.88

For the P-R curve index, the average accuracy AP value is greater than 0.88 and the highest is 0.92 in all experimental results. Compared with Yang's experiments [31], they used the precision-recall (P-R) curve to measure the accuracy-recall trade-off due to unbalanced datasets and reported sensitivity and specificity on the receiver operating characteristic curve. The area under the precise-recall curve showed an average accuracy of 0.874, indicating that the proposed method has good classification performance.

5. Conclusions

This study proposes a two-modal feature layer fusion method based on convolutional neural network, combining the advantages of in vivo biometric features (finger vein) and in vitro biometric features (human face). This method introduces the weight of the self-attention mechanism to update the different features within the feature fusion module, and adopts the residual structure to maximize the effective feature information. The feature extraction was performed on the AlexNet and VGG-19 feature extraction modules, respectively, and then the bimodal feature fusion method proposed in this paper was used. In multimodal experiments, the lowest identification accuracy of the experimental results was 98.84%, and the highest identification accuracy was 99.98%. In addition, the ROC curve and P-R curve of the fusion experiment are plotted, which verifies that the proposed method has very good classification performance. The research in this paper is currently limited to computers with high computing power of GPUs, and cannot be deployed to mobile embedded terminals, which greatly limits the application in practice. Subsequently, the network results of bimodal fusion will be lightweight, and the fusion model will be ported to an embedded terminal within an acceptable range.

Author Contributions: Y.W. was responsible for the model and the experiments. D.S. was responsible for verifying the analysis method. W.Z. was responsible for the overall development direction and planning. All authors participated in the study design and the preparation of the paper. All authors have read and agreed to the published version of the manuscript.

Funding: This research was funded and supported by Tianjin Municipal Education Commission Scientific Research Plan, project 2018KJ102 and Tianjin University "young and middle-aged backbone innovation talent training program" project.

Institutional Review Board Statement: Not applicable.

Informed Consent Statement: Not applicable.

Data Availability Statement: Not applicable.

Conflicts of Interest: The authors declare no conflict of interest.

References

1. Amine, N.-A. (Ed.) *Hidden Biometrics: When Biometric Security Meets Biomedical Engineering*; Springer: Berlin/Heidelberg, Germany, 2019.
2. Bailey, K.O.; Okolica, J.S.; Peterson, G.L. User identification and authentication using multi-modal behavioral biometrics. *Comput. Secur.* **2014**, *43*, 77–89. [CrossRef]
3. Ra'Anan, Z.; Sagi, A.; Wax, Y.; Karplus, I.; Hulata, G.; Kuris, A. Growth, size rank, and maturation of the freshwater prawn, Macrobrachium rosenbergii: Analysis of marked prawns in an experimental population. *Biol. Bull.* **1991**, *181*, 379–386. [CrossRef] [PubMed]
4. Ramírez-Mendoza, R.A.; Lozoya-Santos, J.D.J.; Zavala-Yoé, R.; Alonso-Valerdi, L.M.; Morales-Menendez, R.; Carrión, B.; Cruz, P.P.; Gonzalez-Hernandez, H.G. (Eds.) *Biometry: Technology, Trends and Applications*; CRC Press: Boca Raton, FL, USA, 2022.
5. Jomaa, R.M.; Mathkour, H.; Bazi, Y.; Islam, M.S. End-to-end deep learning fusion of fingerprint and electrocardiogram signals for presentation attack detection. *Sensors* **2020**, *20*, 2085. [CrossRef] [PubMed]
6. Mitra, S.; Gofman, M. (Eds.) *Biometrics in a Data Driven World: Trends, Technologies, and Challenges*; CRC Press: Boca Raton, FL, USA, 2016.
7. Lowe, J. *Ocular Motion Classification for Mobile Device Presentation Attack Detection*; University of Missouri-Kansas City: Kansas City, MO, USA, 2020.
8. Ammour, B.; Boubchir, L.; Bouden, T.; Ramdani, M. Face–iris multimodal biometric identification system. *Electronics* **2020**, *9*, 85. [CrossRef]
9. Zheng, Y.; Blasch, E.; Liu, Z. *Multispectral Image Fusion and Colorization*; SPIE Press: Bellingham, DC, USA, 2018; Volume 481.
10. Zheng, Y.; Blasch, E. An exploration of the impacts of three factors in multimodal biometric score fusion: Score modality, recognition method, and fusion process. *J. Adv. Inf. Fusion* **2015**, *9*, 106–123.
11. Haghighat, M.; Abdel-Mottaleb, M.; Alhalabi, W. Discriminant correlation analysis: Real-time feature level fusion for multimodal biometric recognition. *IEEE Trans. Inf. Forensics Secur.* **2016**, *11*, 1984–1996. [CrossRef]
12. Shaheed, K.; Mao, A.; Qureshi, I.; Kumar, M.; Hussain, S.; Ullah, I.; Zhang, X. DS-CNN: A pre-trained Xception model based on depth-wise separable convolutional neural network for finger vein recognition. *Expert Syst. Appl.* **2022**, *191*, 116288. [CrossRef]
13. Xu, H.; Qi, M.; Lu, Y. Multimodal Biometrics Based on Convolutional Neural Network by Two-Layer Fusion. In Proceedings of the 2019 12th International Congress on Image and Signal Processing, BioMedical Engineering and Informatics (CISP-BMEI), Suzhou, China, 19–21 October 2019; pp. 1–6.
14. Cherrat, E.; Alaoui, R.; Bouzahir, H. Convolutional neural networks approach for multimodal biometric identification system using the fusion of fingerprint, finger-vein and face images. *PeerJ Comput. Sci.* **2020**, *6*, e248. [CrossRef]
15. Abinaya, R.; Indira, D.; Swarup Kumar, J. Multimodal Biometric Person Identification System Based on Speech and Keystroke Dynamics. In *EAI/Springer Innovations in Communication and Computing Book Series (EAISICC), Proceedings of the International Conference on Computing, Communication, Electrical and Biomedical Systems, Online, 28 February 2022*; Springer: Cham, Switzerland, 2022; pp. 285–299.
16. Ding, C.; Tao, D. Robust face recognition via multimodal deep face representation. *IEEE Trans. Multimed.* **2015**, *17*, 2049–2058. [CrossRef]
17. Chawla, B.; Tyagi, S.; Jain, R.; Talegaonkar, A.; Srivastava, S. Finger Vein Recognition Using Deep Learning. In *Formal Ontology in Information Systems, Proceedings of the International Conference on Artificial Intelligence and Applications, Trento, Italy, 6–8 June 2021*; Springer: Singapore, 2021; pp. 69–78.
18. Nada, A.; Al-Baity, H.H. Deep learning approach for multimodal biometric recognition system based on fusion of iris, face, and finger vein traits. *Sensors* **2022**, *20*, 5523.
19. Wan, K.; Song, J.M.; Park, K.R. Multimodal biometric recognition based on convolutional neural network by the fusion of finger-vein and finger shape using near-infrared (NIR) camera sensor. *Sensors* **2018**, *18*, 2296. [CrossRef]
20. Yang, W.; Wang, S.; Hu, J.; Zheng, G.; Valli, C. A fingerprint and finger-vein based cancelable multi-biometric system. *Pattern Recognit.* **2018**, *78*, 242–251. [CrossRef]
21. Soleymani, S.; Dabouei, A.; Kazemi, H.; Dawson, J.; Nasrabadi, N.M. Multi-Level Feature Abstraction from Convolutional Neural Networks for Multimodal Biometric Identification. In Proceedings of the 2018 24th International Conference on Pattern Recognition (ICPR), Beijing, China, 20–24 August 2018; pp. 3469–3476.
22. Al-Waisy, A.S.; Qahwaji, R.; Ipson, S.; Al-Fahdawi, S.; Nagem, T.A. A multi-biometric iris recognition system based on a deep learning approach. *Pattern Anal. Appl.* **2018**, *21*, 783–802.
23. Ren, H.; Sun, L.; Guo, J.; Han, C.; Wu, F. Finger vein recognition system with template protection based on convolutional neural network. *Knowl.-Based Syst.* **2021**, *227*, 107159. [CrossRef]

24. Chollet, F. *Deep Learning with Python*; Simon and Schuster: New York, NY, USA, 2021.
25. Jiao, Y.; Du, P. Performance measures in evaluating machine learning based bioinformatics predictors for classifications. *Quant. Biol.* **2016**, *4*, 320–330. [CrossRef]
26. Simonyan, K.; Andrew, Z. Very deep convolutional networks for large-scale image recognition. *arXiv* **2015**, arXiv:1409.1556.
27. Too, E.C.; Yujian, L.; Njuki, S.; Yingchun, L. A comparative study of fine-tuning deep learning models for plant disease identification. *Comput. Electron. Agric.* **2019**, *161*, 272–279. [CrossRef]
28. Tan, C.; Sun, F.; Kong, T.; Zhang, W.; Yang, C.; Liu, C. A Survey on Deep Transfer Learning. In Proceedings of the International Conference on Artificial Neural Networks, Munich, Germany, 17–19 September 2018; pp. 270–279.
29. Zhou, J.; Jia, X.; Shen, L.; Wen, Z.; Ming, Z. Improved softmax loss for deep learning-based face and expression recognition. *Cogn. Comput. Syst.* **2019**, *1*, 97–102. [CrossRef]
30. Yin, Y.; Liu, L.; Sun, X. SDUMLA-HMT: A multimodal biometric database. In *Chinese Conference on Biometric Recognition*; Springer: Berlin/Heidelberg, Germany, 2011; pp. 260–268.
31. Yang, H.K.; Kim, Y.J.; Sung, J.Y.; Kim, D.H.; Kim, K.G.; Hwang, J.-M. Efficacy for Differentiating Nonglaucomatous versus Glaucomatous Optic Neuropathy Using Deep Learning Systems. *Am. J. Ophthalmol.* **2022**, *216*, 140–146. [CrossRef] [PubMed]

Article

Facial Micro-Expression Recognition Enhanced by Score Fusion and a Hybrid Model from Convolutional LSTM and Vision Transformer

Yufeng Zheng [1,*] and Erik Blasch [2]

1 Department of Data Science, University of Mississippi Medical Center, Jackson, MS 39216, USA
2 MOVEJ Analytics, Fairborn, OH 45324, USA; erik.blasch@gmail.com
* Correspondence: yzheng@umc.edu

Abstract: In the billions of faces that are shaped by thousands of different cultures and ethnicities, one thing remains universal: the way emotions are expressed. To take the next step in human–machine interactions, a machine (e.g., a humanoid robot) must be able to clarify facial emotions. Allowing systems to recognize micro-expressions affords the machine a deeper dive into a person's true feelings, which will take human emotion into account while making optimal decisions. For instance, these machines will be able to detect dangerous situations, alert caregivers to challenges, and provide appropriate responses. Micro-expressions are involuntary and transient facial expressions capable of revealing genuine emotions. We propose a new hybrid neural network (NN) model capable of micro-expression recognition in real-time applications. Several NN models are first compared in this study. Then, a hybrid NN model is created by combining a convolutional neural network (CNN), a recurrent neural network (RNN, e.g., long short-term memory (LSTM)), and a vision transformer. The CNN can extract spatial features (within a neighborhood of an image), whereas the LSTM can summarize temporal features. In addition, a transformer with an attention mechanism can capture sparse spatial relations residing in an image or between frames in a video clip. The inputs of the model are short facial videos, while the outputs are the micro-expressions recognized from the videos. The NN models are trained and tested with publicly available facial micro-expression datasets to recognize different micro-expressions (e.g., happiness, fear, anger, surprise, disgust, sadness). Score fusion and improvement metrics are also presented in our experiments. The results of our proposed models are compared with that of literature-reported methods tested on the same datasets. The proposed hybrid model performs the best, where score fusion can dramatically increase recognition performance.

Keywords: facial micro-expression; human-machine interaction; long short-term memory (LSTM); convolutional neural network (CNN); vision transformer; score fusion; deep learning

Citation: Zheng, Y.; Blasch, E. Facial Micro-Expression Recognition Enhanced by Score Fusion and a Hybrid Model from Convolutional LSTM and Vision Transformer. *Sensors* **2023**, *23*, 5650. https://doi.org/10.3390/s23125650

Academic Editor: Wataru Sato

Received: 15 May 2023
Revised: 2 June 2023
Accepted: 13 June 2023
Published: 16 June 2023

Copyright: © 2023 by the authors. Licensee MDPI, Basel, Switzerland. This article is an open access article distributed under the terms and conditions of the Creative Commons Attribution (CC BY) license (https:// creativecommons.org/licenses/by/ 4.0/).

1. Introduction

Facial expressions serve as a universally understood form of human communication intimately linked to one's mental states, attitudes, and intentions. In addition to the typical facial expressions displayed in daily life, known as *macro-expressions*, there exists a distinct category called *micro-expressions*. These micro-expressions emerge in specific conditions, unveiling people's concealed emotions during high-stakes situations when they strive to mask their true feelings [1,2]. Unlike macro-expressions, micro-expressions are involuntary, spontaneous, subtle, and rapid, lasting typically between 40 and 450 ms. They are instinctive facial movements that react to emotional stimuli [3,4]. While individuals can consciously conceal or restrain their genuine emotions through macro-expressions, micro-expressions are beyond their control, inevitably exposing the authentic emotions they experience [5–8].

Recognizing micro-expressions is a daunting task as they are fleeting, involuntary, and exhibit low intensity. Only extensively trained experts possess the ability to discern these subtle facial cues. However, even with rigorous training, the average human can only identify approximately 47% of micro-expressions [9]. Moreover, human analysis of micro-expressions is error-prone and costly. Therefore, there is a pressing need to develop an automated system for the analysis and recognition of micro-expressions.

Facial expressions result from the intricate interplay of facial skin, connective tissue, and the activation of facial muscles controlled by facial nerve nuclei. These nuclei, in turn, are regulated by cortical and subcortical upper motor neuron circuits. A noteworthy neuropsychological study investigating facial expressions [10] unveiled two distinct neural pathways situated in different brain areas, each playing a role in mediating facial behavior. The cortical circuit, located within the cortical motor strip, primarily governs deliberate, voluntary facial expressions. Conversely, the subcortical circuit, residing in subcortical brain regions, primarily governs spontaneous, involuntary emotional facial expressions. In intense emotional situations where individuals endeavor to conceal or suppress their feelings, both systems are likely to be activated, resulting in fleeting glimpses of genuine emotions through micro-expressions [11]. Consequently, when attempting to mask their emotions, true feelings can swiftly "leak out" and manifest as micro-expressions [5].

Micro-expressions encompass the ability to convey seven universal emotions: disgust, anger, fear, sadness, happiness, surprise, and contempt [12]. Differing from macro-expressions, micro-expressions are characterized by their short duration and pronounced inhibition of facial muscle movement [1,13]. This distinctive feature allows them to authentically reflect a person's true emotions, as they occur involuntarily and are more difficult to control [14]. However, micro-expression recognition is very challenging. According to a literature report [15] tested on the micro- and macro-expression warehouse (MMEW) dataset up to 2021, the best traditional machine learning (ML) method uses Directional Mean Optical Flow (MDMO) features [16], which can achieve 65.7% accuracy. Meanwhile, the best deep learning (DL) method applies a transferring long-term convolutional neural network (TLCNN) to extract frame features [17], which can be as high as 69.4% accuracy.

One of the challenges in micro-expression recognition is how to extract the temporospatial features from facial videos, which contain big and redundant data. Traditional feature extraction methods based on single image analysis [16] are not effective. Regular CNNs [17] are good for capturing spatial features but lack temporal analyses between sequential frames.

Therefore, in this study, to extract facial features from sequential frames (video clips), we propose a hybrid model comprised of a transformer and RNN (LSTM). The transformer can summarize the sparse spatial relations among image blocks [18], whereas the LSTM can analyze the temporal changes among frames. Furthermore, score fusion methods are applied to the multiple scores (from different NN models) in order to improve micro-expression recognition accuracies. The remainder of this paper is organized as follows. Section 2 describes datasets and preprocessing. Section 3 reviews the CNN and RNN. Section 4 presents the experimental results. Section 5 summarizes the paper.

2. Micro-Expression Dataset and Image Preprocessing

2.1. Facial Micro-Expression Dataset

There are two datasets used in this study, which are briefly described in this sub-section.

The MMEW dataset [15] follows the same elicitation paradigm used in other published datasets [19–23], i.e., watching emotional video episodes while attempting to maintain a neutral expression. The full details of the MMEW dataset construction process are presented in Reference [15]. All samples in MMEW were carefully calibrated by experts with onset, apex, and offset frames, and the action unit (AU) [15] annotation from the Facial Action Coding System (FACS) [24] was used to describe the facial muscle movement area. MMEW contains 300 micro-expression samples (image sequences). The samples in MMEW have a large image resolution (1920 × 1080 pixels). Furthermore, MMEW has a facial image size of 400 × 400 pixels. MMEW has seven elaborate emotion classes (see Figure 1), i.e., Happiness (36), Anger (8), Surprise (89), Disgust (72), Fear (16), Sadness (13), and Others (66).

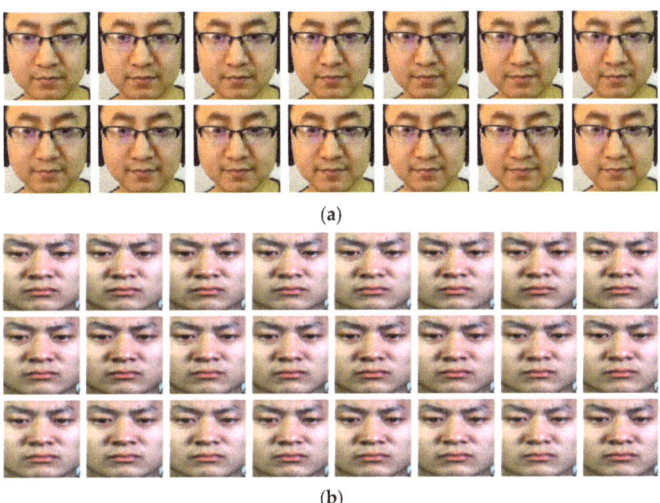

Figure 1. Facial micro-expression samples: (**a**) MMEW dataset—Facial video clip (14 frames shot at 90 FPS) labeled as "Happiness"; (**b**) CASME II dataset—Facial video samples (24 frames shot at 200 FPS) labeled as "Disgust". All facial images are resized to 224 × 224 × 3.

The Chinese Academy of Sciences Micro-expression CASME II [22] dataset was developed in a well-controlled laboratory environment, where four lamps were chosen to provide steady and high-intensity illumination. To elicit micro-expressions, participants were instructed to maintain a neutral facial expression when watching video episodes with high emotional valence. CASME II used a high-speed camera with a sampling rate of 200 fps. There are 247 image sequences, which consist of facial images of 280 × 340 pixels. CASME II contains the samples of five emotion classes (see Figure 1), i.e., Happiness (33 samples), Repression (27), Surprise (25), Disgust (60), and Others (102).

2.2. Facial Image Standardization and Normalization

All facial images have been extracted from both datasets (as shown in Figure 1). Face detection is unnecessary in this study as the dataset contains one face per image.

General standardization is applied to all facial images, which is defined as follows.

$$\mathbf{I}'_N = \mathbf{I}_N - \mathbf{I}_M, \tag{1}$$

$$\mathbf{I}_S = (\mathbf{I}'_N - \mu)/\sigma, \tag{2}$$

where $\mathbf{I}_N$ is the normalized facial image, $\mathbf{I}_M$ is the mean image of all faces in the dataset, and $\mathbf{I}'_N$ is their difference image. $\mathbf{I}_S$ is the standardized image; μ and σ denote the mean and standard deviation of the $\mathbf{I}'_N$ image, respectively.

Image normalization (intensity scaling) is required by neural network models, which is defined as

$$\mathbf{I}_N = (\mathbf{I}_0 - I_{Min}) \frac{L_{Max} - L_{Min}}{I_{Max} - I_{Min}} + L_{Min}, \tag{3}$$

where $\mathbf{I}_N$ is the normalized image, $\mathbf{I}_0$ is the original (input) image; I_{Min} and I_{Max} are the minimum and maximum pixel values in $\mathbf{I}_0$, while L_{Min} and L_{Max} are the desired minimum and maximum pixel values in $\mathbf{I}_N$. For example, we may select $L_{Min} = 0$ and $L_{Max} = 1.0$.

3. Convolutional and Recurrent Neural Networks

3.1. Convolutional Neural Networks

Convolutional neural networks (CNNs) draw inspiration from the biological functioning of the visual cortex, where small groups of cells exhibit sensitivity to specific regions within the visual field. CNNs blend principles from biology, mathematics, and computer science, making them pivotal innovations in the domains of computer vision and artificial intelligence (AI). The year 2012 marked a significant turning point for CNNs when Krizhevsky et al. [25] utilized an eight-layer CNN (comprising five convolutional layers and three fully-connected layers) to secure victory in the ImageNet competition. This groundbreaking achievement, known as *AlexNet*, reduced the classification error rate from 25.8% in 2011 to an impressive 16.4% in 2012, signifying a remarkable improvement at that time. Since then, deep learning CNNs have spurred the development of numerous applications in various domains.

During the training of AlexNet, batch *stochastic gradient descent* (SGD) was employed, incorporating carefully selected momentum and weight decay values. This groundbreaking model achieved exceptional performance on the challenging ImageNet dataset, setting a new record in the competition and solidifying the superior capabilities of CNNs. Later, there are many well-known CNNs developed, such as VGG-19, ResNet-50, Inception-V3, DenseNet-201, etc. Hereby our review begins with ResNet and Xception models, followed by a discussion on vision transformers, and then finishes with RNNs.

3.2. ResNet and Xception Models

The ResNet-50 model [26] is composed of 50 layers, featuring 16 residual blocks with three layers each, in addition to input and output layers. These *residual* blocks introduce identity connections that facilitate incremental or residual learning, enabling effective back-propagation. Through this approach, the identity layers progressively evolve from simple to complex representations. This evolution is particularly beneficial when the parameters of a CNN block start at or near zero. The inclusion of residual blocks helps address the challenging issue of vanishing gradients encountered in training deep neural networks with more than 30 layers.

Recently, an Xception [27] (Extreme Inception) network architecture has been proposed on the following hypothesis: the mapping of cross-channel correlations and spatial correlations in the feature maps of CNNs can be entirely decoupled. Thus, the Inception modules can be replaced with *depthwise separable convolutions*. The feature extraction base of the Xception architecture is constructed with 36 convolutional layers. For image classification, a logistic regression layer follows the convolutional base. Optionally, fully-connected layers can be added before the logistic regression layer. These 36 convolutional layers are organized into 14 modules, with linear residual connections encompassing each module, except for the first and last ones. When compared to Inception V3, Xception exhibits a comparable parameter count while demonstrating slight enhancements in classification performance on the ImageNet dataset.

In the Xception model, a depthwise separable convolution, also known as a separable convolution in deep learning frameworks, such as TensorFlow/Keras, is employed. This approach involves two steps: first, a depthwise convolution is performed independently on each channel of the input, followed by a pointwise convolution, which is a 1×1 convolution. The pointwise convolution projects the output channels from the depthwise convolution into a new channel space. The scenario of separable convolution plus pointwise convolution can significantly reduce the load of convolutional computation in contrast with a regular two-dimensional (2D) or three-dimensional (3D) convolutional layer; thus, it speeds up the CNN model training and the inference process.

3.3. Vision Transformers

Initially, transformers were developed and applied primarily to tasks in natural language processing (NLP), as evidenced by language models, such as BERT (Bidirectional

Encoder Representations from Transformers) [28]. Transformers ascertain the connections between pairs of input tokens, such as words in NLP, through a mechanism called *attention*. However, this approach becomes increasingly computationally expensive with a growing number of tokens. When dealing with images, the fundamental unit of analysis becomes the pixel. Nevertheless, computing relationships between every pair of pixels becomes prohibitively costly in terms of memory and computation. To address this, *Vision Transformers* (ViTs) calculate relationships among smaller image regions, typically 16 × 16 pixels, resulting in reduced computational requirements. These regions, accompanied by positional embeddings, are organized into a sequence. The embeddings represent learnable vectors. Each region is vectorized and multiplied by an embedding matrix. The resulting sequence, along with the positional embeddings, is then fed into the transformer for further processing. The Video ViT (ViViT) model has one additional process, called a video tube (cube, i.e., frames by height by width, e.g., 4 × 16 × 16) positional embedding, while the rest of the ViViT process is the same as ViT.

Self-attention is commonly applied to the vision transformer model. The calculation of self-attention is to create three vectors from each of the encoder's input vectors (in the NLP case, the embedding of each word). So for each word, a *Query* vector, a *Key* vector, and a *Value* vector are created by multiplying the embedding by three matrices that were trained during the training process. Multi-headed attention expands the model's ability to focus on different positions. It gives the attention layer multiple "representation subspaces". The multi-headed attention has multiple sets of Query/Key/Value weight matrices (e.g., a transformer uses eight attention heads, consisting of eight sets of weight matrices for each encoder/decoder). Each of these sets is randomly initialized. Then, after training, each set is used to project the input embeddings (or vectors from lower encoders/decoders) into a different representation subspace.

Two designs of the ViViT architecture are illustrated in Table 1, where the input is the frames of a video clip (shape of MMEW input: (14, 224, 224, 3)), while the output includes seven probabilities corresponding to seven micro-expression classes. The ViViT_FM2 model has an additional CNN block, X_CNN, which is comprised of the first five convolutional layers plus one residual block from the Xception model.

Table 1. Transformer model architectures—ViViT_FM1 and ViViT_FM2. Normalization and Dropout layers are omitted. The batch size (typically shown as None) is omitted in the "Output Shape" column. Attn_FF means attention-based feed-forward network. The numbers shown in "Output Shape" are assumed to be the inputs from the MMEW dataset.

ViViT_FM1 (8.6 M Paras)		ViViT_FM2 (8.8 M Paras)	
Layer (Type)	Output Shape	Layer (Type)	Output Shape
Frame_Input	(14, 224, 224, 3)	Frame_Input	(14, 224, 224, 3)
		time_distributed(X_CNN) (5 × Conv2D, 1 × Res.)	(14, 112, 112, 32)
Tubelet_Embedding (Conv3D → 7 × 16 × 16 patches)	(1792, 64)	Tubelet_Embedding (Conv3D → 7 × 16 × 16 patches)	(1792, 64)
Positional_Encoder	(1792, 64)	Positional_Encoder	(1792, 64)
6 × **Attn_FF**:		6 × **Attn_FF**:	
MultiHeadAttention (heads = 8, key_dim = 64)		MultiHeadAttention (heads = 8, key_dim = 64)	
Feed_Forward_Net (256 → 64)		Feed_Forward_Net (256 → 64)	
Add_Attn_FF_Norm	(1792, 64)	Add_Attn_FF_Norm	(1792, 64)
MaxPooling1D (pool_size = 4, strides = 4)	(448, 64)	MaxPooling1D (pool_size = 4, strides = 4)	(448, 64)

Table 1. *Cont.*

Flatten	(28672)	Flatten	(28672)
Dense	(256)	Dense	(256)
Dense (Output)	(7)	Dense (Output)	(7)

3.4. Recurrent Neural Networks—ConvLSTM models

Recurrent neural networks (RNNs) are designed to leverage sequential information in data. Unlike traditional neural networks that assume inputs and outputs are independent of each other, RNNs recognize the importance of dependencies in tasks such as natural language processing (NLP); for instance, predicting the next word in a sentence benefits from knowledge about the preceding words. RNNs are termed "recurrent" because they execute the same operation for each element of a sequence, with the output relying on previous computations. Another way to conceptualize RNNs is as having a "memory" that retains information about prior calculations. In theory, RNNs can utilize information from arbitrarily long sequences, but in practice, they are typically limited to considering only a few preceding steps. Figure 2 provides an illustration of a typical RNN architecture.

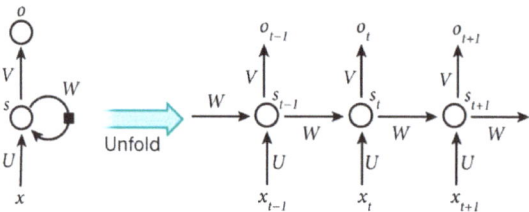

Figure 2. Illustration of a recurrent neural network and the unfolding in time of its forward computation.

RNNs have exhibited remarkable achievements in various NLP tasks and applications involving temporal signals [29]. Among the different types of RNNs, *long short-term memory* (LSTM) networks are widely used and excel in capturing long-term dependencies. LSTMs are essentially similar to RNNs, but they employ a distinct method to compute the hidden state. In LSTMs, memories are referred to as cells, functioning, such as black boxes that take the previous (hidden) state, s_{t-1}, and the current input, x_t, as inputs. These cells internally make decisions about what information to retain or discard from memory. Subsequently, they combine the previous state, the current memory, and the input. Remarkably, these LSTM units have proven highly effective at capturing long-term dependencies.

In some applications (e.g., predicting weather changes), we want to model temporal evolution (e.g., temperature changing over time), ideally using recurrence relations (e.g., LSTM). In facial micro-expression recognition, we need to capture the facial muscle movement over time. At the same time, we also expect to efficiently extract spatial features (e.g., facial muscle movement varying with locations), something that is normally done with convolutional filters. Ideally, then, an architecture includes both recurrent and convolutional mechanisms, which are *convolutional LSTM* (ConvLSTM) layers.

As shown in Table 2, two designs of ConvLSTM models are illustrated for facial micro-expression recognition. The outputs of the 14 × 7 dimension correspond to 14 frames and seven micro-expression classes (MMEW), and the final recognition results of each video sample are the averaged results of 14 frames. The major difference between these two models is that the X_ConvLSTM_FM4 model is comprised of bidirectional ConvLSTM layers. In the context of an NLP model, a unidirectional ConvLSTM block can find some hints (such as the meaning of "it") from future sentences, while a bidirectional ConvLSTM block can find hints from both future sentences and previous sentences. It is not surprising that the number of model parameters in the bidirectional X_ConvLSTM_FM3 model is more than doubled compared with that of a unidirectional X_ConvLSTM_FM4 model.

Table 2. RNN model architectures—X_ConvLSTM_FM3 and X_ConvLSTM_FM4. Normalization and Dropout layers are omitted. The batch size is omitted in the "Output Shape" column. time_distr = time_distributed. R_CNN consists of the first six convolutional layers plus one residual block from the ResNet-50 model.

X_ConvLSTM_FM3 (37.3 M Paras)		X_ConvLSTM_FM4 (17.5 M Paras)	
Layer (Type)	Output Shape	Layer (Type)	Output Shape
Frame_Input	(14, 224, 224, 3)	Frame_Input	(14, 224, 224, 3)
time_distr (R_CNN) (6 × Conv2D, 1 Res)	(14, 56, 56, 64)	time_distr (R_CNN) (6 × Conv2D, 1 Res)	(14, 56, 56, 64)
time_distr (MaxPool2D) (pool_size = (3,3), strides = (2,2))	(14, 28, 28, 64)		
Bidirectional_ConvLSTM2D (filters = 256, kernel_size = (3,3))	(14, 28, 28, 256)	ConvLSTM2D (filters=128, kernel_size=(3,3))	(14, 56, 56, 128)
time_distr (MaxPool2D) (pool_size = (2,2), strides = (2,2))	(14, 14, 14, 256)	time_distr (MaxPool2D) (pool_size = (2,2), strides = (2,2))	(14, 28, 28, 128)
Bidirectional_ConvLSTM2D (filters = 384, kernel_size = (3,3))	(14, 14, 14, 384)	ConvLSTM2D (filters = 256, kernel_size = (3,3))	(14, 28, 28, 256)
time_distr (MaxPool2D) (pool_size = (2,2), strides = (2,2))	(14, 7, 7, 384)	time_distr (MaxPool2D) (pool_size = (2,2), strides = (2,2))	(14, 14, 14, 256)
time_distr (Flatten)	(14, 18816)	time_distr (Flatten)	(14, 50176)
time_distr (Dense)	(14, 256)	time_distr (Dense)	(14, 256)
time_distr (Dense) (Output)	(14, 7)	time_distr (Dense) (Output)	(14, 7)

3.5. Hybrid Models

As described in previous subsections, we know that CNNs can extract spatial features; furthermore, ConvLSTM layers can capture spatial-temporal changes. CNNs are good at modeling neighborhood changes, whereas transformers with attention can grasp sparse spatial relations (e.g., among different image blocks or across frames in a video clip). As shown in Table 3, two designs of hybrid models combine three NN models: CNN, ConvLSTM, and ViViT. The two models differ in the ConvLSTM layer, where the Hybrid_FM5 model has 128 filters and is followed by a pooling layer. The goal is to combine the methods to enhance the performance of a single method.

The Hybrid_FM6 model is illustrated in Figure 3, which consists of three blocks: CNN, ConvLSTM, and Transformer. First, the CNN block (of six convolutional layers) provides local spatial features, where the feature image is the last-layer output randomly selected from 1 of 64 filters and 1 or 14 frames. Second, the ConvLSTM block (of one layer and 64 filters) generates temporal features, where the feature image is randomly selected from 1 of 64 filters and 1 or 14 frames. Third, the Transformer block (of 343 patches and 96 embedding dimensions) presents sparse spatial relations. Fourth, there is a fully connected layer (of 512 filters) prior to the output layer (of seven filters). The Hybrid_FM6 model can predict a micro-expression (e.g., "Anger") with a given facial video (e.g., of 14 frames).

Table 3. Hybrid model architectures—Hybrid_FM5 and Hybrid_FM6 that combine 3 NN models: CNN, ConvLSTM, and ViViT. Normalization and Dropout layers are omitted. The batch size is omitted in the "Output Shape" column. time_distr = time_distributed.

Hybrid_FM5 (20.6 M Paras)		Hybrid_FM6 (20.4 M Paras)	
Layer (Type)	Output Shape	Layer (Type)	Output Shape
Frame_Input	(14, 224, 224, 3)	Frame_Input	(14, 224, 224, 3)
time_distr (R_CNN) (6 × Conv2D, 1 Res)	(14, 56, 56, 64)	time_distr (R_CNN) (6 × Conv2D, 1 Res)	(14, 56, 56, 64)

Table 3. *Cont.*

Hybrid_FM5 (20.6 M Paras)		Hybrid_FM6 (20.4 M Paras)	
Layer (Type)	**Output Shape**	**Layer (Type)**	**Output Shape**
ConvLSTM2D (filters = 128, kernel_size = (3,3))	(14, 56, 56, 128)	ConvLSTM2D (filters = 64, kernel_size = (3,3))	(14, 56, 56, 64)
time_distr (MaxPool2D) (pool_size = (2,2), strides = (2,2))	(14, 28, 28, 128)		
Tubelet_Embedding (Conv3D → 7 × 7 × 7 patches)	(343, 96)	Tubelet_Embedding (Conv3D → 7 × 7 × 7 patches)	(343, 96)
Positional_Encoder	(343, 96)	Positional_Encoder	(343, 96)
6 × **Attn_FF**:		6 × **Attn_FF**:	
MultiHeadAttention (heads = 8, key_dim = 64)		MultiHeadAttention (heads = 8, key_dim = 64)	
Feed_Forward_Net (384 → 96)		Feed_Forward_Net (384 → 96)	
Add_Attn_FF_Norm	(343, 96)	Add_Attn_FF_Norm	(343, 96)
Flatten	(32928)	Flatten	(32928)
Dense	(512)	Dense	(512)
Dense (Output)	(7)	Dense (Output)	(7)

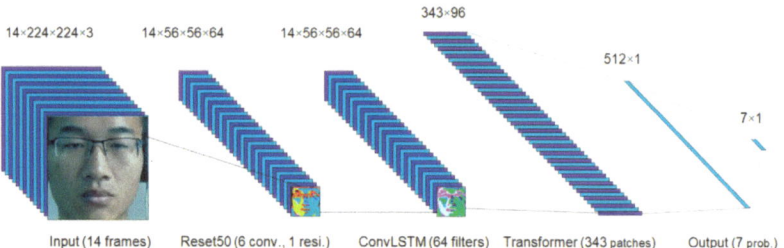

Figure 3. Feature maps of the Hybrid_FM6 model (refer to Table 3): The dimensions of feature maps are shown on the top, whereas the block functions are given at the bottom. The input is a video sample of 14 frames (the 1st MMEW sample), and the output is a vector of 7 probabilities: [0.0, 0.0, 0.0, 0.0, 0.0, 0.0, 1.0]. The index of the predicted class is 6 (max prob.), which corresponds to a labeled micro-expression, "Anger".

4. Experimental Results

Both datasets, MMEW and CASME II, were used in our experiments. The number of frames varies with different video files. Based on manual analyses of the minimal and maximal length of clips and filming speed (FPS), 14 frames are clipped in the MMEW dataset, whereas 24 frames are clipped in the CASME II dataset. If the number of frames in a video file is as m times long as the number of clipped frames (n), then m (typically $m \leq 2$) samples (clips) are clipped from that file. However, there are no overlapped (repeatedly used) frames in the m samples from the same video file. The distributions of facial video samples (clips) from two datasets are shown in Figure 4, where the numbers of clips are larger than the number of video files. During the data splits (k-fold cross-validation) for training and testing, we will make sure that (i) multiple clips from one video file will split into one subset, i.e., either in training or in testing; (ii) the training set includes samples from all classes (stratified split).

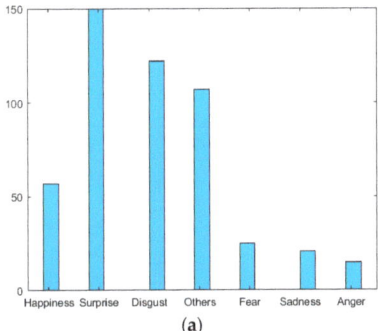
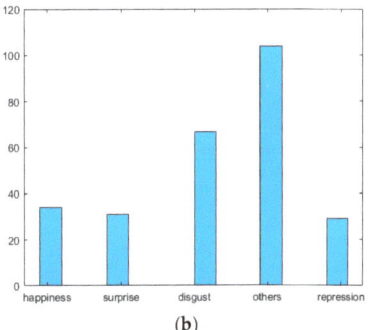

Figure 4. (**a**) Facial video clips (14-frame sequences @ 90 FPS) distributions of seven micro-expressions: a total of 497 samples from 300 original video clips in the MMEW dataset. (**b**) Facial video clips (24-frame sequences @ 200 FPS) distributions of five micro-expressions: a total of 265 samples from 247 original video clips in the CASME II dataset.

All facial images were resized to 224 × 224 pixels, then standardized and normalized (intensity stretched). Ten-fold cross-validation was used in our experiments, and the final classification results were calculated by merging all 10-fold testing scores (instead of averaging 10 testing accuracies).

The first subsection briefly reviews the performance metrics used in our experiments. Then, the classification performances of eight NN models are presented. Score fusion improvement is described and quantitatively measured in the next two subsections. The time costs are reported in the last subsection.

4.1. Classification Performance Metrics

The performance of micro-expression recognition is measured by $F1$ score and accuracy, as shown in Tables 4 and 5, where the two metrics are defined as follows.

Table 4. $F1$ scores of seven micro-expressions, weighted $F1$ scores, and Accuracy values varying over eight different NN models tested on the MMEW dataset (14 frames at 90 FPS in each sample). The tests were conducted using 10-fold cross-validations. The prediction (probability) values from 10 validation folds were merged in one set to calculate the overall $F1$ (weighted average) and *Accuracy* scores. The highest $F1$ score or accuracy in each row is bolded. Notice that the highest accuracy of 0.6940 was reported on this dataset in 2018 [17].

Metric\NN Model	ResNet-50	Xception	ViViT_FM1	X_ViViT_FM2	X_ConvLSTM_FM3	X_ConvLSTM_FM4	Hybrid_FM5	Hybrid_FM6
F1 (Happiness)	0.8622	0.8935	0.9298	0.8947	0.8935	0.9273	0.9298	**0.9298**
F1 (Surprise)	0.9062	0.869	0.9067	0.8467	0.9100	0.8971	0.8667	**0.9133**
F1 (Disgust)	0.8601	0.8946	**0.9262**	0.9016	0.9104	0.9151	0.9098	0.9180
F1 (Others)	0.8284	0.8158	0.7850	0.7664	0.8178	0.8204	0.7944	**0.8318**
F1 (Fear)	0.7371	0.7086	0.6800	0.6400	0.7371	0.7343	**0.7600**	0.7200
F1 (Sadness)	0.7041	0.7551	**0.8095**	0.7143	0.7245	0.7381	0.7619	0.7619
F1 (Anger)	1.0	1.0	1.0	0.933	0.9952	0.9952	1.0	**1.0**
F1 (Weighted Avg.)	0.8601	0.8588	0.8760	0.8365	0.8753	0.8779	0.8675	**0.8855**
Accuracy	0.8588	0.8577	0.8752	0.835	0.8744	0.8765	0.8632	**0.8853**

Table 5. F1 scores of five micro-expressions, weighted F1 scores, and Accuracy values varying over eight different NN models tested on the CASME II dataset (24 frames at 200 FPS in each sample). The highest F1 score or accuracy in each row is bolded. Notice that the best accuracy of 0.6341 was reported in 2014 [29] on this dataset.

Metric\NN Model	ResNet-50	Xception	ViViT_FM1	X_ViViT_FM2	X_ConvLSTM_FM3	X_ConvLSTM_FM4	Hybrid_FM5	Hybrid_FM6
F1 (Happiness)	0.4545	0.4375	0.4706	0.3860	0.3980	0.4595	**0.5067**	0.5000
F1 (Surprise)	0.6809	0.6312	0.5882	0.6301	0.6288	0.6670	**0.6957**	0.6479
F1 (Disgust)	0.6398	0.6335	0.6466	0.5672	0.6364	0.6545	**0.6667**	0.6197
F1 (Others)	0.6910	0.7004	0.7014	0.6564	0.6695	0.7054	0.7058	**0.7281**
F1 (Repression)	0.5905	0.5299	0.6038	0.5588	0.5310	**0.6357**	0.6207	0.5862
F1 (Weighted Avg.)	0.6378	0.6325	0.6544	0.5934	0.6272	0.6627	**0.6739**	0.6543
Accuracy	0.6375	0.6238	0.6301	0.5886	0.6028	0.6468	**0.6565**	0.6452

The *Precision* is the ratio $TP/(TP + FP)$ where TP is the number of true positives, and FP is the number of false positives. False positives are the samples that are predicted as positives but labeled as negatives. The *Precision* is intuitively the ability of the classifier not to predict a negative sample as positive.

The *Recall* is the ratio $TP/(TP + FN)$ where TP is the number of true positives, and FN is the number of false negatives. False negatives are the samples that are predicted as negatives but labeled as positives. The *Recall* is intuitively the ability of the classifier to correctly predict all the positive samples.

The *F1* score can be interpreted as a harmonic mean of the *Precision* and *Recall*, where an F1 score reaches its best value at 1 and worst score at 0. The relative contribution of *Precision* and *Recall* to the F1 score is equal. The formula for the F1 score is

$$F1 = 2 \times (Precision \times Recall)/(Precision + Recall). \tag{4}$$

$$Precision = TP/(TP + FP) \tag{5}$$

$$Recall = TP/(TP + FN) \tag{6}$$

$$Accuracy = (TP + TN)/(TP + FN + TN + FP) \tag{7}$$

In the multi-class and multi-label cases, the average of the F1 score of each class is analyzed with a weighting parameter. The weights for averaging can be calculated by the number of supported samples in each class divided by the total samples. The F1 score is an alternative to the *Accuracy* metric as it does not require one to know the total number of observations (e.g., *TN*). On the other hand, *Accuracy* tells how often we can expect a machine learning model will correctly predict an outcome out of the total number of predictions.

4.2. NN Model Performance on Two Datasets

Table 4 shows the F1 scores of seven micro-expressions, weighted F1 scores, and Accuracy values varying over eight different NN models when tested on the MMEW dataset. Based on the accuracy values, the model of Hybrid_FM6 achieves 0.8853, which is the best on the MMEW dataset. Compared to the literature-reported accuracy of 0.6940 on this dataset (by a CNN model in 2018—the best performance on the same dataset that we could find from the literature) [17], Hybrid_FM6's accuracy is very high. It seems that ConvLSTM models are slightly better than ViViT models, both of which are better than CNN models (ResNet-50, Xception). According to the F1 scores, it looks as though "Anger" is the easiest to be recognized, while "Fear" and "Sadness" are mostly difficult to be detected.

Table 5 shows the F1 scores of seven micro-expressions, weighted F1 scores, and Accuracy values varying with eight different NN models when tested on the CASME II dataset. Based on the accuracy values, the model of Hybrid_FM5 is the best on the MMEW dataset, and its accuracy is 0.6565. Compared to the literature-reported accuracy of 0.6341 on this dataset (by an SVM classifier in 2014—the best performance on the same dataset that we could find from the literature) [29], Hybrid_FM5's accuracy is pretty high. It seems that ConvLSTM models are slightly better than ViViT models, both of which are better than CNN models (ResNet-50, Xception). According to the F1 scores, it looks as though "Happiness" is the most difficult to be detected, while the other four expressions are equally hard to be recognized.

Overall, the hybrid models (combination of ResNet, ConvLSTM, ViViT) overperform non-hybrid models, such as CNN, ViViT, and ConvLSTM models.

4.3. Performance Improvement Using Score Fusion

The performance of facial micro-expression can be improved using score fusion methods, where the multiple scores are from eight different NN models, as presented in Tables 4 and 5. There are several types of score-fusion methods: arithmetic fusion (e.g., average, majority vote) [30], classifier-based fusion, and density-based fusion (e.g., Gaussian mixture model) [31,32]. Based on score fusion performance [33,34], two classification-based score fusion methods are selected and presented in this study: Support-Vector Machine (SVM) and Random Forest (RF). The multiple scores are combined as feature vectors and then fed into a classifier for training (with labeled score vectors) or testing.

The Support-Vector Machine (SVM) is a supervised learning model utilized for non-linear classification and data analysis [35]. In the context of training data with categorized observations, the SVM training algorithm constructs a model that can assign new data points to specific categories. For classification purposes, an SVM establishes a hyperplane (or a set of hyperplanes) as a separating line between data points belonging to different classes. The objective is to find the optimal hyperplane that maximizes the distance between the hyperplane and the closest data points in each class. This approach effectively minimizes the generalization error of the classifier by maximizing the margin between the hyperplane and the nearest data points in each class [36].

Random forest (RF) is a classification model employed in supervised learning tasks. It leverages ensemble learning, which combines multiple models to tackle complex problems rather than relying on a single model. The RF algorithm enhances accuracy by utilizing bagging or bootstrap aggregating. It generates individual decision trees by using random subsets of the training dataset as subsamples. Each decision tree produces its own output or classification. The final output is determined through *majority voting*, where the RF output corresponds to the class chosen by the majority of trees. This approach effectively mitigates the impact of overfitting that may occur in individual decision trees.

In this study, the SVM method employed a Gaussian kernel function and a one-versus-one coding design. This configuration resulted in the utilization of seven (or five) binary learners for the corresponding seven (or five) classes. In the RF model, we trained an ensemble comprising 100 classification trees using the complete training dataset. At each decision split, a random subset of predictors (scores) was utilized. The selection of split predictors aimed to maximize the gain of the split criterion across all possible splits of the predictors. The final classifications were obtained by combining the results from all the trees in the ensemble.

Scores used for fusion are created in two ways: (i) a feature vector consists of eight class indices (e.g., 0, 1, ... $n-1$; $n = 7$ or 5) for the predicted classes (pred-class) from eight NN models; (ii) a feature vector consists of n accumulated probability values (sum-prob) per (onto) its predicted classes ($n = 7$ or 5) from eight different NN models. Three score feature-vector (pred-class or sum-prob) examples from datasets are shown in Tables 6 and 7.

Table 6. Three pred-class feature-vector examples from two datasets (top for MMEW and bottom for CASME II)—predicted class index (0–6 for MMEW, 0–4 for CASME II) and its probability value (between 0 and 1).

Dataset\Model	ResNet-50	Xception	ViViT_FM1	X_ViViT_FM2	X_ConvLSTM_FM3	X_ConvLSTM_FM4	Hybrid_FM5	Hybrid_FM6
MMEW Smpl 1	6 (0.9998)	6 (0.9117)	6 (0.9683)	6 (0.9598)	6 (0.9990)	6 (0.9906)	6 (0.9926)	6 (1.0000)
MMEW Smpl 2	2 (0.9988)	2 (0.8669)	4 (0.4080)	2 (0.5816)	2 (0.9956)	2 (0.9976)	2 (0.9678)	2 (1.0000)
MMEW Smpl 3	1 (0.9998)	1 (0.9317)	1 (0.5577)	4 (0.4170)	1 (0.5820)	1 (0.7767)	2 (0.2843)	2 (0.8847)
CASME II Smpl 1	3 (0.9892)	3 (0.5876)	3 (0.6981)	3 (0.4532)	3 (0.5273)	3 (0.8401)	3 (0.3726)	3 (0.8442)
CASME II Smpl 2	3 (0.9800)	1 (0.5690)	2 (0.3846)	2 (0.8025)	3 (0.4887)	1 (0.8988)	2 (0.4726)	2 (0.6003)
CASME II Smpl 3	2 (0.7515)	0 (0.4152)	0 (0.5328)	4 (0.8950)	2 (0.3420)	3 (0.4979)	4 (0.9895)	4 (0.8853)

Table 7. Three sum-probability feature-vector examples from two datasets (left for MMEW and right for CASME II). For the MMEW Sample 2, the ViViT_FM1 model classified it as "Fear" (4) with a probability of 0.4080 (refer to Table 6), while the rest of the seven models classified it as "Disgust" (2) with the accumulated probability value of 6.4083.

ME\Dataset	MMEW Smpl 1	MMEW Smpl 2	MMEW Smpl 3	CASME II Smpl 1	CASME II Smpl 2	CASME II Smpl 3	Dataset/ME
0 (Happiness)	0.0000	0.0000	0.0000	0.0000	0.0000	0.9480	0 (Happiness)
1 (Surprise)	0.0000	0.0000	3.8479	0.0000	1.4677	0.0000	1 (Surprise)
2 (Disgust)	0.0000	6.4083	1.1689	0.0000	2.2599	1.0936	2 (Disgust)
3 (Others)	0.0000	0.0000	0.0000	5.3123	1.4687	0.4979	3 (Others)
4 (Fear)	0.0000	0.4080	0.4170	0.0000	0.0000	2.7698	4 (Repression)
5 (Sadness)	0.0000	0.0000	0.0000	-	-	-	-
6 (Anger)	7.8217	0.0000	0.0000	-	-	-	-

Table 6 lists six pred-class feature vectors (for six facial video samples in six rows), each of which consists of the predicted class index (0–6 for MMEW, 0–4 for CASME II) across eight NN models, where its probability value (between 0 and 1) is also given (to calculate sum-prob features). Table 7 presents lists six sum-prob feature vectors (in six columns), each of which consists of the accumulated probability values (sum-prob) with regard to its predicted classes across n micro-expresses (n = 7 or 5) for each facial video sample. For CASME II Sample 2, two models classified it as "Surprise" (1) with sum-prob = 1.4677, four models classified it as "Disgust" (2) with sum-prob = 2.2599, and the other two models classified it as "Others" (3) with sum-prob = 1.4687. The majority-voted result is "Disgust".

In sum, each pred-class feature vector is the concatenated classification results (indices of classes) from different models, while each sum-prob feature vector is the summed classification probabilities from different models unfolded along with various micro-expressions.

The score fusion experiments are conducted using 10-fold cross-validation, and the final results (as shown in Tables 6 and 7) are calculated with the merged 10-fold prediction outcomes. In the MMEW dataset, the RF method with pred-class features achieves the best overall. The accuracy of facial micro-expression recognition is improved to 0.9684 from 0.8853, which is a very good improvement. In the CASME II dataset, overall, the best method is still the RF method with pred-class features, which reaches 0.9112 in contrast with the best single NN model accuracy, 0.6565.

It seems the RF method with sum-prob features is better at recognizing some facial micro-expressions, such as Happiness, Surprise, and Disgust.

4.4. Metric for Fusion Improvement—Relative Rate Increase (RRI)

The performance improvement using score fusion (SF) cannot be properly measured by using the absolute difference of two accuracy rates (R_V). For example, improving

R_V from 80% to 90% seems to be more difficult than the improvement from 98% to 99%. Generally speaking, the improvement of R_V via SF becomes increasingly difficult when the original rate approaches 100%. Thus, it is proposed to use the Relative Rate Increase (RRI) [29] to evaluate the fusion improvement, where

$$\text{RRI} = \frac{\text{ARI}}{1 - \overline{R_V}} = \frac{R_F - \overline{R_V}}{1 - \overline{R_V}}, \quad (8)$$

R_F is the accuracy rate via SF and $\overline{R_V}$ is the mean of the accuracy rates from all classification models. If $\overline{R_V} = 1$ (no need to improve the accuracy via SF), then set RRI = 1. The *absolute rate increase* ARI $= R_F - \overline{R_V}$, may not precisely measure the performance improvement as stated earlier. RRI $\in$ (0, 1], where a higher value is better. According to the RRI definition, two fusion improvements—from 80% to 90% and from 98% to 99%—are equivalent, and both RRI values are 0.50. The two improvements are equivalent in the sense of their difficulty levels and the extent of the effort to implement them. Many metrics (e.g., F1, Precision, Recall) can be devised, wherein the RRI metric seeks to measure the actual improvement against the total amount of possible improvement.

The RRI values from the best SF results are listed in the right-most columns in Tables 8 and 9. In Table 8, RRI [F_1(Anger)] = 1.0 means the SF improvement is perfectly done (cannot be better). RRI [F_1(Fear)] = 0.8977 (the second best) means that the SF rate of 0.9708 is 89.77% improved in contrast with the mean rate of 0.7146. In Table 9, the best RRI [F_1(Repression)] = 0.8914 represents that the SF rate of 0.9546 has an 89.14% improvement from the mean rate of 0.5821. The second best RRI [F_1(Surprise)] = 0.8527 means an 85.27% SF increase on the basis of the averaged performance of individual models.

Table 8. Improved performance via score fusion (SF)—F1 scores of seven micro-expressions, weighted F1 scores, and Accuracy values varying over two fusion methods vs. two combined scores originating from the MMEW dataset. The Mean Rate and Max Rate are computed (or extracted) from Table 4. The Best SF RRI values are calculated with the best SF rates (bolded, from Sum-Prob RF column and Pred-Class RF column) and the mean rates using Equation (8). The top two RRI values are highlighted with a shaded background.

Metric\NN Model	Mean Rate	Max Rate Hybrid_FM6	Pred-Class SVM	Pred-Class RF	Sum-Prob SVM	Sum-Prob RF	Best SF RRI
F1 (Happiness)	0.9076	0.9298	0.934	0.9739	0.9145	**0.9823**	0.8085
F1 (Surprise)	0.8895	0.9133	0.9103	0.9722	0.9126	**0.9743**	0.7675
F1 (Disgust)	0.9045	0.9180	0.9044	**0.9732**	0.9351	0.9691	0.7194
F1 (Others)	0.8075	0.8318	0.8774	**0.9619**	0.8637	0.9545	0.8021
F1 (Fear)	0.7146	0.7200	0.7750	**0.9708**	0.8336	0.9514	0.8977
F1 (Sadness)	0.7462	0.7619	0.7561	**0.9048**	0.8105	0.9017	0.6249
F1 (Anger)	0.9904	1.0	0.9952	**1.0**	1.0	0.9976	1.0000
F1 (Weighted Avg.)	0.8672	0.8855	0.8951	**0.9686**	0.9033	0.9664	0.7636
Accuracy	0.8658	0.8853	0.8948	**0.9684**	0.9024	0.9662	0.7646

Table 9. Improved performance via score fusion—F1 scores of seven micro-expressions, weighted F1 scores, and Accuracy values varying over two fusion methods vs. two combined scores originating from the CASME II dataset. The Mean Rate and Max Rate are computed (or extracted) from Table 5. The highest F1 score or accuracy in each row is bolded. The top two RRI values are highlighted with a shaded background.

Metric\NN Model	Mean Rate	Max Rate Hybrid_FM5	Pred-Class SVM	Pred-Class RF	Sum-Prob SVM	Sum-Prob RF	Best SF RRI
F1 (Happiness)	0.4516	0.5067	0.2338	0.8653	0.4681	**0.8726**	0.7677
F1 (Disgust)	0.6331	0.6667	0.6041	0.8826	0.6689	**0.8871**	0.6923
F1 (Others)	0.6947	0.7058	0.7112	**0.9246**	0.7330	0.9161	0.7530
F1 (Surprise)	0.6462	0.6957	0.5635	0.9377	0.7507	**0.9479**	0.8527
F1 (Repression)	0.5821	0.6207	0.6646	**0.9546**	0.6043	0.9383	0.8914
F1 (Weighted Avg.)	0.6420	0.6739	0.6290	**0.9143**	0.6707	0.9120	0.7606
Accuracy	0.6289	0.6565	0.6164	**0.9112**	0.6733	0.9093	0.7607

4.5. Time Costs of NN Models

All models were implemented with Tensorflow 2.10 and ran in Jupyterlab (Version 3.4.4) on a desktop computer, HP Omen, with the following configuration: Intel i7-10700KF CPUs 3.8 GHz, 32 GB RAM, 1 TB hard disk, Ubuntu 20.04; NVIDIA GeForce RTX 3090 Graphics Board with 24 GB video memory (onboard) and 10,496 CUDA cores.

The number of model parameters and time costs of models are presented in Table 10 (on MMEW) and Table 11 (on CASME II). Time costs are related to the NN model (number of parameters) and data size (number of frames and samples). Model training is typically completed offline. In a real application, model inferencing (predicting) only processes one set of given frames or images. For example, using the Hybrid_FM6 model takes approx. 20 milliseconds per sample (14 frames) for predicting micro-expression, which is fast enough for real-time applications. The time costs on the CASME II dataset are longer due to processing more frame data (24 frames per video sample).

Table 10. Model parameters, training time (seconds per epoch), and testing time (milliseconds per sample) vary with NN models tested on the MMEW dataset. Time costs slightly change at different runs due to data caching and optimization.

Metric\NN Model	ResNet-50	Xception	ViViT_FM1	X_ViViT_FM2	X_ConvLSTM_FM3	X_ConvLSTM_FM4	Hybrid_FM5	Hybrid_FM6
Number of Parameters	25,693,063	22,966,831	33,934,215	13,029,487	28,597,319	17,578,183	20,643,719	20,446,727
Training Time (s/epoch)	17	22	17	101	34	22	15	17
Testing Time (ms/sample)	1	1	18	14	26	18	22	20

Table 11. Model parameters, training time (seconds per epoch), and testing time (milliseconds per sample) vary with NN models tested on the CASME II dataset. Time costs slightly change at different runs due to data caching and optimization.

Metric\NN Model	ResNet-50	Xception	ViViT_FM1	X_ViViT_FM2	X_ConvLSTM_FM3	X_ConvLSTM_FM4	Hybrid_FM5	Hybrid_FM6
Number of Parameters	25,691,013	22,964,781	29,823,877	13,569,133	28,596,805	17,577,669	18,603,941	18,800,165
Training Time (s/epoch)	17	21	8	165	32	21	14	19
Testing Time (ms/sample)	1	1	26	27	46	31	39	36

5. Summary and Discussion

In this study, we compared eight different neural network models in recognizing facial micro-expressions based on two datasets, where 6 of 8 models were newly designed for micro-expression recognition. The performance of the NN model in terms of accuracy (from high to low) is as follows: hybrid, ConvLSTM, vision transformer, and CNN. Overall we suggest the hybrid models that achieve the highest accuracy and yet are fast enough for real applications. The hybrid models are created by combining the fundamental building blocks from CNN, ConvLSTM, and vision transformer models, which are capable of extracting spatial features (in image neighborhood by CNN), summarizing temporal features (among video frames by LSTM), and capturing sparse spatial relations (among image blocks and video frames by transformer).

Score fusion can significantly increase facial micro-expression recognition rate. For example, "Fear" was only recognized at a low rate of 0.7146 (on the MMEW dataset). Random forest fusion improved the rate up to 0.9708, which is an 89.77% improvement according to the Relative Rate Increase (RRI) metric. The best overall accuracies from the hybrid models are 0.8853 (on the MMEW dataset) and 0.6565 (on the CASME II dataset), while score fusion can boost them up to 0.9684 and 0.9112. In addition, score fusion utilizes the outputs (e.g., predicted classes or probabilities) from multiple classifiers and has no additional hardware costs.

Information fusion can increase the recognition accuracy by combining the classification scores from different NN models and from different imaging modalities (e.g., infrared camera). With a large-scale dataset, the recognition reliability will also be improved. The inference latency may be further reduced with highly configured hardware (GPUs or multiple GPUs).

Our experimental results shed light on a new method for real-time micro-expression recognition. Also, score fusion can further improve the recognition system performance without extra hardware costs. Real-time micro-expression recognition can be implemented and integrated into mobile devices or humanoid robots, which will enable a friendly human–machine interface taking micro-expressions into account for better decision-making.

Author Contributions: Conceptualization, Y.Z. and E.B.; Methodology, Y.Z.; Validation, E.B.; Investigation, Y.Z.; Writing—original draft, Y.Z.; Writing—review & editing, E.B. All authors have read and agreed to the published version of the manuscript.

Funding: This research received no external funding.

Institutional Review Board Statement: Not applicable.

Informed Consent Statement: Not applicable.

Data Availability Statement: Not applicable.

Acknowledgments: The authors would like to thank the dataset providers who shared their precious MMEW dataset [15] and CASME II [20] dataset. We also thank Richard Zheng, who pioneered the work of facial micro-express recognition in this study.

Conflicts of Interest: The authors declare no conflict of interest.

References

1. Ekman, P. Darwin, deception, and facial expression. *Ann. N.Y. Acad. Sci.* **2003**, *1000*, 205–221. [CrossRef]
2. Zhang, L.; Arandjelović, O. Review of Automatic Microexpression Recognition in the Past Decade. *Mach. Learn. Knowl. Extr.* **2021**, *3*, 21. [CrossRef]
3. Ekman, P.; Friesen, W.V. Constants across cultures in the face and emotion. *J. Pers. Soc. Psychol.* **1971**, *17*, 124–129. [CrossRef]
4. Ekman, P. Lie Catching and Microexpressions. In *The Philosophy of Deception*; Oxford University Press: New York, NY, USA, 2009; pp. 118–133. [CrossRef]
5. Ekman, P.; Friesen, W.V. Nonverbal leakage and clues to deception. *Psychiatry* **1969**, *32*, 88–106. [CrossRef]
6. Bhushan, B. Study of facial micro-expressions in psychology. In *Understanding Facial Expressions in Communication*; Springer: Berlin/Heidelberg, Germany, 2015; pp. 265–286.
7. Ekman, P. *Telling Lies: Clues to Deceit in the Marketplace, Politics, and Marriage (Revised Edition)*; WW Norton & Company: New York, NY, USA, 2009.
8. Porter, S.; Brinke, L.T. Reading between the lies: Identifying concealed and falsified emotions in universal facial expressions. *Psychol. Sci.* **2008**, *19*, 508–514. [CrossRef] [PubMed]

9. Frank, M.; Herbasz, M.; Sinuk, K.; Keller, A.; Nolan, C. I see how you feel: Training laypeople and professionals to recognize fleeting emotions. In *The Annual Meeting of the International Communication Association*; Sheraton: New York, NY, USA, 2009.
10. Rinn, W.E. The neuropsychology of facial expression: A review of the neurological and psychological mechanisms for producing facial expressions. *Psychol. Bull.* **1984**, *95*, 52–77. [CrossRef]
11. Matsumoto, D.; Hwang, H.S. Evidence for training the ability to read microexpressions of emotion. *Motiv. Emot.* **2011**, *35*, 181–191. [CrossRef]
12. Polikovsky, S.; Kameda, Y.; Ohta, Y. Facial micro-expressions recognition using high speed camera and 3D-gradient descriptor. In Proceedings of the 3rd International Conference on Imaging for Crime Detection and Prevention (ICDP 2009), London, UK, 3 December 2009.
13. Ekman, P.; O'Sullivan, M. From flawed self-assessment to blatant whoppers: The utility of voluntary and involuntary behavior in detecting deception. *Behav. Sci. Law* **2006**, *24*, 673–686. [CrossRef]
14. Hurley, C.M.; Frank, M.G. Executing facial control during deception situations. *J. Nonverbal Behav.* **2011**, *35*, 119–131. [CrossRef]
15. Ben, X.; Ren, Y.; Zhang, J.; Wang, S.-J.; Kpalma, K.; Meng, W.; Liu, Y.-J. Video-based Facial Micro-Expression Analysis: A Survey of Datasets, Features and Algorithms. *IEEE Trans. Pattern Anal. Mach. Intell.* **2021**, *44*, 5826–5846. [CrossRef]
16. Liu, Y.-J.; Zhang, J.-K.; Yan, W.-J.; Wang, S.-J.; Zhao, G.; Fu, X. A Main Directional Mean Optical Flow Feature for Spontaneous Micro-Expression Recognition. *IEEE Trans. Affect. Comput.* **2015**, *7*, 299–310. [CrossRef]
17. Wang, S.-J.; Li, B.-J.; Liu, Y.-J.; Yan, W.-J.; Ou, X.; Huang, X.; Xu, F.; Fu, X. Micro-expression recognition with small sample size by transferring long-term convolutional neural network. *Neurocomputing* **2018**, *312*, 251–262. [CrossRef]
18. Zhang, L.; Hong, X.; Arandjelovic, O.; Zhao, G. Short and Long Range Relation Based Spatio-Temporal Transformer for Micro-Expression Recognition. *IEEE Trans. Affect. Comput.* **2022**, *13*, 1973–1985. [CrossRef]
19. Davison, A.K.; Lansley, C.; Costen, N.; Tan, K.; Yap, M.H. SAMM: A Spontaneous Micro-Facial Movement Dataset. *IEEE Trans. Affect. Comput.* **2016**, *9*, 116–129. [CrossRef]
20. Li, X.; Pfister, T.; Huang, X.; Zhao, G. A spontaneous microexpression database: Inducement, collection and baseline. In Proceedings of the 2013 10th IEEE International Conference and Workshops on Automatic Face and Gesture Recognition (FG), Shanghai, China, 22–26 April 2013; pp. 1–6.
21. Yan, W.-J.; Wu, Q.; Liu, Y.-J.; Wang, S.-J.; Fu, X. CASME database: A dataset of spontaneous micro-expressions collected from neutralized faces. In Proceedings of the 2013 10th IEEE International Conference and Workshops on Automatic Face and Gesture Recognition (FG), Shanghai, China, 22–26 April 2013; pp. 1–7. [CrossRef]
22. Yan, W.-J.; Li, X.; Wang, S.-J.; Zhao, G.; Liu, Y.-J.; Chen, Y.-H.; Fu, X. CASME II: An Improved Spontaneous Micro-Expression Database and the Baseline Evaluation. *PLoS ONE* **2014**, *9*, e86041. [CrossRef] [PubMed]
23. Qu, F.; Wang, S.-J.; Yan, W.-J.; Li, H.; Wu, S.; Fu, X. CAS(ME): A Database for Spontaneous Macro-Expression and Micro-Expression Spotting and Recognition. *IEEE Trans. Affect. Comput.* **2017**, *9*, 424–436. [CrossRef]
24. Ekman, P.; Friesen, W.V. Facial action coding system (FACS): A technique for the measurement of facial actions. *Riv. Psichiatr.* **1978**, *47*, 126–138.
25. Krizhevsky, A.; Sutskever, I.; Hinton, G.E. Imagenet classification with deep convolutional neural networks. In Proceedings of the 25th International Conference on Neural Information Processing Systems, Lake Tahoe, NV, USA, 3–6 December 2012; Volume 1; pp. 1097–1105.
26. He, K.; Zhang, X.; Ren, S.; Sun, J. Deep Residual Learning for Image Recognition. In Proceedings of the 2016 IEEE Conference on Computer Vision and Pattern Recognition (CVPR), Las Vegas, NV, USA, 26 June–1 July 2016; pp. 770–778.
27. Chollet, F. Xception: Deep Learning with Depthwise Separable Convolutions. In Proceedings of the 2017 IEEE Conference on Computer Vision and Pattern Recognition (CVPR), Honolulu, HI, USA, 21–26 July 2017; pp. 1800–1807.
28. Devlin, J.; Chang, M.-W.; Lee, K.; Toutanova, K. BERT: Pre-training of Deep Bidirectional Transformers for Language Understanding. *arXiv* **2018**, arXiv:1810.04805v2.
29. Wensel, J.; Ullah, H.; Munir, A. ViT-ReT: Vision and Recurrent Transformer Neural Networks for Human Activity Recognition in Videos. 2022. Available online: https://arxiv.org/pdf/2208.07929.pdf (accessed on 16 August 2022).
30. Kuncheva, L. A theoretical study on six classifier fusion strategies. *IEEE Trans. Pattern Anal. Mach. Intell.* **2002**, *24*, 281–286. [CrossRef]
31. Prabhakar, S.; Jain, A.K. Decision-level fusion in fingerprint verification. *Pattern Recognit.* **2002**, *35*, 861–874. [CrossRef]
32. Ulery, B.; Hicklin, A.; Watson, C.; Fellner, W.; Hallinan, P. Studies of Biometric Fusion. NIST Interagency Report. 2006. Available online: https://tsapps.nist.gov/publication/get_pdf.cfm?pub_id=50722 (accessed on 1 September 2006).
33. Zheng, Y.; Blasch, E. An Exploration of the Impacts of Three Factors in Multimodal Biometric Score Fusion: Score Modality, Recognition Method, and Fusion Process. *J. Adv. Inf. Fusion* **2015**, *9*, 106–123.
34. Zheng, Y.; Blasch, E.; Liu, Z. *Multispectral Image Fusion and Colorization*; SPIE Press: Bellingham, WA, USA, 2018.

35. Burges, C.J. A tutorial on support vector machines for pattern recognition. *Data Min. Knowl. Discov.* **1998**, *2*, 121–167. [CrossRef]
36. Hastie, T.; Tibshirani, R.; Friedman, J.H. *The Elements of Statistical Learning: Data Mining, Inference, and Prediction*; Springer: Berlin/Heidelberg, Germany, 2009.

Disclaimer/Publisher's Note: The statements, opinions and data contained in all publications are solely those of the individual author(s) and contributor(s) and not of MDPI and/or the editor(s). MDPI and/or the editor(s) disclaim responsibility for any injury to people or property resulting from any ideas, methods, instructions or products referred to in the content.

Article

Remote Photoplethysmography and Motion Tracking Convolutional Neural Network with Bidirectional Long Short-Term Memory: Non-Invasive Fatigue Detection Method Based on Multi-Modal Fusion

Lingjian Kong [1], Kai Xie [1,*], Kaixuan Niu [1], Jianbiao He [2] and Wei Zhang [3]

1. School of Electronic Information and Electrical Engineering, Yangtze University, Jingzhou 434023, China; 202003899@yangtzeu.edu.cn (L.K.); 202101374@yangtzeu.edu.cn (K.N.)
2. School of Computer Science, Central South University, Changsha 410083, China; jbhe@mail.csu.edu.cn
3. School of Electronic Information, Central South University, Changsha 410083, China; csuzwzbn@csu.edu.cn
* Correspondence: xiekai@yangtzeu.edu.cn; Tel.: +86-136-9731-5482

Citation: Kong, L.; Xie, K.; Niu, K.; He, J.; Zhang, W. Remote Photoplethysmography and Motion Tracking Convolutional Neural Network with Bidirectional Long Short-Term Memory: Non-Invasive Fatigue Detection Method Based on Multi-Modal Fusion. *Sensors* 2024, 24, 455. https://doi.org/10.3390/s24020455

Academic Editors: Yufeng Zheng and Erik Blasch

Received: 17 December 2023
Revised: 4 January 2024
Accepted: 8 January 2024
Published: 11 January 2024

Copyright: © 2024 by the authors. Licensee MDPI, Basel, Switzerland. This article is an open access article distributed under the terms and conditions of the Creative Commons Attribution (CC BY) license (https://creativecommons.org/licenses/by/4.0/).

Abstract: Existing vision-based fatigue detection methods commonly utilize RGB cameras to extract facial and physiological features for monitoring driver fatigue. These features often include single indicators such as eyelid movement, yawning frequency, and heart rate. However, the accuracy of RGB cameras can be affected by factors like varying lighting conditions and motion. To address these challenges, we propose a non-invasive method for multi-modal fusion fatigue detection called RPPMT-CNN-BiLSTM. This method incorporates a feature extraction enhancement module based on the improved Pan–Tompkins algorithm and 1D-MTCNN. This enhances the accuracy of heart rate signal extraction and eyelid features. Furthermore, we use one-dimensional neural networks to construct two models based on heart rate and PERCLOS values, forming a fatigue detection model. To enhance the robustness and accuracy of fatigue detection, the trained model data results are input into the BiLSTM network. This generates a time-fitting relationship between the data extracted from the CNN, allowing for effective dynamic modeling and achieving multi-modal fusion fatigue detection. Numerous experiments validate the effectiveness of the proposed method, achieving an accuracy of 98.2% on the self-made MDAD (Multi-Modal Driver Alertness Dataset). This underscores the feasibility of the algorithm. In comparison with traditional methods, our approach demonstrates higher accuracy and positively contributes to maintaining traffic safety, thereby advancing the field of smart transportation.

Keywords: intelligent traffic; fatigue detection; multi-modal feature fusion; heart rate; bidirectional LSTM

1. Introduction

Recently, with the rapid expansion of the transportation industry and the widespread use of vehicles, instances of traffic accidents resulting from fatigue driving have become increasingly common. Prolonged periods of driving or insufficient sleep can induce fatigue in drivers, significantly elevating the risk of accidents. Research indicates that driver drowsiness and sleep deprivation are primary contributors to road traffic accidents [1], accounting for approximately 25% to 30% of such incidents [2]. The ramifications of traffic accidents extend beyond individual safety and property loss, permeating into the broader stability of both a nation and society. Consequently, the timely detection of fatigue in drivers and the provision of alerts to prompt breaks are critical measures for upholding traffic safety and ensuring secure travel. In light of these considerations, addressing the issue of fatigue-related accidents assumes paramount importance. Developing effective methods for detecting and mitigating driver fatigue can substantially contribute to reducing accident rates, thereby safeguarding lives, property, and the overall stability of society.

Currently, fatigue driving detection methods can be broadly categorized into advantage detection, single-mode feature detection, and multi-mode feature detection [3]. Advantages are primarily assessed with public questionnaires and advantage scales. However, these individual methods exhibit significant differences, and their time-consuming nature renders them insufficient for real-time detection and prevention. This article addresses fatigue characteristics in two other dimensions.

Single-modal feature detection relies on individual features to assess fatigue. Among the current methods focusing on single-modal features, utilizing facial features has proven to be effective in determining a driver's fatigue status. Facial features encompass expressions, eye states, head posture, etc., extracted from a driver's facial images or videos. For instance, Zhuang [4] introduced an efficient fatigue detection method based on eye status, utilizing pupil and iris segmentation. Yang et al. [5] proposed a yawn detection method based on subtle facial action recognition, utilizing 3D convolution and bidirectional long short-term memory networks to detect a driver's fatigue state. Liu [6] presented a fatigue detection algorithm based on facial expression analysis. Xing [7] applied a convolutional neural network to face recognition, implementing a straightforward eye state judgment method using the PERCLOS algorithm to determine a driver's fatigue state, with experimental results demonstrating an 87.5% fatigue recognition rate. Moujahid [8] introduced a face monitoring system based on compact facial texture descriptors, capable of encompassing the most discriminative drowsy features. Bai [9] utilized the facial landmark detection method to extract a driver's facial landmarks from real-time videos, subsequently obtaining driver drowsiness detection results with 2s-STGCN and significantly improving driver drowsiness detection. Ahmed [10] proposed an ensemble deep learning architecture operating on merged features of eye and mouth subsamples, along with decision structures, to ascertain driver fitness.

The exploration of fatigue driving detection based on physiological characteristics has evolved into a significant research direction. In recent years, traditional heart rate detection has predominantly relied on wearable devices utilizing electroencephalogram (EEG) or electrocardiogram (ECG). For instance, Zhu [11] proposed a wearable EEG-based vehicle driver drowsiness detection method using a convolutional neural network (CNN). Gao [12] developed a novel EEG-based spatiotemporal convolutional neural network (ESTCNN) for driver fatigue detection, achieving a high classification accuracy of 97.37%. Despite their accuracy, traditional methods are hindered by issues such as expensive equipment and inconvenient wearing. In response to these challenges, non-contact physiological feature extraction has emerged as a research hotspot. Heart rate (HR) and heart rate variability (HRV) are crucial vital signs, with their changes directly or indirectly reflecting information on the physiological state of the human body. Research indicates that heart rate and HRV can objectively indicate fatigue. For instance, Dobbs [13] used a portable device to conveniently record HRV, showing a small absolute error compared with electrocardiography. Monitoring changes in heart rate and HRV is crucial for determining driver fatigue. Lu [14] emphasized the significance of HRV as a physiological marker for detecting driver fatigue, measurable during real-life driving. Systematic reviews, such as the one conducted by Persson [15], explore the relationship between HRV measurements and driver fatigue, as well as the performance of HRV-based fatigue detection systems. In medical contexts, Allado et al. [16] evaluated the accuracy of imaging photoplethysmography compared to existing contact point measurement methods in clinical settings, demonstrating that rPPG can accurately and reliably assess heart rate. Cao [17] et al. introduced a drowsiness detection system using low-cost photoplethysmography (PPG) sensors and motion sensors integrated into wrist-worn devices. Comas [18] proposed a lightweight neural model for remote heart rate estimation, focusing on efficient spatiotemporal learning of facial photoplethysmography (PPG). Patel [19] introduced an artificial intelligence-based system designed to detect early driver fatigue by leveraging heart rate variability (HRV) as a physiological measurement. Experimental results demonstrated that this HRV-based fatigue detection technology served as an effective countermeasure against fatigue. Gao [20] proposed a novel remote heart estimation algorithm incorporating a signal quality attention mechanism and a long short-term memory (LSTM) network. Experiments indicated that the LSTM with an attention mechanism accurately estimated heart rate from corrupted rPPG signals, performing well across cross-subject and cross-dataset tasks. Additionally, the signal quality model's predicted scores were found to be valuable for extracting reliable heart rates. The accuracy of existing heart rate detection based on RGB cameras is susceptible to various factors such as lighting conditions and motion, leading to challenges in achieving precise heart rate estimation. Recent advancements have addressed these challenges. For instance, Yin [21] and colleagues proposed a new multi-task learning model combining the strengths of signal-based methods and deep learning methods to achieve accurate heart rate estimation, even in scenarios with changing lighting and head movement. Given the dynamic lighting changes typical in vehicle cabins, heart rate measurement in automotive contexts presents specific challenges. To tackle these issues, Ming [22] and collaborators introduced a method named Illumination Variation Robust Remote Photoplethysmography (Ivrr-PPG) for monitoring a driver's heart rate during road driving. Rao [23] proposed a distracted driving recognition method based on a deep convolutional neural network using in-vehicle camera-captured driving image data. Experimental analysis indicated an accuracy of 97.31%, surpassing existing machine learning algorithms. Consequently, methods based on deep convolutional neural networks prove effective in enhancing the accuracy of distracted driving identification. Addressing challenges related to dramatic lighting changes and significant driver head movements during driving, Nowara [24] demonstrated that narrowband near-infrared (NIR) video recordings can mitigate external light variations and yield reliable heart rate estimates. Rajesh [25] utilized the Pan–Tompkins method for R-peak detection to identify irregularities in human heart rate (DIIHR), achieving an average accuracy of 96.

The above-mentioned fatigue driving detection methods mainly rely on single modal data, which limits the adaptability to various scenarios and the reliability of model processing. Each parameter has its advantages and disadvantages. Therefore, identifying how to effectively combine and utilize multiple driver characteristics is an important research direction for real-time and accurate driver detection.

There are currently some methods that combine multi-modal data together for fatigue detection, which involve multi-modal feature fusion models. Most of the existing multi-modal fusion is implemented based on decision fusion and feature fusion methods of RGB cameras. For example, Kassem [26] proposed a low-cost driver fatigue level prediction framework (DFLP) for detecting driver fatigue at the earliest stage. Experimental results show that this method can predict the driver fatigue level with an overall accuracy of 93.3%. Du [27] proposed a novel non-invasive method for driver multi-modal fusion fatigue detection by extracting eyelid features and heart rate signals from RGB videos. The results show that the multi-modal feature fusion method can significantly improve the accuracy of fatigue detection. Dua and colleagues [28] proposed a driver drowsiness detection system in their paper. They use the driver's RGB video as input to help detect drowsiness. The results show that the accuracy of the system reaches 85%. Liu [29] focused on RGB-D cameras and deep learning generative adversarial networks and utilized multi-channel schemes to improve fatigue detection performance. Research indicates that fatigue features extracted with convolutional neural networks outperform traditional manual fatigue features. However, relying on a single feature may not guarantee robustness. Du [30] and colleagues used a single RGB-D camera to extract three fatigue features: heart rate, eye-opening, and mouth-opening. They proposed a novel multi-modal fusion recurrent neural network (MFRNN) that integrates these three features to enhance the accuracy of driver fatigue detection. To address issues such as poor comfort, susceptibility to external factors, and poor real-time performance in existing fatigue driving detection algorithms, Jia [31] designed a system for detecting driver facial features (FFD-System) and an algorithm for judging driver fatigue status (MF-Algorithm). Akrout [32] proposed a fusion system based on yawn detection, drowsiness detection, and 3D head pose estimation.

Traditional fatigue detection methods often require the connection of inconvenient sensors (such as EEG and ECG) or use video camera systems sensitive to light, compromising privacy. Akrout suggests accounting for changes in lighting conditions during the day and night to avoid limiting the fusion system. Using an infrared camera could be a potential solution. Zhang [33] introduced Ubi-Fatigue, a non-contact fatigue monitoring system combining vital signs and facial features to achieve reliable fatigue detection. The results demonstrated that Fatigue-Radio's detection accuracy reached 81.4%, surpassing ECG or visual fatigue detection systems. Ouzar [34] and colleagues compared the performance of a single-modal approach using facial expressions or physiological data with a multi-modal system fusing facial expressions with video-based physiological cues. The multi-modal fusion model improved emotion recognition accuracy, with the fusion of facial expression features and iPPG signals achieving the best accuracy of 71.90%. This underscores the efficacy of multi-modal fusion, particularly in combining facial expression features with iPPG signals for enhanced emotion recognition accuracy.

To summarize, the existing fatigue driving detection systems face limitations in equipment deployment, environmental changes, and real-time monitoring. Addressing these challenges represents a crucial research direction for the future development of driving fatigue detection systems [35]. Consequently, this article will concentrate on resolving the following three problems:

1. The problem of the low fatigue detection accuracy of a single feature. Traditional vision-based fatigue detection methods usually only use a single feature, such as facial features, physiological features, etc., resulting in low fatigue detection accuracy.
2. The problem of low feature extraction accuracy. Existing multi-modal fusion is mostly implemented based on RGB camera methods, and its detection accuracy will be affected by different lighting conditions, motion, etc., resulting in the inability to correctly detect a driver's fatigue state.
3. The problem of the poor robustness and temporal nature of detection models. In the actual driving environment, a driver's fatigue state changes dynamically, and the fatigue state is continuous time series data. The existing methods focus on processing the characteristics of a certain moment while ignoring the changes in fatigue characteristics over time, which affects the robustness of the detection model.

It can be seen that it is very important to design a multi-modal fatigue driving detection system with high accuracy, strong robustness, portability, and real-time performance.

To address the aforementioned challenges, we propose a non-invasive method for multi-modal fusion fatigue detection based on heart rate features and eye and face features. Our approach involves the use of an infrared camera in conjunction with rPPG and MTCNN to extract a driver's physiological features and eye and face features, respectively. This combination aims to reduce errors in extracting physiological signals and facial features caused by varying lighting conditions during the day and night. To enhance feature extraction accuracy, we implemented feature extraction enhancement modules based on an improved Pan–Tompkins algorithm and 1D-MTCNN. These modules aim to more accurately extract heart rate signals and eyelid features. Subsequently, we utilize one-dimensional convolutional neural networks (1D CNNs) to establish two models based on PERCLOS values and heart rate signals for fatigue detection. Heart rate signals and PERCLOS are critical analysis objects, and their accurate extraction is pivotal for driver fatigue detection. For the extraction of heart rate signals, we use singular spectrum analysis (SSA) and filtering technology to process rPPG physiological signals. This process aims to extract relatively pure heart rate signals and enhance detection accuracy. The heart rate signal is then analyzed in the time–frequency domain, and the time–frequency domain temporal feature matrix related to fatigue is extracted. This matrix is input into a one-dimensional convolutional neural network (1D CNN) to establish a fatigue detection model based on heart rate. For PERCLOS extraction, 1D-MTCNN is utilized to calculate the PERCLOS value. Specifically, the MTCNN algorithm is used for face detection and key point positioning, offering faster and more accurate results compared with traditional algorithms while minimizing the impact of varying lighting conditions. Finally, the trained data results from the two models are input into the BiLSTM network, and the outputs of the two models are weighted to achieve multi-modal fusion fatigue detection.

2. Principles and Methods

This paper proposes a non-invasive method for multi-modal fusion fatigue detection based on heart rate features and eye and face features: RPPMT-CNN-BiLSTM. The overall framework of the multi-modal fusion fatigue driving detection model can be seen in Figure 1.

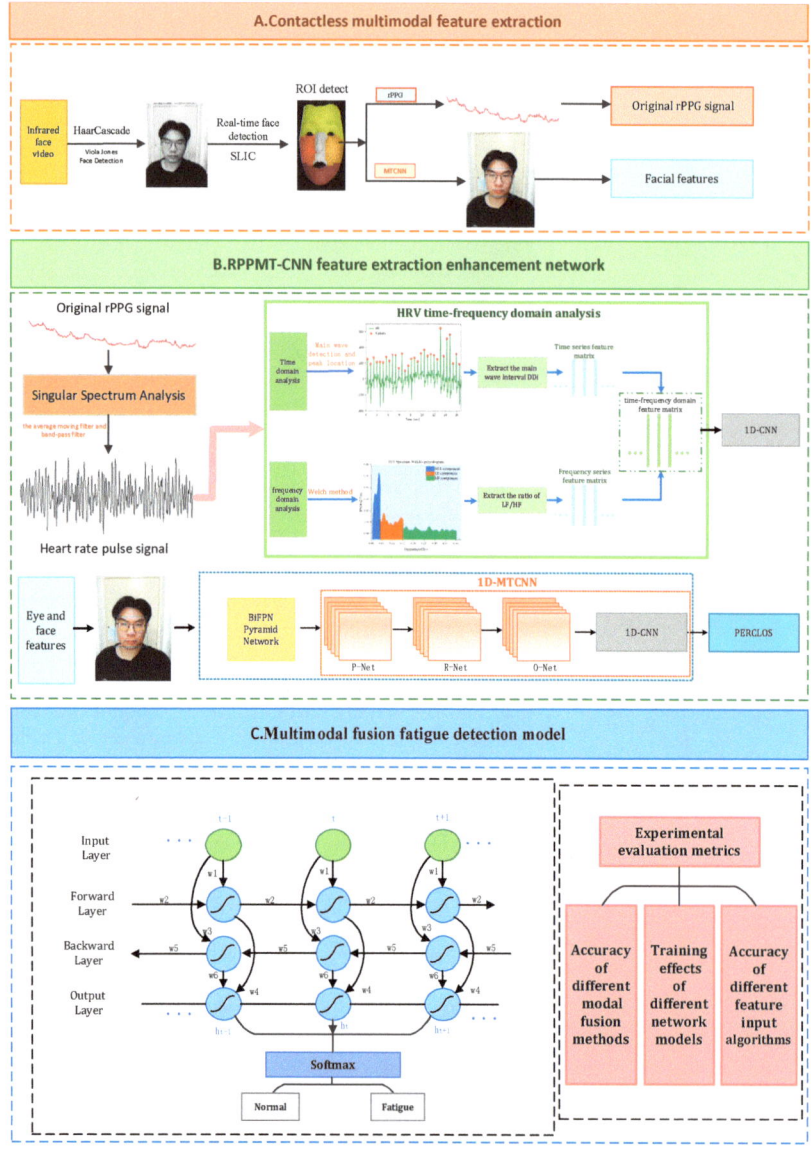

Figure 1. Multi-modal fusion fatigue driving detection model.

The model mainly includes the following 3 parts:

(1) Non-contact multi-modal feature extraction. We apply real-time face detection and ROI area tracking, utilizing infrared cameras in conjunction with rPPG and MTCNN combined with the BiFPN pyramid network to extract a driver's physiological characteristics and facial features, thereby reducing errors in extracting the driver's physiological signals and facial features caused by lighting changes during the day and night.

(2) An RPPMT-CNN feature extraction enhanced network. We introduce an infrared-based enhanced network for RPPMT-CNN feature extraction. In this paper, we establish an improved feature extraction enhancement module based on Pan–Tompkins and 1D-MTCNN. This module aims to extract heart rate signals and eyelid features more accurately. Subsequently, we create two fatigue detection models based on heart rate and PERCLOS values, utilizing one-dimensional convolutional neural networks (1D CNNs) for each model, respectively.

(3) A multi-modal feature fusion fatigue driving recognition model. To enhance the robustness and timeliness of fatigue detection, we introduce a multi-modal feature fusion fatigue driving recognition model. The outcomes of the trained model data are fed into the Bidirectional Long Short-Term Memory (BiLSTM) network. This allows the BiLSTM network to learn the temporal relationships between the data extracted from the 1D CNN, facilitating effective dynamic modeling of the input and output data. Ultimately, the outputs of the two models are weighted to achieve multi-modal fusion fatigue detection.

2.1. Non-Contact Multi-Modal Feature Extraction

2.1.1. Face Detection and ROI Area Tracking

When collecting a real-time driver video, accounting for the driver's head movement is crucial. Fixed-face Region of Interest (ROI) areas may inadvertently include non-skin areas alongside the actual skin area, thereby compromising the quality of subsequently extracted remote photoplethysmography (rPPG) signals. To address this, we use the Haar-Cascade face detector to identify faces in all frames of the video stream. Subsequently, we utilize the SLIC algorithm for superpixel skin segmentation on the detected face areas. This process determines the face ROI area for each frame in the picture, ensuring its precise position. The input video stream is segmented into multiple regions called superpixels. Superpixels corresponding to the cheek region, with the highest achromaticity in the forehead region, are selected as ROI. rPPG is then calculated for these selected superpixels, and the remaining superpixels are eliminated. This approach significantly reduces computation time. The method guarantees that during the extraction of physiological signals, the ROI area exclusively encompasses facial skin, thereby minimizing interference from motion artifacts.

2.1.2. Physiological Feature Extraction

Remote photoplethysmography (rPPG) is a non-contact method for extracting human physiological signals, developed based on the traditional photoplethysmography (PPG) principle. This approach leverages the periodic changes in blood flow induced by the human heartbeat within the skin capillaries, causing the absorption or reflection of periodic light signals. While these periodic signals are not directly observable by the human eye, high-definition cameras can capture facial data, enabling the analysis and monitoring of human physiological characteristics.

The advantage of rPPG technology lies in its non-invasive nature, as it eliminates the need for subjects to wear sensors, thereby avoiding interference with the human body. Additionally, the widespread availability and use of ordinary high-definition cameras have significantly reduced the cost of implementing rPPG technology, making it highly promising for various applications. For instance, in the context of driving fatigue monitoring, rPPG technology can be used to monitor a driver's heart rate and heart rate variability in real time. This real-time monitoring allows for the determination of the degree of fatigue, enabling timely reminders for the driver to take necessary rest measures and ensuring overall driving safety.

For rPPG signal extraction, the approach involves calculating the average of the pixel intensity values within the Region of Interest (ROI) area. In each frame of the facial video, assumed to correspond to time t, all pixels within the infrared single channel in the selected ROI area are spatially averaged. The spatial average value of the ROI area at time t can be expressed as:

$$a_t = \sum_{i=0}^{M} \frac{a_i}{M} \tag{1}$$

where is the total pixels in the selected ROI area and is the value of the $i - th$ pixel in the ROI area.

A 30s video (a total of 900 frames) is collected starting from time t at a frame rate of 30 frames/second. The sequence of skin areas in consecutive image frames can be expressed as:

$$A_t = [a_t, a_{t+\tau}, a_{t+2\tau}, \cdots, a_{t+(N-1)\tau}], N = 900 \tag{2}$$

where is the time interval used to obtain one frame of video, which is the reciprocal of the sampling frequency.

The obtained signal is defined as the original input rPPG physiological signal at time t, and every 1 s (30 frames) thereafter, the original input rPPG physiological signal starting from the next second is obtained.

2.1.3. Facial Feature Extraction

To extract facial features from drivers in a fatigued state during driving, this article uses the MTCNN algorithm in conjunction with the BiFPN pyramid network as the core of the facial feature extraction module. The MTCNN algorithm comprises three cascaded networks (P-Net, R-Net, O-Net) and is utilized for face detection and key point localization. However, in complex driving environments, factors such as lighting changes, facial postures, gender, and partial occlusion may impact its performance. To enhance the algorithm's robustness, the BiFPN pyramid network is introduced, which better captures multi-scale features and improves adaptability to illumination changes. The output of BiFPN is connected with the cascade network of MTCNN to form a comprehensive facial feature extraction module. This approach yields a richer and more accurate representation of facial features. The algorithm demonstrates faster and more accurate performance than traditional methods, reducing the impact of varying lighting conditions. It maintains accurate face and key point detection even when a face is tilted, pitched, or partially obscured. Consequently, it is highly suitable for driver detection during driving. The structure of the BiFPN pyramid network is illustrated in Figure 2.

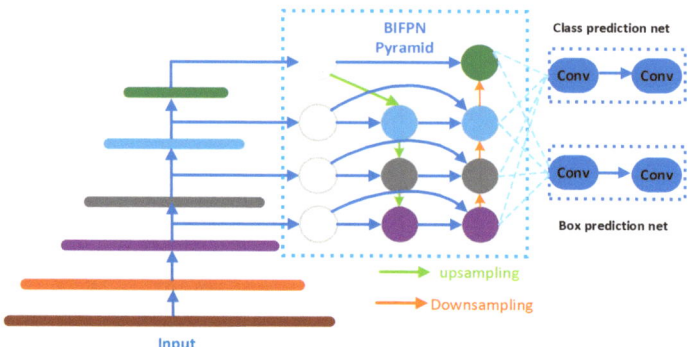

Figure 2. The BiFPN pyramid network.

The input N videos are sampled at 30 frames/s to obtain n groups of frame images. The size of these images is reduced to 0.5 times the original images and formed into n sample sets. The sample set is used as the input of the MTCNN network and is calculated as follows:

$$L_i(\text{det}, box, landmark) = \overline{MTCNN}(X_i) \tag{3}$$

where $i \in [1, n]$ $L_i(\text{det}, box, landmark)$ represents the face candidate frames and key points X_i obtained using the network output. Among them, $L_i(\text{det}, box, landmark)$ includes the coordinate values of 5 key points on the face (left eye, right eye, nose, and left and right corners of the lips).

2.2. The RPPMT-CNN Feature Extraction Enhancement Network

After acquiring the initial facial information and original physiological signals, we incorporated the RPPMT-CNN feature extraction enhancement network. This method is grounded in an improved algorithm and devises a 1D CNN (convolutional neural network) structure tailored for facial feature processing, enabling the capture of spatiotemporal relationships inherent in facial features. Concurrently, a distinct 1D CNN structure was formulated for processing physiological features, aiming to more precisely capture the time–frequency domain characteristics of physiological signals. Following the separate optimization of facial and physiological features, their characteristic information can be maximally captured. The combination of facial and physiological information yields more accurate and comprehensive features, establishing the groundwork for subsequent comprehensive analysis and application.

2.2.1. Singular Spectrum Analysis

Due to the non-orthogonal characteristics of physiological sources, the usual blind source separation method cannot directly extract the heart rate pulse signal from the original rPPG signal. Therefore, based on singular spectrum analysis, we propose the following method to separate the target signal. The data matrix A of each time series A_t of length N can be expressed as:

$$A = \begin{pmatrix} a_t & a_{t+\tau} & \cdots & a_{t+(K-1)\tau} \\ a_{t+\tau} & a_{t+2\tau} & \cdots & a_{t+K\tau} \\ \vdots & \cdots & \ddots & \vdots \\ a_{t+(M-1)\tau} & a_{t+M\tau} & \cdots & a_{t+(N-1)\tau} \end{pmatrix} \tag{4}$$

where $K = N - M + 1$.

Then, we perform singular value decomposition (SVD) on the data matrix to solve the characteristic matrix of A. Its singular value decomposition expression is:

$$A = U \sum V^T \tag{5}$$

$$A = \sum_{i=1}^{M} U_i P_i^T \tag{6}$$

$$P_i = \sqrt{\lambda_i} V_i \tag{7}$$

where U and V are the two orthogonal bases representing the left singular matrix and right singular matrix, respectively. The diagonal matrix $\sum$ is composed of singular values σ_i. It satisfies the relationship with the eigenvalue λ of AAT (covariance matrix) in eigenvalue decomposition (EVD): $\sigma_i = \sqrt{\lambda_i}$.

After singular value decomposition, the data matrix A is decomposed into M components. Then, we extract the heart rate pulse signal Ri from the M independent components, where $R_i = P_i U_i^T (i < r)$. Finally, we recover the output time series $g_i(t)$ from Ri using anti-angle averaging.

$$g_i(t) = \begin{cases} \frac{1}{m}\sum_{h=1}^{m} R^i_{h,t+1-h}, (t \leq K) \\ \frac{1}{m}\sum_{h=1}^{m} R^i_{t+h-K,K-h+1}, (t > K) \end{cases} \quad (8)$$

Due to significant noise corruption in the heart rate pulse signal obtained with singular spectrum analysis, further filtering is necessary. In this case, a moving average filter is used for low-pass filtering to eliminate low-frequency interference caused by factors such as breathing. The original sampled data forms a one-dimensional queue of length N, and a sliding window of length L is applied to it. The average value of the data within the window is computed as the output of the filter at the current moment. The window progresses in the positive direction of the time axis, generating filter outputs for subsequent moments until all the data points are covered. The calculation formula for the moving average filter at the $i - th$ moment is given by:

$$G(i) = \frac{1}{L} \sum_{j=i}^{j=i+L-1} g(j)(i = 1, 2, 3, \ldots, N - L + 1) \quad (9)$$

Subsequently, a Hamming window bandpass filter with a passband frequency of 0.8~4 Hz is applied to eliminate high-frequency and low-frequency noise outside the heart rate range, aiming to minimize noise interference.

2.2.2. Time Domain Analysis of Heart Rate Signals

Building upon [10], this paper uses the enhanced Pan–Tompkins algorithm for primary wave detection and localization. The main wave detection involves a combination of Shannon energy and adaptive dual threshold methods to accurately identify the main wave and pinpoint its peak for extracting the target signal. The detailed algorithmic flow is illustrated in Figure 3.

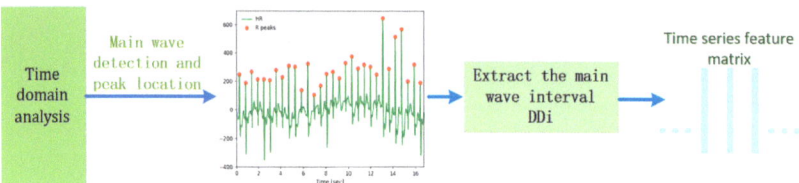

Figure 3. Time domain analysis of heart rate signals.

The time difference between two adjacent main wave peaks is called the DD interval, denoted as $DD_i (i = 1, 2, 3 \ldots\ldots)$. By integrating the physiological characteristics of the heart rate signal with pertinent medical insights, we designate the interval between the peaks of the main waves as the duration of one heartbeat, constituting a single cardiac cycle. According to existing research, the standard deviation of the cardiac cycle in the human body tends to notably increase as fatigue intensifies. Hence, this paper uses the standard deviation (SD) of the RR interval as the time domain analysis index, with its calculation formula as follows:

$$MEAN = \sum_{i=1}^{N} \frac{DD_i}{N} \quad (10)$$

$$SD = \sqrt{\frac{1}{N}\sum_{i=1}^{N}(DD_i - MEAN)^2} \quad (11)$$

The flow chart and specific implementation process of the improved algorithm are shown in Figure 4.

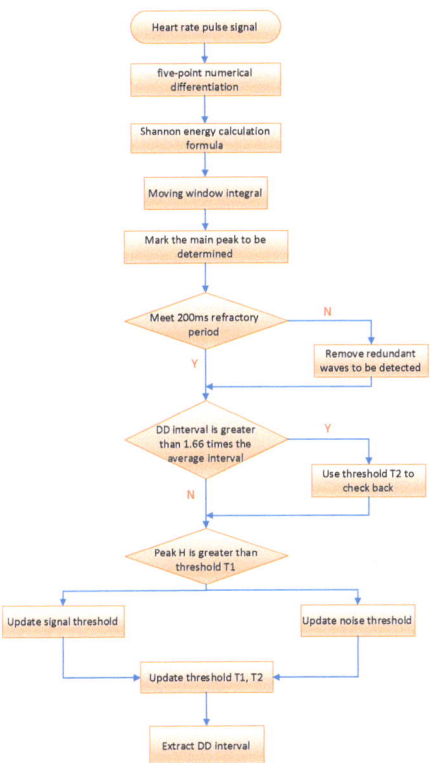

Figure 4. Flow of the HRV time domain analysis algorithm.

We initially differentiate the filtered heart rate pulse signal to extract waveform slope information, using the five-point numerical differentiation formula:

$$y\prime(i) = x[i-2:i+2] \cdot [\frac{1}{12}, -\frac{8}{12}, 0, \frac{8}{12}, -\frac{1}{12}]^T \qquad (12)$$

where $y'(i)$, $y'(i)$ represents the slope of the heart rate pulse signal at the $i-th$ time point, the symbol ":" is used to represent an array or vector, and the symbol "·" is used to represent matrix or vector multiplication. Before performing the Shannon energy calculation, the differentiated data are standardized as follows:

$$\widetilde{y(n)} = y(n)/\max_i(|y(n)|) \qquad (13)$$

Then, the output of the derivative undergoes nonlinear amplification using the Shannon energy formula. This process ensures that all data points become positive, accentuates high- and medium-intensity components, and attenuates other intensity values. This enhancement aids in better locating the main wave and detecting its peak. The Shannon energy formula is as follows:

$$y(nT) = -[x(nT)]^2 \ln([x(nT)]^2) \qquad (14)$$

Following the calculation of Shannon energy, numerous closely spaced and small wave peaks are obtained. To enhance the concentration of energy, a moving window integration is applied to smooth the waveform. The choice of window size is crucial for main wave detection. If the selected window is too small, the resulting signal waveform after moving integration may lack smoothness, hindering main wave peak detection and potentially leading to false detections. Conversely, if the window is too large, the energy of the main wave in the signal may be dispersed, increasing the risk of missed detection.

Typically, the size of the moving window integral after Shannon energy processing is correlated with the sampling frequency. The window size is generally chosen as 0.18 times the sampling frequency of 0.18 fs. For instance, with a sampling rate of 200 samples/s, the window width is set to 30 samples (150 MS).

The rising edge peak of the signal waveform obtained after moving window integration is marked as the main wave peak to be detected, and it is then adjusted using adaptive dual-threshold technology to determine the true main wave peak. If the peak value DP to be detected is greater than the threshold $T1$, it is the main wave peak value; otherwise, it is the noise peak value. The driver's heart rate signal extracted in the first 3 s is selected as the initial data, one-third of the maximum detected peak value is used as the initial signal threshold (ST), and half of the average value of all detected peak values is used as the initial noise threshold (NT). The adaptive dual threshold adjustment process is as follows:

If DP is the peak of the main wave:
If H is the main wave peak:

$$ST = \frac{1}{8}DP + \frac{7}{8}ST \tag{15}$$

If DP is the noise peak:

$$NT = \frac{1}{8}DP + \frac{7}{8}NT \tag{16}$$

Our dual thresholds, denoted as $T1$ and $T2$ for discrimination, vary with ST and NT. As ST and NT change, $T1$ and $T2$ dynamically adjust accordingly. This relationship can be expressed by the following formula:

$$T1 = NT + \frac{1}{4}(ST - NT) \tag{17}$$

$$T2 = \frac{1}{2}T1 \tag{18}$$

Considering the refractory period between two adjacent main waves and the physiological characteristics of the human heartbeat, we set the refractory period to 200 MS. During this period, redundant detection points are removed to prevent errors.

The average of the last eight DD intervals serves as the reference for the average interval. If the presently detected DD interval exceeds 1.66 times the average interval, indicating a potential detection miss, we initiate a backcheck using threshold $T2$ and update the signal threshold as follows:

$$ST = \frac{1}{4}DP + \frac{3}{4}ST \tag{19}$$

2.2.3. Frequency Domain Analysis of Heart Rate Signals

Frequency domain analysis is used to depict the fundamental information regarding the changes in signal energy concerning frequency. The frequency domain component of the heart rate variability signal is intricately linked to the physiological state of the human body. Notably, high-frequency power mirrors the regulatory influence of the vagus nerve on the heart rate, while low-frequency power reflects the intricate interplay between sympathetic and parasympathetic nerves in the heart rate regulation process. The LF/HF ratio is a metric used to quantify the balance between sympathetic and parasympathetic tension. When the body is fatigued, sympathetic tension tends to dominate. Studies have indicated that the power spectral ratio of low-frequency power values (LF) and LF/HF to the heart rate variability signal significantly increases during fatigue, while the high-frequency power value diminishes. The LF/HF index serves as a crucial indicator of driver sleepiness and fatigue status. Therefore, LF/HF is utilized as the frequency domain analysis index for the target signal. The specific algorithm flow is illustrated in Figure 5.

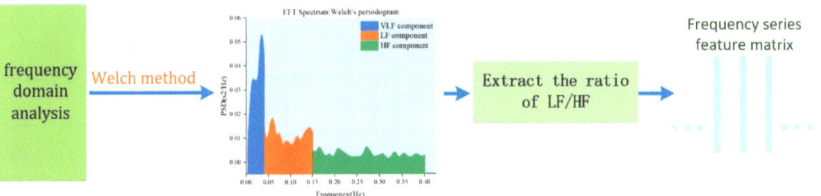

Figure 5. Frequency domain analysis of heart rate signals.

Given the target signal's inherent variability associated with a driver's heart rate and its substantial randomness, we use the Welch method to estimate the power spectrum of the target signal. The Welch method is an enhanced periodogram power spectral density estimation technique that is well-suited for rapid Fourier calculations. This method involves selecting window data, segmentally obtaining the power spectrum, and subsequently averaging it. The specific steps for our frequency domain analysis are outlined as follows.

First, we sample the obtained target signal to obtain the discretized signal $y(n)$ ($0 \leq n \leq N$). The window size is positioned as L, and y(n) is divided into J segments when a quarter overlap is allowed, $J = (N - L/4)/(L/4)$. For the data in paragraph i:

$$y_i(m) = y[m + \frac{(i-1)L}{4}] \tag{20}$$

where $0 \leq m \leq L - 1, 1 \leq i \leq J$.

Consider the $i - th$ segment as an example to calculate the power spectrum of each segment of the data:

$$\hat{Y}_i(w) = \frac{1}{LU} |\sum_{m=0}^{L-1} y_i(m) D(m) e^{-jwm}|^2 \tag{21}$$

where, in this formula, $U = \frac{1}{L}\sum_{m=0}^{L-1} D^2(m)$ is the normalization factor, which ensures that the obtained spectrum is an asymptotically unbiased estimate, and $D(m)$ is the added window function. Next, we add the power spectra of all segments and take the average value to obtain the power spectrum y(n):

$$\widetilde{Y(w)} = \frac{1}{LUJ} \sum_{i=1}^{J} |\sum_{m=0}^{L-1} y_i(m) D(m) e^{-jwm}|^2 \tag{22}$$

The extracted time–frequency domain feature matrix is input into 1D CNN for processing. The 1D CNN method is effective in capturing the correlation between time–frequency domain features using convolution and pooling operations. HRV (heart rate variability) refers to the change in the heart rate over a period of time. The current HRV-based fatigue detection method typically obtains an ECG signal by attaching electrodes to the subject's skin and then converts the signal into HRV. However, obtaining HRV directly from the heart rate is not feasible. Since our goal is to learn how the heart rate signal changes over time, considering the heart rate of the sliding window over time can aid in achieving fatigue driving detection. Therefore, we designed a method to extract heart rate changes at adjacent moments during fatigue activities and established a 1D CNN-based model.

We utilize the 1D CNN for fatigue detection, as illustrated in Figure 6. The network comprises an input layer, three convolutional layers, and two fully connected layers. The input size is 1024 × 1. There are three convolutional layers with a filter length of 32, each utilizing the ReLU activation function. The convolution kernel sizes for layer 1, layer 2, and layer 3 are 16 × 1, 8 × 1, and 4 × 1, respectively. Following the convolutional layers, two fully connected layers (FCLs) with 256 and 128 neurons, respectively, are added for classification. To prevent overfitting, a dropout layer is introduced after the fully connected layer. Finally, the SoftMax classifier calculates the probability of two fatigue states.

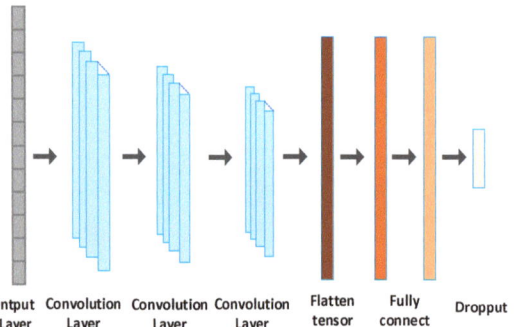

Figure 6. A 1D CNN based on fatigue detection.

2.2.4. Using 1D-MTCNN to Extract PERCOLS

Leveraging insights from [25], our approach acquires accurate key point coordinates with MTCNN. Subsequently, these key point coordinates are used to extract images of the eye and mouth regions. The extracted eye areas serve as input for the 1D CNN model to extract features. The detailed implementation process is depicted in Figure 7.

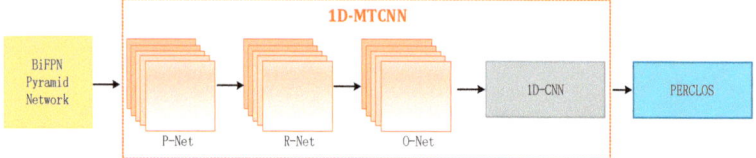

Figure 7. Using 1D-MTCNN to extract PERCOLS.

Local patterns and correlation information in sequence data can be captured using 1D CNN; therefore, it is suitable for processing video data from consecutive frames. The eye and mouth status of the region image is then classified. First, we connect the center points of the left and right eyes to obtain line a and define the angle between the connecting line a and the horizontal line as α. The width and height of the eye area frame are defined as w and $h = w/2$, respectively. Then, we connect the left and right corners of the lips to obtain line b and draw a vertical line b from the key point of the nose to the connecting line c. The vertical distance is defined as d; then, the upper edge of the mouth area frame is $d/2$, and the lower edge is the vertical line c. At the extension line $5d/3$, after obtaining the eye and mouth area frames, we perform two classifications. With an interval of 60 s, there are a total of 1200 frames of images. Based on the PECLOS criterion and prior knowledge, we can state that:

$$P = \frac{\text{Eyes closed frames}}{\text{The total number of frames in the detection period}} \times 100\% \quad (23)$$

$$L = \frac{\text{yawn frames}}{\text{The total number of frames in the detection period}} \times 100\% \quad (24)$$

The PERCOLS algorithm has been proven to be able to accurately determine driver fatigue in real time. At the same time, based on prior knowledge, it can be determined whether the driver is in a fatigue state by detecting the number of times the subject yawns per minute. Therefore, the p value and the L value are selected as facial features.

Our proposed one-dimensional CNN architecture for fatigue detection is depicted in Figure 3. The architecture is composed of an input layer, three convolutional layers, and two fully connected layers. The input size is specified as 600 × 1. The convolutional layers feature filter lengths of 24, using the ReLU activation function. The size of the convolution kernel is set at 10 × 1 for convolution layer 1, 5 × 1 for layer 2, and 3 × 1 for convolution layer 3. To finalize the network, two fully connected layers (FCLs) are added, containing 128 and 64 neurons, respectively, for classification. In order to mitigate overfitting, the SoftMax classifier is utilized to compute the probability of two fatigue states. The initialization of the network's values is accomplished by assigning random values. Given that PERCLOS and heart rate features are represented by one-dimensional signals, and considering the time and performance advantages of one-dimensional CNN in processing such signals, our choice of one-dimensional CNN for fatigue detection is well-founded.

2.3. Multi-modal Feature Fusion Fatigue Driving Identification Model

The LSTM network has shown unique advantages in the fields of text generation, machine translation, speech recognition, generated image description, and video tagging, demonstrating its powerful functions in processing and searching for spatio-temporal data. The schematic representation of the LSTM cell structure is shown in Figure 8.

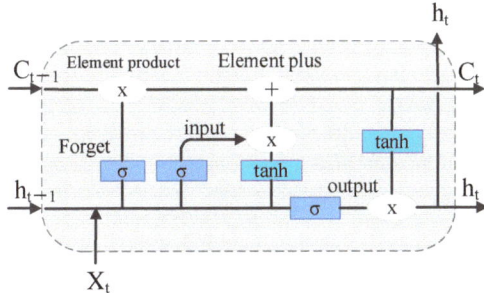

Figure 8. Schematic diagram of the LSTM cell structure.

The control gate mechanism of LSTM solves the problem of gradient disappearance when the RNN is processing a long sequence. The working principles of the three control gates are as follows:

The forget gate:
$$\Gamma_f = \sigma(W_f[h_{t-1}, x_t] + b_f) \tag{25}$$

the input gate:
$$\Gamma_u = \sigma(W_u[h_{t-1}, x_t] + b_u) \tag{26}$$

$$\tilde{C}_t = \tanh(W_c[h_{t-1}, x_t] + b_c) \tag{27}$$

$$C_t = \Gamma u \times \tilde{C}_t + \Gamma_f \times C_{t-1} \tag{28}$$

and the output gate:
$$\Gamma_o = \sigma(W_o[h_{t-1}, x_t] + b_o) \tag{29}$$

$$h_t = \Gamma_o \times C_t \tag{30}$$

where x_t is the input at time t. $\Gamma_u, \Gamma_f, \Gamma_o$ are the input, forget, and output gates at time t, respectively. The output gate passes the activation function values W_u, W_f, W_c, W_o, and b_u, b_f, b_o are the weights and deviations of the gates, respectively. $\tilde{C}_t$ is the state of the memory element at time t, and h_t is the final output.

A driver's driving state is a dynamic process, with the driver's physiological signal data changing over time, representing standard time series data. When utilizing LSTM to process the time–frequency domain time series feature matrix reflecting the driver's state, only past information is considered, and future information is disregarded [20], potentially impacting the accurate assessment of the current state. To address this limitation, Bi-LSTM emerges as a solution. Bi-LSTM comprises two LSTMs operating in opposite directions. One LSTM processes information in a forward pass to retain past information, while the other LSTM processes information in a backward pass to incorporate future information. The outputs of these two LSTMs are then combined to derive the driver's status judgment using Softmax. The network structure of Bi-LSTM is illustrated in Figure 9. In driver fatigue detection, the time–frequency domain time series feature matrix reflecting the driver's state is fed into Bi-LSTM in real time. The forward and backward propagation layers in Bi-LSTM work together to precisely determine the driver's status by incorporating both past and future information.

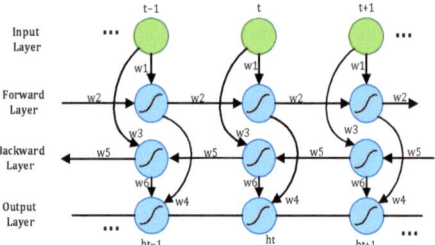

Figure 9. Bidirectional LSTM.

3. Experimental Results and Discussion

This section encompasses an introduction to the experimental environment, collected in-vehicle driving datasets, evaluation indicators, currently prevalent models, and a conclusive analysis of the experimental results. The experimental flow chart is depicted in Figure 10.

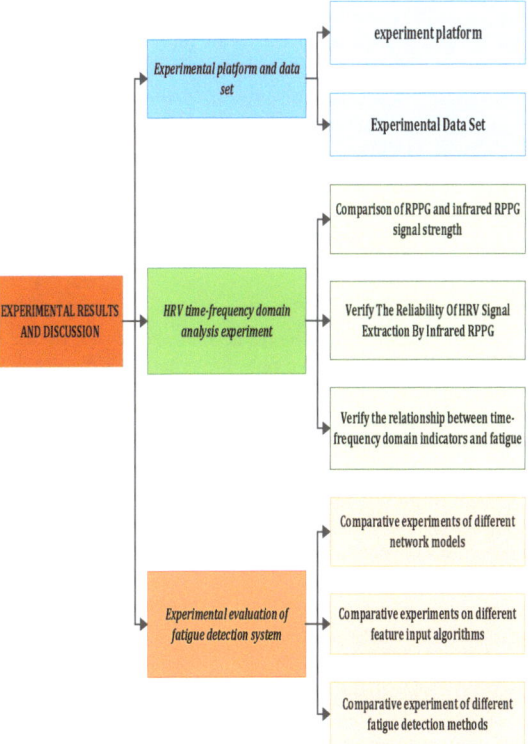

Figure 10. Experimental framework.

In this section, we elaborate on the details of the experiment. Figure 10 illustrates the entire experimental process, which is segmented into four parts. Initially, we introduce some fundamental experimental settings and datasets. Subsequently, utilizing the ECG signal as a reference, we measure the error induced by infrared rPPG, analyze heart rate variability, and calculate its correlation. Following that, the Bi-LSTM and LSTM networks are trained using time–frequency domain feature indicators, emphasizing the distinctive advantages of Bi-LSTM in detecting driver fatigue through comparison. Finally, we evaluate and analyze our model against existing physiological feature-based driver fatigue detection models.

3.1. Experiment Platform

3.1.1. Experimental Equipment

The experimental equipment and environment include an infrared camera, Windows11 64-bit operating system, Intel i9 2.20 GHz processor, 16 GB memory, NVIDIA RTX 4060 (GPU), Python (3.7), and the Keras (Tensorflow2.1) framework.

In this research, the infrared camera serves as the pivotal hardware component for extracting the infrared rPPG signal. The original infrared time signal is derived from the video data collected with the infrared camera and utilized as the input for the rPPG signal. Consequently, the quality of the infrared camera profoundly influences the accuracy and reliability of heart rate signal extraction.

In selecting the appropriate infrared camera, several factors were taken into consideration, encompassing signal quality, performance, and cost. To guarantee the extraction of high-quality infrared rPPG signals, a cost-effective yet high-performance infrared camera with a superior signal-to-noise ratio was chosen, namely, the Oni S500 model. The

Oni S500 infrared camera was selected for its advantageous cost-performance ratio and notable signal-to-noise ratio. This attribute proves pivotal in extracting delicate biological signals, as a high signal-to-noise ratio aids in diminishing interference and noise, ultimately enhancing the accuracy and stability of heart rate signal extraction.

The utilization of the Oni S500 infrared camera, known for its high performance and cost-effectiveness, allowed us to acquire high-quality infrared time signals for our research. This serves as a robust foundation for the subsequent processing and analysis of rPPG signals. Additionally, considering cost implications, the selection of the Oni S500 presents an economical hardware solution for our research, ensuring that this study can yield accurate and reliable experimental results.

3.1.2. Experimental Dataset

In this study, due to the limited availability of public RGB and infrared multi-modal fatigue driving datasets, we opted to create our dataset, named the MDAD (Multi-Modal Driver Alertness Dataset). The dataset is illustrated in Figure 11. Our data collection used the Oni S500 binocular infrared camera, equipped with color RGB and infrared IR sensors. This binocular camera, capable of flexible installation on the rearview mirror or dashboard, simultaneously captures RGB and infrared images at a sampling rate of 30 Hz, with an image resolution of 640 × 480. Despite a minor offset in the positions of the RGB and IR cameras, it has been validated that this difference insignificantly impacts the effectiveness of data collection.

To ensure the diversity of the dataset, we collected real driving scenarios involving different types of vehicles (private cars, taxis, trucks, etc.) during both the daytime and nighttime. Throughout the data collection phase, a total of 52 participants (28 men and 24 women) with ages ranging from 22 to 53 years were involved. Each participant drove once on different road segments, engaging in typical driving activities. The entire driving session lasted approximately 2 h. Various real-world complexities, including changes in lighting conditions, driver head deflection, and partial occlusion, were intentionally introduced during the data collection to guarantee the diversity and randomness of the samples.

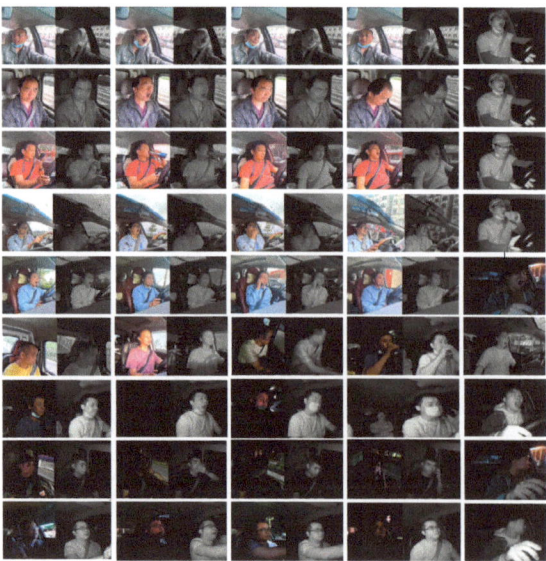

Figure 11. MDAD sample image.

Following the completion of data collection, we organized and segmented the videos into 60 s short video clips. These videos were then precisely labeled as either awake or fatigued based on the Karolinska Sleepiness Scale (KSS) in chronological order. The entire dataset comprises a total of 4000 1200 × 650 RGB-IR driving videos. Among these, 2530 are labeled as awake, while 1470 are labeled as fatigued.

The self-made MDAD serves as a crucial experimental foundation and resource for this research. Encompassing a diverse range of real driving scenarios, the dataset facilitates robust testing and evaluation of fatigue driving detection algorithms. This initiative contributes to enhancing the model's resilience and generalization. The utilization of this dataset is anticipated to drive further advancements in the field of fatigue driving detection.

3.2. HRV Time–Frequency Domain Analysis Experiment

In this section, we perform experiments to evaluate the accuracy and stability of our proposed HRV time–frequency domain analysis method for assessing driver fatigue. For the comparative analysis, we selected a widely used and advanced contact electrocardiogram (ECG) monitoring system. Fifteen participants, randomly chosen from a total of 52 subjects, were involved in the experiment. Subjects numbered 1–7 conducted the experiment at 12:00 noon, while subjects numbered 8–15 conducted it at 23:00 in the evening. This timing variation allowed us to accurately extract their heart rate (HR) and heart rate variability (HRV), including low frequency (LF) and high frequency (HF) data. Subjects were required to wear ECG monitors, ensuring correct placement for accurate signal extraction. The collected electrocardiogram and infrared video signals were then processed to extract heartbeat intervals for calculating heart rate variability.

3.2.1. Comparison of the RPPG and Infrared RPPG Signal Intensities

In the case of the long-distance extracted remote photoplethysmography (rPPG) signal, its signal strength is weakened due to the extended transmission distance, making it susceptible to environmental factors like lighting. The extraction of rPPG signals is crucial for our fatigue recognition task, particularly in relation to the extraction of heart rate variability (HRV) signals. If the obtained HRV signal is too weak or significantly affected by noise interference, it can directly impact our subsequent data analysis, ultimately reducing the accuracy of driver fatigue assessment.

To mitigate this issue, we used both infrared rPPG and traditional rPPG methods for heart rate extraction. The experimental results, depicted in Figure 12, indicate that in comparison with the traditional rPPG signal, the heart rate signal obtained with infrared rPPG is less influenced by noise. Moreover, the heart rate signal acquired with infrared rPPG exhibits enhanced anti-interference capabilities. This implies that the rPPG signal, after minimizing environmental interference, is more robust, thus aiding in the extraction of accurate heart rate signals during subsequent data processing. Consequently, this contributes to an improved accuracy in assessing the driver's fatigue state.

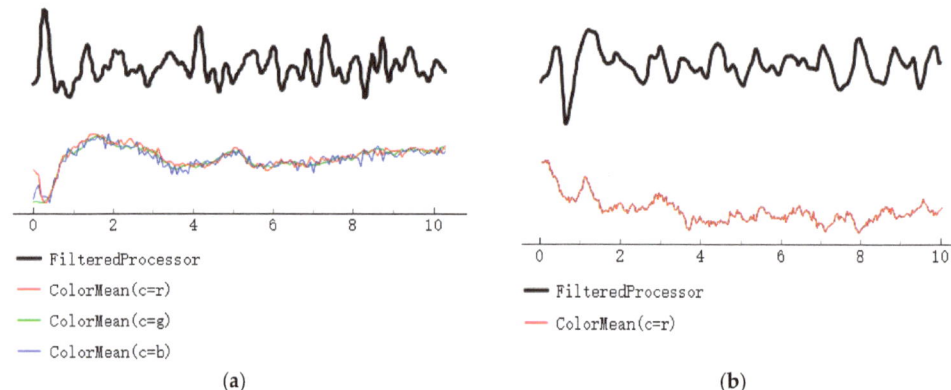

Figure 12. A comparison of the RPPG and infrared RPPG signal strengths. (**a**) Remote photoplethysmography; (**b**) Infrared remote photoplethysmography.

3.2.2. Verification of the Reliability of HRV Signal Extraction with Infrared RPPG

We assessed the reliability of HRV signal extraction using infrared rPPG by comparing the time–frequency domain index values of the ECG signal and the infrared rPPG signal, as illustrated in Figure 13. A comparison with the literature [16] reveals a high degree of overlap in the upper points of the polyline in Figure 13a,b for the SD value obtained from the infrared rPPG signal and the LF/HF ratio, which is consistent with the ECG signal. This suggests that the HRV signal can be effectively extracted from the infrared rPPG signal. Using electrocardiogram results as the standard, the accuracy of infrared rPPG, measured after 25 rounds of experiments, is approximately 95%.

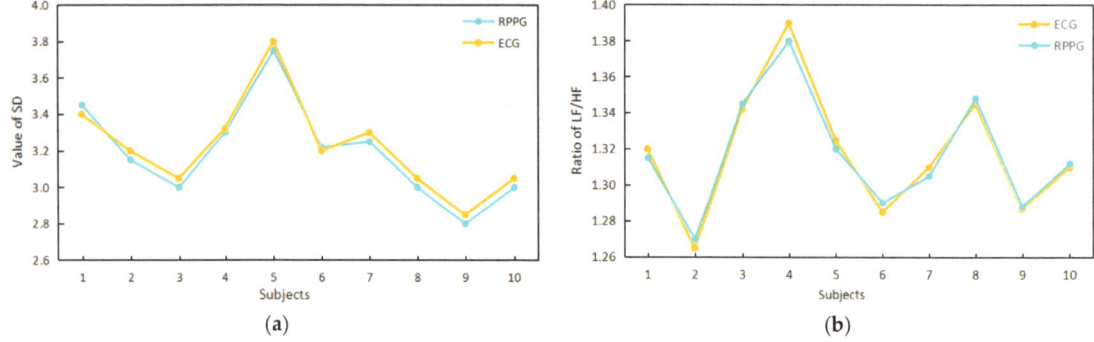

Figure 13. A comparison of the time–frequency domain index values of the ECG signals and infrared RPPG signals and a comparison of the RPPG and infrared RPPG signal intensities. (**a**) Time domain index comparison. (**b**) Frequency domain index comparison.

3.2.3. Verification of the Relationship between Time–Frequency Domain Indicators and Driver Fatigue

We conducted tests and verified, as indicated in the literature [14], that when transitioning from a normal state to a fatigue state, there are changes in time domain features such as SD and frequency domain features like the LF/HF ratio. This validates the feasibility of utilizing HRV time–frequency domain indices to detect driver fatigue.

The observations in Figure 14 indicate that when subjects transition from a normal state to a fatigue state, there are notable changes in both the SD value and the LF/HF value. With increasing fatigue, both SD and LF/HF values are higher than in the normal state, exhibiting a clear upward trend.

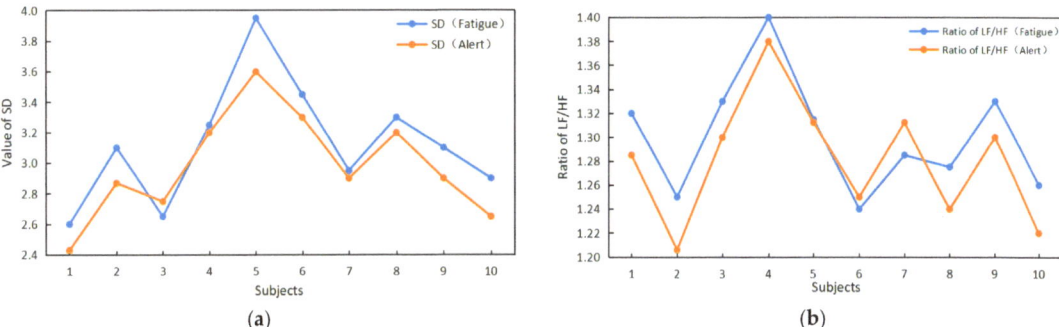

(a) (b)

Figure 14. Verification of the relationship between time–frequency domain indicators and driver fatigue. (**a**) The value of SD changes between alarm and fatigue. (**b**) The LF/HF ratio changes between alarm and fatigue.

3.3. Experimental Evaluation of the Fatigue Driving Detection System

3.3.1. Comparison of the Effects of BI-LSTM and LSTM Network Training

The experiment aimed to assess the training effect of the Bi-LSTM network. Thirty sets of 30 s facial videos were utilized to train both the Bi-LSTM and LSTM network models. A comparison of the loss function and accuracy between the two networks is presented in Figure 10. The results reveal that in most cases, Bi-LSTM outperforms LSTM, demonstrating superiority in fatigue detection. This suggests that integrating both past and future physiological signals of the driver enhances the judgment process, leading to improved results. As shown in Figure 15, after the 15th training cycle, the loss function approaches 0, and around the 20th cycle, the accuracy of Bi-LSTM approaches 98%.

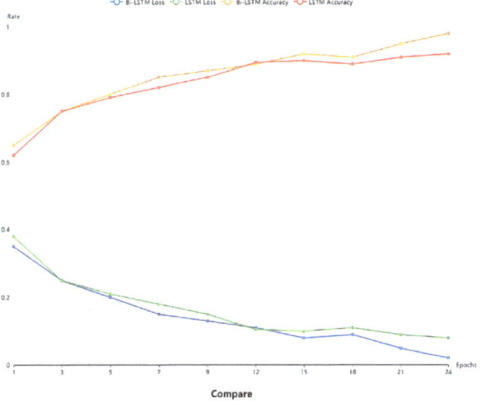

Figure 15. Comparison of the LSTM and Bi-LSTM models.

3.3.2. Model Evaluation

The sensitivity (*Se*), true positive rate (+*P*), and accuracy rate (*Acc*) were used to evaluate the overall performance of the proposed algorithm. The calculation formulas are as follows:

$$Se = \frac{TP}{TP + FN} \tag{31}$$

$$+P = \frac{TP}{TP + FP} \tag{32}$$

$$Acc = \frac{TP}{TP + FP + FN} \quad (33)$$

where true positive (*TP*) represents the number of correctly detected fatigue states or non-fatigue states, false positive (*FP*) represents the number of non-fatigue states judged as fatigue states, and false negative (*FN*) indicates the number of fatigue states judged as non-fatigue.

The average absolute error is the average of the sum of the absolute differences between the *DD* interval values of all test samples in the test sample and the *RR* interval control value, which can be expressed as:

$$\overline{Error} = \frac{1}{N}\sum_{i=1}^{N}|DD_i - RR_i| \quad (34)$$

where *N* represents the total number of test samples.

3.3.3. Comparative Experiments on Different Modal Feature Input Algorithms

By comparing the literature [34], this study aims to more effectively evaluate the classification effect after multi-modal data fusion and the single-modal classification effect. To ensure consistency in the experimental environment and subjects, different modal features were input into various algorithms for comparison. The feature categories were divided into four groups: physiological features, visual features, mixed visual features, and mixed features. These four groups of feature types were categorized into a fatigue state and an awake state based on given labels. They were then input into traditional SVM, DBN, random forest, and CNN networks simultaneously to compare the obtained classification accuracy with the method proposed in this paper. The experimental results are shown in Table 1.

Table 1. Comparison of the accuracy rates of different algorithms input with different feature types.

Category	Feature	Index	SVM	Random Forest	DBN	CNN	Ours
Physiological characteristics	ECG	HR, HRV	0.792	0.827	0.853	0.874	**0.894**
	rPPG	HRV	0.778	0.817	0.827	0.864	**0.895**
Visual characteristics	Eyes	PERCLOS	0.785	0.822	0.849	0.868	**0.877**
	Mouth	Yawn frequency	0.757	0.783	0.806	0.855	**0.895**
Mixed visual features	Eyes + Mouth	PERCLOS, yawn frequency	0.814	0.836	0.869	0.896	**0.948**
Mixing multi-class features	Eyes + Mouth + rPPG	PERCLOS, yawn frequency, HRV	0.774	0.842	0.871	0.949	**0.972**

Note: Bold is the best result.

According to Table 1, the proposed method in this paper achieves the highest classification accuracy across all feature types. Concerning physiological characteristics, our method demonstrates relatively high accuracy rates when utilizing both ECG and rPPG, achieving 0.894 and 0.895, respectively. Regarding visual features, the accuracy of our method using eye and mouth features is 0.877 and 0.895, respectively, also surpassing the traditional classification algorithms. Notably, the combination of visual features (Eyes + Mouth) performs exceptionally well in our method, achieving an accuracy of 0.948, significantly outperforming other algorithms. Additionally, we observe that compared with traditional algorithms (SVM and random forest), the deep learning models (DBN and CNN) exhibit superior performance, affirming the advantages of deep learning in processing multi-modal data. In the case of mixed multi-class features (Eyes + Mouth + rPPG), our method excels with an accuracy of 0.972, surpassing the other traditional algorithms.

In summary, based on the data results from the comparative experiments, we can confidently conclude that the multi-modal fusion fatigue detection method proposed in this article demonstrates high accuracy across various feature types. The optimal performance is achieved when combining multiple types of features, establishing the effectiveness and promising application prospects of this method in fatigue detection.

3.3.4. Comparative Experiments on Different Fatigue Driving Detection Methods

In order to enhance the generalization and reliability of the experimental results, we conducted a comparison between the algorithm proposed in this paper and the existing mainstream fatigue driving detection algorithms. This comparison includes the multi-class support vector machine (MCSVM) presented in [20] and the multi-granularity deep convolution model (RF-DCM) introduced in [27]. The dataset utilized for experimentation is our self-compiled MDAD, comprising continuous facial videos of drivers navigating diverse and challenging driving scenarios. The accuracy of each algorithm in fatigue detection was evaluated, and the results are presented in Table 2 below.

Table 2. Comparison of the accuracy of fatigue driving detection methods in complex environments.

Method	Acc (%)
MCSVM [36]	87.3%
RF-DCM [37]	94.6%
Drowsiness detection system [28]	85%
Fatigue-Radio [33]	81.4%
FFD system [31]	97.8%
MTCNN + InceptionV3 [10]	91.1%
HDDD [38]	88.9%
CNN + DF-LSTM [39]	88.9%
Fusion system [32]	97.3%
Ours	**98.2%**

Note: Bold is the best result.

4. Conclusions

To address the challenges associated with the low accuracy and poor robustness of traditional fatigue detection methods, particularly under varying lighting conditions, we propose a non-invasive approach for multi-modal fusion fatigue detection that integrates heart rate features with eye and face features. Leveraging an infrared camera in conjunction with rPPG and MTCNN combined with the BiFPN pyramid network, we extract a driver's physiological characteristics and facial features, mitigating errors introduced by day-to-night lighting changes and enhancing the stability and reliability of fatigue detection. Furthermore, we introduce a feature extraction enhancement module based on an improved Pan–Tompkins algorithm and 1D-MTCNN to more accurately extract heart rate signals and eyelid features. These enhancement modules contribute to elevating the quality and precision of the data, forming a solid foundation for subsequent fatigue detection models. Two independent fatigue detection models are established using a one-dimensional convolutional neural network (1D CNN), where one model is based on the PERCLOS value, and the other is based on the heart rate signal. After training these models, accurate detection and classification of different fatigue characteristics are achieved. To enhance robustness, real-time performance, and accuracy, the BiLSTM network is used to input the data results from the two trained models, allowing it to learn the temporal relationships between the data. This dynamic modeling approach effectively processes forecast time series input data, improves the model's applicability in real-world scenarios, and realizes multi-modal fatigue detection. Extensive experiments and comparisons of the self-compiled MDAD demonstrate that the proposed model exhibits excellent fatigue detection performance, boasting high accuracy and robustness across different scenarios. The multi-modal fusion fatigue detection method presented in this paper provides an effective solution for achieving accurate and reliable fatigue driving detection.

Author Contributions: Conceptualization, L.K. and K.X.; methodology, L.K.; software, K.N.; validation, K.N.; formal analysis, L.K.; investigation, J.H.; resources.; data curation, K.X.; writing—original draft preparation, L.K.; writing—review and editing, K.N.; visualization, J.H.; supervision, K.X.; project administration, W.Z.; funding acquisition, K.X. All authors have read and agreed to the published version of the manuscript.

Funding: This work was supported by the National Natural Science Foundation of China (No. 62373372 and 62272485), under whose cooperation, this work was successfully established and conducted. This work was mainly supported by a project sponsored by the Undergraduate Training Programs for Innovation and Entrepreneurship of Yangtze University under Grant Yz2022062. This project enabled us to purchase and build a non-invasive fatigue detection system based on multimodal fusion and conduct the experiments in this paper to complete the identification of fatigue states under different lighting conditions.

Institutional Review Board Statement: Not applicable. The digital products mentioned in this article are for illustrative and explanatory purposes only and do not imply any advertising or marketing intent.

Informed Consent Statement: Informed consent was obtained from all subjects involved in this study.

Data Availability Statement: The original contributions presented in the study are included in the article, further inquiries can be directed to the corresponding author.

Acknowledgments: We express our sincere gratitude to all the drivers who willingly participated in the dataset collection. Their support and cooperation have been invaluable to the success of this research.

Conflicts of Interest: The authors declare no conflicts of interest.

References

1. Hussein, M.K.; Salman, T.M.; Miry, A.H.; Subhi, M.A. Driver drowsiness detection techniques: A survey. In Proceedings of the 2021 1st Babylon International Conference on Information Technology and Science (BICITS), Babil, Iraq, 28–29 April 2021; pp. 45–51.
2. Chacon-Murguia, M.I.; Prieto-Resendiz, C. Detecting driver drowsiness: A survey of system designs and technology. *IEEE Consum. Electron. Mag.* **2015**, *4*, 107–119. [CrossRef]
3. Shi, S.Y.; Tang, W.Z.; Wang, Y.Y. A review on fatigue driving detection. *ITM Web Conf.* **2017**, *12*, 01019. [CrossRef]
4. Zhuang, Q.; Kehua, Z.; Wang, J.; Chen, Q. Driver fatigue detection method based on eye states with pupil and iris segmentation. *IEEE Access* **2020**, *8*, 173440–173449. [CrossRef]
5. Yang, H.; Liu, L.; Min, W.; Yang, X.; Xiong, X. Driver yawning detection based on subtle facial action recognition. *IEEE Trans. Multimed.* **2020**, *23*, 572–583. [CrossRef]
6. Liu, Z.; Peng, Y.; Hu, W. Driver fatigue detection based on deeply-learned facial expression representation. *J. Vis. Commun. Image Represent.* **2020**, *71*, 102723. [CrossRef]
7. Xing, J.; Fang, G.; Zhong, J.; Li, J. Application of face recognition based on CNN in fatigue driving detection. In Proceedings of the 2019 International Conference on Artificial Intelligence and Advanced Manufacturing, Dublin, Ireland, 17–19 October 2019.
8. Moujahid, A.; Dornaika, F.; Arganda-Carreras, I.; Reta, J. Efficient and compact face descriptor for driver drowsiness detection. *Expert Syst. Appl.* **2021**, *168*, 114334. [CrossRef]
9. Bai, J.; Yu, W.; Xiao, Z.; Havyarimana, V.; Regan, A.C.; Jiang, H.; Jiao, L. Two-stream spatial–temporal graph convolutional networks for driver drowsiness detection. *IEEE Trans. Cybern.* **2021**, *52*, 13821–13833. [CrossRef] [PubMed]
10. Ahmed, M.; Masood, S.; Ahmad, M.; Abd El-Latif, A.A. Intelligent driver drowsiness detection for traffic safety based on multi CNN deep model and facial subsampling. *IEEE Trans. Intell. Transp. Syst.* **2021**, *23*, 19743–19752. [CrossRef]
11. Zhu, M.; Chen, J.; Li, H.; Liang, F.; Han, L.; Zhang, Z. Vehicle driver drowsiness detection method using wearable EEG based on convolution neural network. *Neural Comput. Appl.* **2021**, *33*, 13965–13980. [CrossRef]

12. Gao, Z.; Wang, X.; Yang, Y.; Mu, C.; Cai, Q.; Dang, W.; Zuo, S. EEG-based spatio–temporal convolutional neural network for driver fatigue evaluation. *IEEE Trans. Neural Netw. Learn. Syst.* **2019**, *30*, 2755–2763. [CrossRef]
13. Dobbs, W.C.; Fedewa, M.V.; MacDonald, H.V.; Holmes, C.J.; Cicone, Z.S.; Plews, D.J.; Esco, M.R. The accuracy of acquiring heart rate variability from portable devices: A systematic review and meta-analysis. *Sports Med.* **2019**, *49*, 417–435. [CrossRef] [PubMed]
14. Lu, K.; Dahlman, A.S.; Karlsson, J.; Candefjord, S. Detecting driver fatigue using heart rate variability: A systematic review. *Accid. Anal. Prev.* **2022**, *178*, 106830. [CrossRef] [PubMed]
15. Persson, A.; Jonasson, H.; Fredriksson, I.; Wiklund, U.; Ahlström, C. Heart rate variability for classification of alert versus sleep deprived drivers in real road driving conditions. *IEEE Trans. Intell. Transp. Syst.* **2020**, *22*, 3316–3325. [CrossRef]
16. Allado, E.; Poussel, M.; Moussu, A.; Hily, O.; Temperelli, M.; Cherifi, A.; Saunier, V.; Bernard, Y.; Albuisson, E.; Chenuel, B. Accurate and Reliable Assessment of Heart Rate in Real-Life Clinical Settings Using an Imaging Photoplethysmography. *J. Clin. Med.* **2022**, *11*, 6101. [CrossRef] [PubMed]
17. Cao, Y.; Li, F.; Liu, X.; Yang, S.; Wang, Y. Towards reliable driver drowsiness detection leveraging wearables. *ACM Trans. Sens. Netw.* **2023**, *19*, 1–23. [CrossRef]
18. Comas, J.; Ruiz, A.; Sukno, F. Efficient remote photoplethysmography with temporal derivative modules and time-shift invariant loss. In Proceedings of the IEEE/CVF Conference on Computer Vision and Pattern Recognition, New Orleans, LA, USA, 19–20 June 2022; pp. 2182–2191.
19. Patel, M.; Lal SK, L.; Kavanagh, D.; Rossiter, P. Applying neural network analysis on heart rate variability data to assess driver fatigue. *Expert Syst. Appl.* **2011**, *38*, 7235–7242. [CrossRef]
20. Gao, H.; Wu, X.; Geng, J.; Lv, Y. Remote heart rate estimation by signal quality attention network. In Proceedings of the IEEE/CVF Conference on Computer Vision and Pattern Recognition, New Orleans, LA, USA, 19–20 June 2022; pp. 2122–2129.
21. Yin, R.-N.; Jia, R.-S.; Cui, Z.; Sun, H.-M. PulseNet: A multitask learning network for remote heart rate estimation. *Knowl.-Based Syst.* **2022**, *239*, 108048. [CrossRef]
22. Xu, M.; Zeng, G.; Song, Y.; Cao, Y.; Liu, Z.; He, X. Ivrr-PPG: An Illumination Variation Robust Remote-PPG Algorithm for Monitoring Heart Rate of Drivers. *IEEE Trans. Instrum. Meas.* **2023**, *72*, 3515510. [CrossRef]
23. Rao, X.; Lin, F.; Chen, Z.; Zhao, J. Distracted driving recognition method based on deep convolutional neural network. *J. Ambient. Intell. Humaniz. Comput.* **2021**, *12*, 193–200. [CrossRef]
24. Nowara, E.M.; Marks, T.K.; Mansour, H.; Veeraraghavan, A. Near-infrared imaging photoplethysmography during driving. *IEEE Trans. Intell. Transp. Syst.* **2020**, *23*, 3589–3600. [CrossRef]
25. Rajesh, N.; Ramachandra, A.C.; Prathibha, A. Detection and Identification of Irregularities in Human Heart Rate. In Proceedings of the 2021 International Conference on Intelligent Technologies (CONIT), Hubli, India, 25–27 June 2021; pp. 1–5.
26. Kassem, H.A.; Chowdhury, M.; Abawajy, J.H. Drivers fatigue level prediction using facial, and head behavior information. *IEEE Access* **2021**, *9*, 121686–121697. [CrossRef]
27. Du, G.; Zhang, L.; Su, K.; Wang, X.; Teng, S.; Li, P.X. A multimodal fusion fatigue driving detection method based on heart rate and PERCLOS. *IEEE Trans. Intell. Transp. Syst.* **2022**, *23*, 21810–21820. [CrossRef]
28. Dua, M.; Shakshi Singla, R.; Raj, S.; Jangra, A. Deep CNN models-based ensemble approach to driver drowsiness detection. *Neural Comput. Appl.* **2021**, *33*, 3155–3168. [CrossRef]
29. Liu, F.; Chen, D.; Zhou, J.; Xu, F. A review of driver fatigue detection and its advances on the use of RGB-D camera and deep learning. *Eng. Appl. Artif. Intell.* **2022**, *116*, 105399. [CrossRef]
30. Du, G.; Li, T.; Li, C.; Liu, P.X.; Li, D. Vision-based fatigue driving recognition method integrating heart rate and facial features. *IEEE Trans. Intell. Transp. Syst.* **2020**, *22*, 3089–3100. [CrossRef]
31. Jia, H.; Xiao, Z.; Ji, P. Real-time fatigue driving detection system based on multi-module fusion. *Comput. Graph.* **2022**, *108*, 22–33. [CrossRef]
32. Akrout, B.; Mahdi, W. A novel approach for driver fatigue detection based on visual characteristics analysis. *J. Ambient. Intell. Humaniz. Comput.* **2023**, *14*, 527–552. [CrossRef]
33. Zhang, J.; Wu, Y.; Chen, Y.; Wang, J.; Huang, J.; Zhang, Q. Ubi-Fatigue: Toward Ubiquitous Fatigue Detection via Contactless Sensing. *IEEE Internet Things J.* **2022**, *9*, 14103–14115. [CrossRef]
34. Ouzar, Y.; Bousefsaf, F.; Djeldjli, D.; Maaoui, C. Video-based multimodal spontaneous emotion recognition using facial expressions and physiological signals. In Proceedings of the IEEE/CVF Conference on Computer Vision and Pattern Recognition, New Orleans, LA, USA, 19–20 June 2022; pp. 2460–2469.
35. Zhang, Z.; Ning, H.; Zhou, F. A systematic survey of driving fatigue monitoring. *IEEE Trans. Intell. Transp. Syst.* **2022**, *23*, 19999–20020. [CrossRef]
36. Sun, W.; Zhang, X.; Peeta, S.; He, X.; Li, Y. A real-time fatigue driving recognition method incorporating contextual features and two fusion levels. *IEEE Trans. Intell. Transp. Syst.* **2017**, *18*, 3408–3420. [CrossRef]
37. Huang, R.; Wang, Y.; Li, Z.; Lei, Z.; Xu, Y. RF-DCM: Multi granularity deep convolutional model based on feature recalibration and fusion for driver fatigue detection. *IEEE Trans. Intell. Transp. Syst.* **2020**, *23*, 630–640. [CrossRef]

38. Jamshidi, S.; Azmi, R.; Sharghi, M.; Soryani, M. Hierarchical deep neural networks to detect driver drowsiness. *Multimed. Tools Appl.* **2021**, *80*, 16045–16058. [CrossRef]
39. Liu, W.; Qian, J.; Yao, Z.; Jiao, X.; Pan, J. Convolutional two-stream network using multi-facial feature fusion for driver fatigue detection. *Future Internet* **2019**, *11*, 115. [CrossRef]

Disclaimer/Publisher's Note: The statements, opinions and data contained in all publications are solely those of the individual author(s) and contributor(s) and not of MDPI and/or the editor(s). MDPI and/or the editor(s) disclaim responsibility for any injury to people or property resulting from any ideas, methods, instructions or products referred to in the content.

Article

Deep Learning-Based Child Handwritten Arabic Character Recognition and Handwriting Discrimination

Maram Saleh Alwagdani * and Emad Sami Jaha

Department of Computer Science, Faculty of Computing and Information Technology, King Abdulaziz University, Jeddah 21589, Saudi Arabia; ejaha@kau.edu.sa
* Correspondence: msalokidin@kau.edu.sa

Abstract: Handwritten Arabic character recognition has received increasing research interest in recent years. However, as of yet, the majority of the existing handwriting recognition systems have only focused on adult handwriting. In contrast, there have not been many studies conducted on child handwriting, nor has it been regarded as a major research issue yet. Compared to adults' handwriting, children's handwriting is more challenging since it often has lower quality, higher variation, and larger distortions. Furthermore, most of these designed and currently used systems for adult data have not been trained or tested for child data recognition purposes or applications. This paper presents a new convolution neural network (CNN) model for recognizing children's handwritten isolated Arabic letters. Several experiments are conducted here to investigate and analyze the influence when training the model with different datasets of children, adults, and both to measure and compare performance in recognizing children's handwritten characters and discriminating their handwriting from adult handwriting. In addition, a number of supplementary features are proposed based on empirical study and observations and are combined with CNN-extracted features to augment the child and adult writer-group classification. Lastly, the performance of the extracted deep and supplementary features is evaluated and compared using different classifiers, comprising Softmax, support vector machine (SVM), k-nearest neighbor (KNN), and random forest (RF), as well as different dataset combinations from Hijja for child data and AHCD for adult data. Our findings highlight that the training strategy is crucial, and the inclusion of adult data is influential in achieving an increased accuracy of up to around 93% in child handwritten character recognition. Moreover, the fusion of the proposed supplementary features with the deep features attains an improved performance in child handwriting discrimination by up to around 94%.

Keywords: child handwriting; handwritten character recognition; writer-group classification; convolutional neural network; deep learning; machine learning

Citation: Alwagdani, M.S.; Jaha, E.S. Deep Learning-Based Child Handwritten Arabic Character Recognition and Handwriting Discrimination. *Sensors* **2023**, *23*, 6774. https://doi.org/10.3390/s23156774

Academic Editors: Erik Blasch, Zhe-Ming Lu and Yufeng Zheng

Received: 16 May 2023
Revised: 12 July 2023
Accepted: 26 July 2023
Published: 28 July 2023

Copyright: © 2023 by the authors. Licensee MDPI, Basel, Switzerland. This article is an open access article distributed under the terms and conditions of the Creative Commons Attribution (CC BY) license (https://creativecommons.org/licenses/by/4.0/).

1. Introduction

Despite significant advances in technology, the textual compositions of many people are still handwritten [1]. Thus, using automated recognition techniques for handwritten data in many applications is crucial. These techniques convert handwritten data (e.g., texts, words, characters, or digits) into corresponding digital representations, which can be accurately processed offline, such as scanned handwritten documents, or online, such as handwriting data input via electronic pen tip [2–4]. Developing automatic handwriting recognition systems is a difficult task in computer vision due to the wide variety of handwriting sizes and styles, besides the characteristics of the language to be recognized [4]. Handwritten character recognition is one of the most challenging research fields in document image processing. Most investigations in this field have been conducted on different languages (e.g., English, French, and Chinese), but only a little work has been conducted on other languages like Arabic [3].

In recent years, handwritten Arabic character recognition has gained considerable research interest. This is due to the importance of the Arabic language, which is considered one of the five most widely spoken languages worldwide and used for reading and writing by hundreds of millions of people from hundreds of nations [5]. However, it is considered a challenging task in pattern recognition and computer vision, as it still requires significant effort to construct generalized systems capable of handling various recognition problems and achieving highly feasible accuracy [6]. These challenges are due to the unique characteristics of the Arabic script, e.g., cursive nature, the existence of diacritics and dots, diagonal strokes, different alternative character shapes in the middle of words, and many characteristics [2,3,5]. Moreover, there are high diversities in handwriting styles across individuals; even at the individual level, a person's handwriting may change significantly or slightly every time, which may make it difficult for a system to recognize the letters from their own handwriting [4].

The majority of research on handwritten Arabic character recognition has focused on adult handwriting, as the findings have revealed the effectiveness of their systems in achieving accuracy rates of up to 99% using deep learning and machine learning techniques [7–15]. Furthermore, a few researchers have recently focused on children's handwriting data for recognizing Arabic letters due to its great significance for many applications and different purposes [4,16–20]. Employing character recognition capabilities in child-related applications such as education [21], interactive learning, physical or mental health assessment, or other possible practical purposes is critical for many future research areas. However, it poses a further challenge due to many differences between the nature of children's and adults' handwriting in several different aspects, including generally being of lesser quality, having more variances, and having more considerable distortions [16].

Handwritten Arabic character recognition technologies have evolved rapidly and achieved progress dramatically using different algorithms, such as support vector machines (SVMs), k-nearest neighbor (KNN), artificial neural networks (ANNs), and, later, convolutional neural networks (CNNs). CNNs have recently outperformed machine learning (ML) techniques that require manually generated features, while CNNs automatically detect and extract distinctive and representative features from the analyzed images [18]. Furthermore, building handwriting recognition hybrid systems using CNNs as a feature extractor and ML algorithms as a classifier has yielded effective results in several handwritten Arabic character datasets [15,18].

In this paper, we develop a novel CNN architecture for recognizing children's handwritten isolated Arabic characters using the Hijja dataset [4] to compose a child data subset for testing. As previously stated, most recognition systems nowadays were neither trained nor tested on children's handwriting character datasets. However, they were exclusively on adult datasets, although character recognition accuracy may be significantly improved by carefully considering the strategy in selecting the training dataset on which the model is to be trained. Therefore, we investigate and analyze the effect of training the suggested model using different datasets of handwritten Arabic characters by either children, adults, or both to assess and compare the fluctuations in model performance in recognizing children's handwritten characters. Moreover, we use three popular machine learning SVM, KNN, and RF techniques as classifiers to assess the automatically derived features from the suggested trained CNN-based deep learning model and compare their performance variation; then, we observe how well these models perform in the classification process compared with the Softmax classifier.

For writer-group classification, Shin et al. [22] proposed a machine learning-based method to automatically classify individuals as adults or children based on their handwritten data, including Japanese scripts and drawn patterns. To the best of our knowledge, no similar research has focused on differentiating between children's and adults' handwriting for Arabic characters. Establishing this capability in this research could open new horizons for other research fields serving multiple purposes, such as fraud or forgery detection and prevention, recognizing and discriminating handwriting more accurately and working on improving skills, comprehending similarities and differences in ways of writing, and further estimating age groups. Finally, after analyzing handwriting, we propose some appropriate supplementary features that can be used along with the extracted deep features of the proposed CNN model to improve the accuracy of child and adult writer-group classification.

The main contributions of this study can be summarized as follows:

- Developing an effective CNN model for recognizing children's handwritten Arabic characters.
- Investigating and analyzing the effect on child handwritten Arabic character recognition performance when training the proposed CNN model on a variety of datasets that either belong to children, adults, or both.
- Examining the capability of the suggested CNN model to classify the writers of Arabic characters into two writer groups, either children or adults.
- Suggesting some supplementary features that contribute to distinguishing between children's and adults' handwriting and augment the performance of the suggested CNN model.
- Extended performance analysis, evaluation, and comparison of the extracted deep features learned by the proposed CNN model and the proposed supplementary features using SVM, KNN, RF, and Softmax classifiers.

The rest of this paper is organized as follows: Section 2 discusses the previous studies on handwritten Arabic character recognition for adults and children data samples. Section 3 describes the used datasets and the proposed research methodology. The experimental work is demonstrated in Section 4, and the findings are provided in Section 5. Section 6 discusses the results and compares the proposed approach to other approaches from the literature. Finally, Section 7 concludes the proposed research work alongside a few ideas for future works.

2. Related Work

In this section, the previous work in the literature is reviewed, presenting various approaches using machine learning and deep learning techniques for adults' and children's handwritten Arabic character recognition. Most of the latest previous studies relevant to our work mainly focused on proposing different approaches to solve this challenging task using CNN-based models.

2.1. Handwritten Arabic Character Recognition for Adult Writers

Most researchers have focused on adult handwriting in Arabic character recognition. In 2017, El-Sawy et al. [7] developed a novel CNN model that was trained and tested on their own dataset, AHCD, which contains 16,800 handwritten Arabic characters collected from 60 persons aged between 19 and 40 years and divided into 28 classes, where their model achieved an accuracy of 94.9%. Another research by Younis [8] introduced a deep model using CNN to recognize handwritten Arabic letters, and it was improved by applying multiple optimization strategies to avoid overfitting. The results demonstrated that their model could classify letters using two datasets, AIA9k and AHCD, achieving 94.8% and 97.6% accuracy, respectively. In 2021, another new handwritten Arabic character dataset named HMBD was introduced by Balaha et al. [9]. They also suggested two CNN-based architectures known as HMB1 and HMB2. They investigated the effect of changing the complexity of these architectures using overfitting reduction strategies on various datasets, including HMBD, AIA9k, and CMATER, to increase recognition accuracy. The uniform weight initializer and the AdaDelta optimizer scored the highest accuracies, where the performance was improved via data augmentation using the HMB1 model, achieving the top overall performance of 90.7%, 98.4%, and 97.3% on AIA9k, HMBD, and CMATER datasets, respectively.

In [10], De Sousa suggested VGG12 and REGU deep models for recognizing handwritten Arabic letters and numbers. Both models were trained twice, once with data augmentation and once without. Then, an ensemble of the four models was created by averaging the predictions of each model. The highest accuracy of their ensemble model was 98.42% for AHCD and 99.47% for MADbase. Boufenar et al. [11] also built a DCNN model similar to Alexnet architecture. They investigated the role of preprocessing data samples in enhancing their model performance using three learning strategies: training the model from scratch, utilizing a transfer-learning technique, and fine-tuning the CNN. Overall, their experimental findings revealed that the first technique outperformed the others, either way, with and without preprocessing, achieving an average of 100% and 99.98% accuracy on OIHACDB-40 and AHCD, respectively. Moreover, Ullah et al. [12] investigated the dropout technique's effect on their built CNN model. They noticed a considerable difference in performance when the model was trained with and without dropout, indicating that dropout regularization could effectively prevent model overfitting. The model reported a test accuracy of 96.78% on the AHCD dataset using dropout. Alyahya et al. [13] studied how the ResNet-18 architecture could be effective in recognizing handwritten Arabic characters. They suggested four ensemble models: the first two were the original ResNet-18 and the updated ResNet-18, using one fully connected layer with or without a dropout layer. The last two models were the original ResNet-18 and the updated ResNet18, but they included two fully connected layers with or without a dropout layer. The original ResNet-18 model achieved the highest test score of 98.30% from other ensemble models on the AHCD dataset. In [14], a CNN model was developed to recognize Arabic letters written by hand. The model was trained and tested using an AHCD dataset. Their experiment has shown that the suggested method achieved a recognition rate of 97.2%. Meanwhile, once data augmentation techniques were used, their model's accuracy rose to 97.7%. Ali et al. [15] designed a CNN-based SVM model with a dropout technique utilizing two deep neural networks and evaluated it on various datasets, including AHDB, AHCD, HACDB, and IFN/ENIT, for recognizing handwritten Arabic letters. The authors reported improved performance of the suggested model compared to previous models created for the same domain by obtaining the accuracies of 99%, 99.71%, 99.85%, and 98.58% on AHDB, AHCD, HACDB, and IFN/ENIT, respectively. Table 1 summarizes these handwritten Arabic character recognition studies using adults' data.

Table 1. A summary of related work on handwritten Arabic character recognition for adult writers.

Ref.	Year	Feature Extractor	Classifier	Dataset	Type	Size	Accuracy
[7]	2017	CNN	Softmax	AHCD	Characters	16,800	94.9%
[8]	2017	CNN	Softmax	AIA9k	Characters	9000	94.8%
				AHCD	Characters	16,800	97.6%
[10]	2018	CNN	Softmax	AHCD	Characters	16,800	98.42%
				MADbase	Digits	70,000	99.47%
[11]	2018	CNN	Softmax	OIHACD	Characters	30,000	100%
				AHCD	Characters	16,800	99.98%
[9]	2020	CNN	Softmax	HMBD	Characters	54,115	90.7%
				AIA9k	Characters	9000	98.4%
				CMATER	Digits	3000	97.3%
[13]	2020	CNN	Softmax	AHCD	Characters	16,800	98.30%
[14]	2021	CNN	Softmax	AHCD	Characters	16,800	97.7%
[15]	2021	CNN	SVM	AHDB	Words and Texts	15,084	99%
				AHCD	Characters	16,800	99.71%
				HACDB	Characters	6600	99.85%
				IFN/ENIT	Words	26,459	98.58%
[12]	2022	CNN	Softmax	AHCD	Characters	16,800	96.78%

2.2. Handwritten Arabic Character Recognition for Child Writers

A few efforts have been made to address the issue of children's Arabic handwriting recognition. In 2020, unlike earlier research, Altwaijry et al. [4] concentrated on recognizing Arabic letters for children's writing. They collected a new dataset named Hijja, consisting of 47,434 disconnected and connected Arabic characters written by children aged 7 to 12 years. They also developed a functional CNN-based model to study and evaluate its performance on their dataset. They compared the performance of their model with the model suggested in El-Sawy's paper [7] on both datasets, Hijja and AHCD. According to the experiment findings, their model outperformed the other compared model, achieving an accuracy of 88% and 97% on the Hijja and AHCD datasets, respectively. Alkhateeb et al. [16] also proposed a deep learning-based system for recognizing handwritten Arabic letters using CNN and three separate datasets, AHCR, AHCD, and Hijja, to validate the proposed system. Based on their experimental results, the suggested approach achieved accuracies of 89.8%, 95.4%, and 92.5% on the AHCR, AHCD, and Hijja datasets, respectively. Another study proposed by Nayef et al. [17] discussed using CNN models to recognize handwritten Arabic characters with an improved Leaky-ReLU activation function. To evaluate the performance of their compared models, they used four datasets, AHCD, HIJJA, and MNIST, in addition to their own dataset containing 38,100 handwritten Arabic characters, categorized into 28 classes that were collected from elementary school students in grades one to three. The proposed CNN model with Leaky-ReLU optimization outperformed the other compared model of [8] with an accuracy of 99%, 95%, and 90% on AHCD, their dataset, and Hijja, respectively.

Alrobah et al. [18] employed a different approach, merging CNN deep-learning models for feature extraction with SVM and XGBoost machine-learning models for classification to build a hybrid model. They used the two CNN architectures presented in [9], namely HMB1 and HMB2. The study attained an accuracy of 96.3% using the HMB1 model and the SVM classifier on the Hijja dataset, highlighting their hybrid model's efficiency. In 2022, Wagaa et al. [19] presented a new CNN architecture that achieved 98.48% and 91.24% accuracies on the AHCD and Hijja datasets, respectively, by applying rotation and shifting data augmentation techniques and using the Nadam optimizer. They also investigated the impact of mixing the two AHCD and Hijja datasets of handwritten Arabic characters in varying proportions on the model's performance during the training and testing phases using different data augmentation approaches. Their results showed that using the Nadam optimizer together with rotation and shifting data augmentation techniques gave their highest test accuracy of 98.32% among other choices when mixed with 80% of AHCD and 20% of Hijja for training along with 20% of AHCD and 10% of Hijja for testing. Bouchriha et al. [20] also presented a novel CNN model for recognizing handwritten Arabic characters. They focused on unique characteristics of Arabic text, particularly the difference in the shape of letters according to their location in the word, and by using the Hijja dataset, they attained an accuracy of 95%. Table 2 summarizes these handwritten Arabic character recognition studies on children's data.

Table 2. A summary of related work on handwritten Arabic character recognition for child writers.

Ref.	Year	Feature Extractor	Classifier	Dataset	Type	Size	Accuracy
[4]	2020	CNN	Softmax	Hijja	Characters	47,434	88%
				AHCD	Characters	16,800	97%
[16]	2020	CNN	Softmax	AHCR	Characters	28,000	89.8%
				AHCD	Characters	16,800	95.4%
				Hijja	Characters	47,434	92.5%

Table 2. Cont.

Ref.	Year	Feature Extractor	Classifier	Dataset	Type	Size	Accuracy
[17]	2021	CNN	Softmax	AHCD Proposed dataset Hijja MNIST	Characters Characters Characters Digits	16,800 38,100 47,434 70,000	99% 95.4% 90% 99%
[18]	2021	CNN	Softmax SVM XGBoost	Hijja	Characters	47,434	89% 96.3% 95.7%
[19]	2022	CNN	Softmax	AHCD Hijja	Characters Characters	16,800 47,434	98.48% 91.24%
[20]	2022	CNN	Softmax	Hijja	Characters	47,434	95%

3. Proposed Methodology

In this study, we conducted two different tasks, handwritten character recognition and writer-group classification. Figure 1 shows the framework designed to achieve the suggested approach for recognizing children's handwritten Arabic characters and classifying them into a child or an adult writer group. The proposed approach is divided into four phases: data preprocessing, feature extraction using CNN and other supplementary features, classification using three additional popular ML-based classifiers, and evaluation of the results and model performance using standard assessment measurement techniques. The following subsections provide more information on each of these four stages.

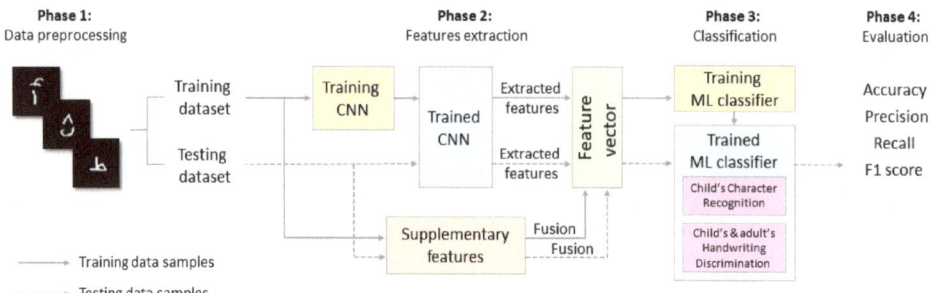

Figure 1. Overview of the framework of the proposed methodology.

3.1. Data Preprocessing Phase

Data preprocessing is an important prior step meant to facilitate the extraction of significant features and improve classification accuracy. This section describes the datasets used to conduct the experimental work and how they were preprocessed using several methods.

3.1.1. Datasets Description

This study uses two publicly available datasets of handwritten Arabic characters to conduct all experiments: Hijja dataset for child writers and the Arabic handwritten characters dataset (AHCD) for adult writers. The Hijja dataset [4] comprises 47,434 letter samples of size 32 × 32 written by 591 children aged 7 to 12 years. It has 108 subclasses arranged into 29 main classes, 28 classes for Arabic letters, and one more class for the "Hamza" character (ء). Each of the 28 classes contains up to four additional subclasses, categorized into connected (beginning, middle, and end of a word) and disconnected characters. Moreover, it was divided into 80% (37,933 samples) for training and 20% (9501 samples) for testing. Note that we only used the disconnected characters totaling 12,355 character samples for conducting the experimental work here. We divided them by the same proportion into two groups, 80% (9884 characters) for training and 20% (2471 characters) for testing.

The AHCD dataset [7] contains 16,800 character samples of size 32×32 written by 60 people aged 19 to 40 years, and it includes 28 classes for isolated (disconnected) Arabic characters. Each participant wrote each of the 28 characters ten times, from the character "Alf" (ا) to "Yaa" (ي). It was similarly divided into 80% for training with 13,440 samples (480 per class) and 20% for testing with 3360 samples (120 per class). Table 3 statistically describes the used child and adult datasets.

Table 3. Description of the used datasets.

Dataset Characteristic	Hijja	AHCD
Number of writers	591	60
Total samples per character for each writer	1	10
Total character samples per writer	28	280
Total samples per character	400~500	600
Total isolated character samples	12,355	16,800
Category of writers	Children	Adults

3.1.2. Character Image Preprocessing

As mentioned above, we used all images in the Hijja dataset that only contain the character in its separate (disconnected) form. The preprocessing stage has included a number of procedures that help the proposed system achieve the highest possible accuracy. Firstly, these images were converted into grayscale images and then inverted to set the foreground as bright pixels and the background as dark pixels. Secondly, because some of the grayscale inverted images were too low-contrast and blurry, the contrast was adjusted to increase the intensity values of the foreground components and reduce the pixel values of the background to appear as dark as possible. After that, the brightness was raised by 2%. Thirdly, after empirically testing different image threshold values, the resulting pixels were thresholded by considering all values less than 90 as background pixels and resetting their values to zero. Finally, the foreground pixels were centered by drawing a rectangle around the character pixels and then cropped, after which zeros were added around the character to be the size 32×32.

For the AHCD dataset, the same thresholding was applied to all images and the characters were then centered. Moreover, the Hijja dataset has a different number of images for each class, which may negatively impact the efficiency of adequately training our deep model on all classes, especially in the comparison between different training strategies using child, adult, and both datasets. To solve this problem, we just increased the number of samples for each class in the Hijja training dataset using different data augmentation methods to match the number of samples for each class in the AHCD training dataset, which is 480, resulting in a sum of 13,440 character samples in the new augmented Hijja training dataset. The augmentation techniques used were zoom range, height shift range, and width shift range, all of which are equal to 0.1, and a rotation range of 5.

In addition, we combined both the Hijja and AHCD training and testing datasets to create a new dataset consisting of 26,880 characters for training and 5831 characters for testing. The combined training dataset was used in training both tasks of character recognition and writer-group classification, while the combined testing dataset was only used for probing the second task of writer-group classification. It is worth noting that, for writer-group classification, all images were further converted into binary images, unlike for character recognition using grayscale images.

3.2. Feature Extraction Phase

In this phase, the proposed features for handwritten character images and the classification process are extracted. This section explains the suggested CNN architecture and supplementary features in detail.

3.2.1. Proposed CNN Architecture

CNNs have been proven successful and effective in recognizing handwritten characters [23]. A CNN is a multi-layered hierarchical model composed of convolution, pooling, and fully connected layers (FCLs). The purpose of convolution layers is to extract essential features from input images and generate feature maps using several filters. Pooling layers are used to minimize the dimensions of feature maps and to retrain the most critical features. Eventually, FCLs receive the high-level features from the preceding layers as input (formed as flat feature vectors) and yield several output classes, each with a value that indicates the class probability [23].

As shown in Figure 2, the suggested CNN model to extract features has ten layers, comprising four convolution layers, four max-pooling layers, and two fully connected layers. The input is a grayscale image for the child's character recognition task and a binary image for the child's handwriting discrimination task from the adult's handwriting as a writer's group, both images with a size of 32 × 32. All convolution layers use a 3 × 3 kernel, one stride, padding equal to the same input size, and a ReLU activation function that converts x (an input feature value) less than zero to zero, as defined in Equation (1).

$$\text{ReLU}(x) = max(0, x) \tag{1}$$

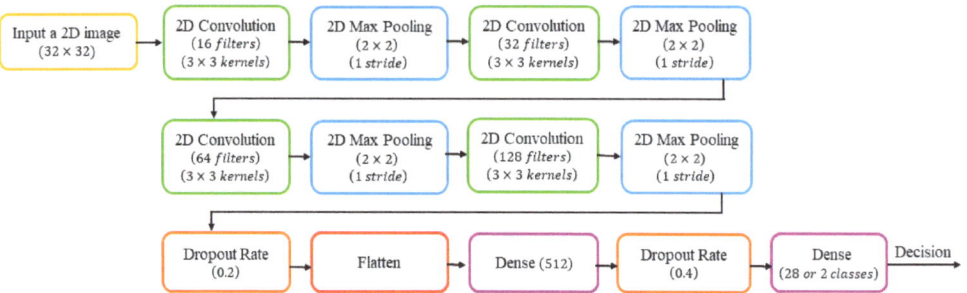

Figure 2. The proposed CNN architecture.

The number of filters used in each convolution layer varies, such that the first convolution has 16 filters, the second has 32 filters, the third has 64 filters, and the fourth has 128 filters. Each convolution layer is followed by a max-pooling layer with a size of 2 × 2 and a stride of 1. A dropout rate of 0.2 is used after all convolution and max-pooling layers. Moreover, there are two dense layers in the last two fully connected layers. The first contains 512 neurons with a ReLU activation. In contrast, the second has 28 neurons for the character recognition task and two for the writer-group classification task with a Softmax activation, as defined in Equation (2), where x_i is the output feature vector from CNN, e is a mathematical constant known as Euler's number, and N is the number of output classes. After the first dense layer, a dropout rate of 0.4 was applied.

$$\text{Softmax}(x_i) = \frac{e^{x_i}}{\sum_{k=1}^{N} e^{x_k}} \tag{2}$$

3.2.2. Proposed Supplementary Features

The purpose of these features is to supplement the CNN-based features and improve the discrimination accuracy between children's and adults' handwriting. In this study, we used a histogram of oriented gradient (HOG)-based features and other statistical-based features as supplementary features to help distinguish between child and adult writers.

- Histogram of Oriented Gradient (HOG)-based Features

An HOG generates descriptive features for an object's shape and appearance in an image by calculating gradients distribution or contour directions [24]. We used an HOG to extract features of the distinctive shape aspects of handwritten characters to distinguish children's writing typical style from that of adults. To extract HOG features, the gradients for each pixel in the image were first computed in both the vertical and horizontal directions using the following Equations (3) and (4):

$$d_x = I(x+1, y) - I(x, y) \quad (3)$$

$$d_y = I(x, y+1) - I(x, y) \quad (4)$$

where d_x and d_y represent the horizontal and vertical gradient directions, and $I(x, y)$ is the pixel value at (x, y). Hence, the gradient magnitude, $|d|$, and orientation, θ, were then calculated by Equations (5) and (6):

$$|d| = \sqrt{d_x^2 + d_y^2} \quad (5)$$

$$\theta(x, y) = tan^{-1}\frac{d_y}{d_x} \quad (6)$$

In the next step, the gradient image was divided into small cells of 8×8 pixels to calculate the histogram of gradient direction for each pixel inside the cell and place them into a nine-bin histogram. These histograms were then combined to represent HOG features. For better results, these histograms were then normalized by taking overlapping 3×3 blocks and applying L2-Hys normalization. Finally, the gradient histograms inside each cell between each block were then added together to obtain the final HOG feature vector of size 1×324.

- Statistical-Based Features

We proposed some statistical-based features that assisted in differentiating between the handwriting of children and adults after analyzing their character data samples. Figure 3 shows the differences between adult and children's handwriting for some characters such as Kha (خ), Alif (ا), Thaa (ث), Qaaf (ق), Tha (ط), and Shiin (ش). Statistical features are based on the analysis of the spatial distribution of pixels and basic dimensions of a character sample [25]. The total of these features is twelve, as illustrated in Table 4, which are divided into two main groups as follows:

1. Ratio of Height to Width:

This feature depends on the main dimensions of a character, where the ratio of height h to width w (F1) is calculated for the bounding box of an unnormalized character sample (only the main body of the character, no "hamza" or "dots") [25,26]. This feature is useful in differentiating between the sizes of letters written by children and adults since most letters written by children usually have common sizing characteristics, which can be utilized for differentiating them from those written by adults, and vice versa.

2. Ratios of Pixel Distribution:

These features depend on the spatial distribution of pixels in an image. We derived eleven features through the distribution of foreground fg (white) pixels and background bg (black) pixels for the bounding box of an unnormalized character sample. Firstly, we computed (F2) as the ratio of all foreground pixels to all background pixels for the whole character image [25]. Secondly, we divided the character image into four equal quadrants: upper-left (UL), upper-right (UR), bottom-left (BL), and bottom-right (BR) to calculate the ratio of the number of foreground pixels to the number of background pixels in each quarter (F3–F6). Finally, as inspired by [25], we computed the ratio of background pixels in each pairwise combination of the four quarters computed as $\binom{4}{2}$, resulting in six features (F7–F12). These features are helpful for distinguishing pen strokes and font width between child and adult handwriting, as most of the children's handwriting was intermittent, pen-down, pen-up actions, and displayed hesitancy, and it was somewhat light, while the adults' handwriting was mostly uninterrupted and bold, indicating more confidence, convenience, and consistency.

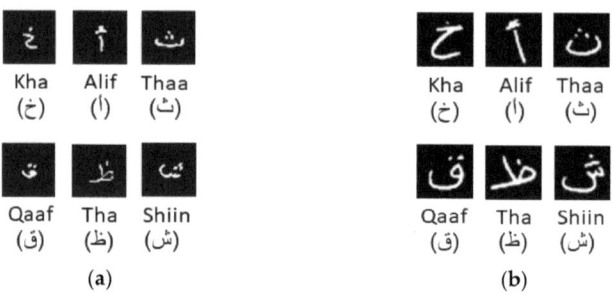

Figure 3. Some preprocessed Hijja and AHCD character data samples: (**a**) Child writers' samples; (**b**) Adult writers' samples.

Table 4. A summary of statistical-based features.

Feature	Formula	Feature	Formula	Feature	Formula
F1	h/w	F5	$(fg/bg)_{BL}$	F9	bg_{UL}/bg_{BL}
F2	$(fg/bg)_{All}$	F6	$(fg/bg)_{BR}$	F10	bg_{UR}/bg_{BR}
F3	$(fg/bg)_{UL}$	F7	bg_{UL}/bg_{UR}	F11	bg_{UR}/bg_{BL}
F4	$(fg/bg)_{UR}$	F8	bg_{UL}/bg_{BR}	F12	bg_{BL}/bg_{BR}

3.3. Classification Phase

After the CNN model was trained and tested, we used it as a major feature extractor by replacing the final output FCL (Softmax classifier) with three well-known ML-based SVM, KNN, and RF classifiers for performance variation measurement and comparison purposes across all experiments. The feature vector obtained from the trained CNN consists of 512 features when trained on child data, adult data, and both for the children's character recognition. For the writer-group classification task, we used the feature fusion method to supplement CNN-extracted features with statistical-based features, with HOG-based features, and with both. The feature vector obtained from the statistical-based feature extractor constitutes twelve features, while the one obtained from the HOG-based feature extractor comprises 324 features. All these extracted features were normalized to range from 0 to 1 using min-max normalization. The fused feature vector was initially trained and evaluated using the Softmax classifier by constructing a feed-forward neural network (FFNN) with six layers. The first layer is the input layer showing the number of features in each feature vector, while the remaining layers are illustrated in Figure 4. All these extracted features were then trained by the ML classifiers. After training, the trained ML classifiers were used for testing in the classification phase for children's handwritten character recognition and children and adult handwriting discrimination with/without supplementary features.

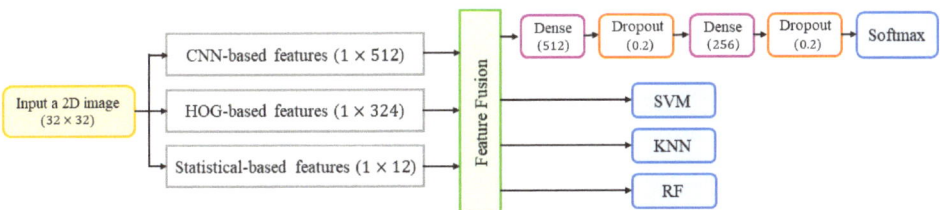

Figure 4. The FFNN-based feature fusion model for the writer-group classification task.

3.3.1. Support Vector Machine (SVM) Classifier

SVM is an effective supervised learning technique used for classification and regression tasks. It works on training data examples as plotted points in a high-dimensional feature space. The classification process is then performed by finding an optimal hyperplane that separates between classes correctly when achieving the maximum possible margin between them [1,27,28]. The nascent SVM performance significantly relies on the three primary hyperparameters: kernel function, regularization (usually defined as C), and *gamma* [15]. In this work, we used a nonlinear SVM classifier that can be defined as shown in Equation (7).

$$f(x) = \sum_{i=1}^{l} w_i \Phi_i(x) + b \qquad (7)$$

where $\Phi(x)$ represents a feature map and w refers to the corresponding weights. Φ means transform x input vector from input space into a higher dimensional feature space using kernel functions. Kernel functions have two main parameters: *C* and *gamma*. We examined here multiple nonlinear SVM kernel functions with several values of *C* and *gamma* to find the optimal values that yield the best possible classification accuracy.

3.3.2. K-Nearest Neighbor (KNN) Classifier

KNN is the simplest supervised learning classifier, requiring no previous intensive training process or probabilistic classification. It works by finding *k* nearest samples and their class labels in the training dataset to predict the class of a new sample in the testing dataset. The classification process is performed by measuring the distance between feature vectors of training and testing samples in feature space. The *k*-nearest samples with their class labels are then retrieved to choose the predominant class label as a class for the test sample [29]. In this work, we tested different distance metrics with different odd k numbers, where the distance measures used in KNN are Euclidean and Manhattan distances, which can be defined as given in Equations (8) and (9), respectively [30].

$$d(x,y) = \sqrt{\sum_{i=1}^{n}(x_i - y_i)^2} \qquad (8)$$

$$d(x,y) = \sum_{i=1}^{n} |x_i - y_i| \qquad (9)$$

where x and y are the feature vectors in the feature space, and x_i and y_i refer to their *i*-th feature of the total n features.

3.3.3. Random Forest (RF) Classifier

RF is an ensemble machine learning algorithm used for classification and regression problems [31]. It is composed of multiple decision trees that are generated in parallel using a subset of randomly selected training data samples, each of which works as an independent classifier. Their predictions are then aggregated to determine the final outcome by calculating the majority vote for the results of each output decision tree. RF enables fast learning even with high-dimensional features. Moreover, the random selection of training data makes it robust against noise [22]. This work tested different numbers of trees and their maximum depth.

3.4. Evaluation Phase

The overall performance of the proposed model was evaluated using the *accuracy*, *precision*, *recall*, and *F1-score* metrics inferred via the four distributions, true positive (*TP*), false negative (*FP*), true negative (*TN*), and false negative (*FN*), as follows:

- *Accuracy (A)* is the ratio of correctly predicted characters to the total of all predicted characters. Equation (10) shows the accuracy evaluation metric.

$$Accuracy = \frac{TP + TN}{TP + FP + FN + TN} \quad (10)$$

- *Precision (P)* is the ratio of correctly predicted positive characters to the total number of correctly and incorrectly predicted positive characters. Equation (11) shows the precision classification rate.

$$Precision = \frac{TP}{TP + FP} \quad (11)$$

- *Recall (R)* is the ratio of correctly predicted positive characters to the total number of positive characters, calculated using Equation (12).

$$Recall = \frac{TP}{TP + FN} \quad (12)$$

- *F1-score (F1)* combines the recall and precision measures, as shown in Equation (13).

$$F1 - score = \frac{2(Recall \times Precision)}{Recall \times Precision} \quad (13)$$

4. Experiments

This section describes the environment used, experimental setup, and design of experiments with implementation details, hyperparameter tuning, and data augmentation.

4.1. Experimental Setup

All experiments were conducted using the Google Colab environment. In addition, several open source Python libraries were used, such as Kares to build and train the CNN model, Scikit-learn to address ML classifiers and print evaluation measurement tools, CSV to read Excel data files, TensorFlow to implement and evaluate the CNN model, and others.

4.2. Experiments Design

In this research, we conducted five experiments with different scenarios. The first three experiments are related to testing the proposed CNN model in recognizing children's handwritten Arabic letter data (Hijja) by training the model on children's data (Hijja), on adult data (AHCD), and on both types of data (combined Hijja and AHCD). The last two experiments are associated with discriminating between adult and child handwriting of Arabic letters by training and testing the model on both data samples (combined Hijja and AHCD) with and without the proposed supplementary features. The extracted features by the CNN and supplementary features are trained and evaluated using Softmax, SVM, KNN, and RF classifiers. Table 5 briefly describes the objective of each experiment. The three datasets used for experimental work were prepared and rearranged by dividing them into 80% for training and 20% for testing for all classifiers. To tune the CNN model's hyperparameters, the training dataset was divided into 60% for training and 20% for validation. Table 6 shows for each experiment the number of images and the image type of each of the training, validation, and test datasets for tuning the proposed CNN model's hyperparameters.

Table 5. An overview of conducted experimental work.

Experiment No.	Task	Training Dataset	Testing Dataset
Experiment 1		Hijja	Hijja
Experiment 2	Character Recognition	AHCD	Hijja
Experiment 3		Combined Hijja and AHCD	Hijja
Experiment 4	Writer-Group Classification *without* Supplementary Features	Combined Hijja and AHCD	Combined Hijja and AHCD
Experiment 5	Writer-Group Classification *with* Supplementary Features	Combined Hijja and AHCD	Combined Hijja and AHCD

Table 6. Statistics of the used datasets.

Dataset	Training Dataset	Validation Dataset	Testing Dataset	Normalized Image Type
Hijja	10,752	2688	2471	
AHCD	10,752	2688	2471	Grayscale
Combined Hijja and AHCD	21,504	5376	2471	
Combined Hijja and AHCD	21,504	5376	5831	Binary

4.3. Hyperparameters Tuning and Data Augmentation

To tune the proposed CNN model's hyperparameters, we examined three different optimizers and three weight initializers in all experiments using the validation dataset to find the optimal hyperparameters for the training dataset in order to make the model generalized and as not overfitted as possible. The examined optimizers are Adam, Nadam, and RMSProp, while the weight initializers are Normal, Uniform, and He Normal. Nadam optimizer and He Normal weight initializer are used to optimize our model since they gave better results than the others. In addition, categorical cross-entropy was used to calculate the loss for the child's character recognition and binary cross-entropy for the writer-group classification, where accuracy was assigned as the metric. The model was also trained using a batch size equal to 80 and an epoch number set to 100. Moreover, we used the ReduceonLRPPlateau approach that periodically reduces the learning rate in the Kares library, beginning from 0.001 until 0.00001 when multiplied by a factor equal to 0.1. For the FFNN model, it was also trained using 80 batch sizes and 100 epochs, with the Nadam optimizer and the binary cross-entropy.

We set the following hyperparameters for the SVM classifier: C = (1, 10, 100, 1000), kernel = ['poly', 'sigmoid', 'rbf'], and *gamma* = (0.01, 0.001, 0.0001), whereas for the KNN classifier we set the following: k = (5, 7, 9, 11), weights = ['distance'], and metric = ['Euclidean', 'Manhattan']. We also set the hyperparameters for the RF classifier as n_estimators = (50, 100, 200, 300, 400) and max_depth = (5, 10, 15, 20, 25, 30). We tuned the hyperparameters of these classifiers using the grid search method and then determined the optimal hyperparameters that provide the highest possible classification accuracy. Finally, we used the same data augmentation techniques applied to the Hijja training dataset to be balanced in training the CNN model by increasing the overall size of the Hijja and AHCD datasets, with a view to overcome the overfitting problem and improve the model's performance.

5. Results

The results obtained from the five conducted experiments are reported in this section to evaluate and compare the proposed model's performance using different classifiers for recognizing children's handwritten Arabic characters and distinguishing between child and adult handwriting. It is worth noting that the data split for all five experiments using different classifiers was 80% for training and 20% for testing.

Experiment 1 was conducted to show how the proposed model performed after being trained and tested on the Hijja dataset alone. The results of this experiment are presented in

Table 7. Furthermore, the accuracy and loss curves of the training and validation are shown in Figure 5. The model achieved the best performance with an accuracy of 91.95% using the SVM classifier with radial basis function (SVM-RBF) kernel values set to ($C = 100$ and $gamma = 0.001$). The RF classifier reported the second-highest accuracy at 91.87%, while Softmax and KNN achieved the lowest performance compared to the others. It is worth noting that the hyperparameter of the KNN was set to Manhattan distance and $k = 5$, and the RF was set to n_estimators = 300 and max_depth = 30.

Table 7. Child character recognition results of Experiment 1, using Hijja for training and testing.

Classifier	Accuracy	Precision	Recall	F1-Score
Softmax	91.78%	91.87%	91.76%	91.76%
SVM	**91.95%**	**92.07%**	**91.91%**	**91.93%**
KNN	91.50%	91.62%	91.46%	91.47%
RF	91.87%	91.93%	91.82%	91.81%

Results in bold indicate the highest scores achieved among the different classifiers.

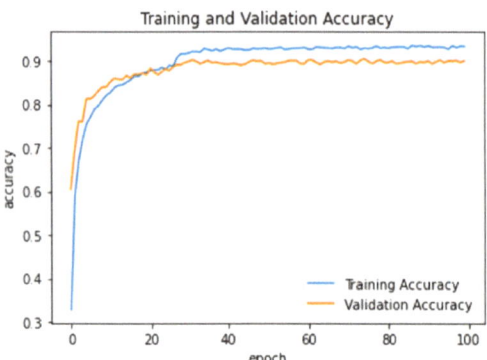 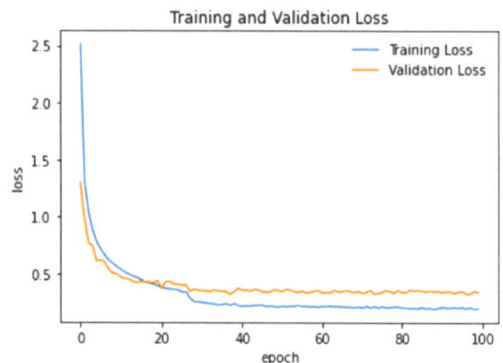

Figure 5. Learning accuracy and loss performance of Experiment 1.

Experiment 2 investigated the effect of training the proposed model using the adult handwriting dataset (AHCD) alone on testing the child handwriting dataset (Hijja). Table 8 shows the results obtained in Experiment 2, and Figure 6 displays its accuracy and loss curves over the training epochs. Here, the highest accuracy reported in this experiment was 80.17%, achieved by the SVM classifier, while KNN, RF, and Softmax received lower accuracies of 79.24%, 79.16, and 78.67%, respectively. Noting that the hyperparameter of the SVM-RBF kernel was set to $C = 10$ and $gamma = 0.01$, KNN was set to Manhattan distance, with $k = 9$, and RF was set to n_estimators = 200 and max_depth = 20.

Table 8. Child character recognition results of Experiment 2, using AHCD for training and Hijja for testing.

Classifier	Accuracy	Precision	Recall	F1-Score
Softmax	78.67%	80.54%	78.66%	78.87%
SVM	**80.17%**	**81.87%**	**80.12%**	**80.28%**
KNN	79.24%	81.12%	79.21%	79.40%
RF	79.16%	80.62%	79.12%	79.15%

Results in bold indicate the highest scores achieved among the different classifiers.

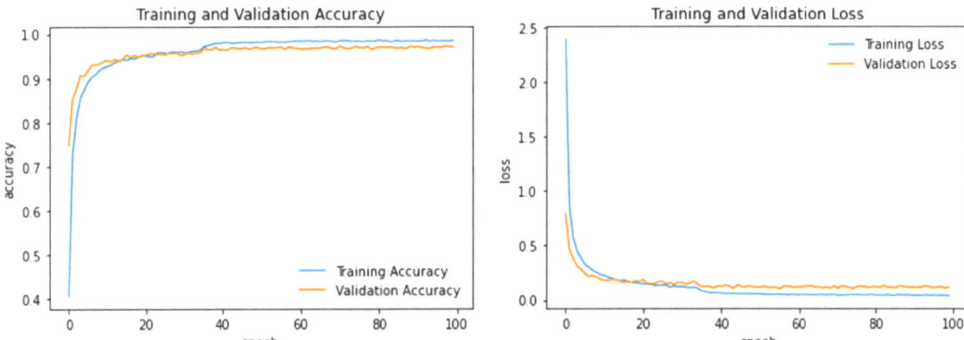

Figure 6. Learning accuracy and loss performance of Experiment 2.

Experiment 3 was carried out to see whether the proposed model could improve recognition accuracy when trained on both child and adult data samples (combined Hijja and AHCD) to recognize the Hijja testing dataset. Table 9 summarizes all the recognition results of the Hijja testing dataset when child and adult datasets were combined during the training phase. Also, the accuracy and loss curves are shown in Figure 7. Interestingly, this experiment achieved a higher accuracy of 92.96% for both SVM and Softmax and 92.72% for RF and 92.47% for KNN than the prior two experiments trained only on either the child or adult dataset in isolation. It is worth noting that hyperparameter of the SVM-RBF kernel was set to $C = 10$ and *gamma* = 0.01, KNN was set to Manhattan distance and $k = 9$, and RF was set to n_estimators = 400 and max_depth = 25. Table 10 summarizes and compares the performance results of the three experiments along with their average performance of the different classifiers used.

Table 9. Child character recognition results of Experiment 3, using combined Hijja and AHCD for training and Hijja for testing.

Classifier	Accuracy	Precision	Recall	F1-Score
Softmax	**92.96%**	92.99%	**92.92%**	92.92%
SVM	**92.96%**	93.14%	92.91%	**92.94%**
KNN	92.47%	92.52%	92.44%	92.42%
RF	92.72%	92.81%	92.68%	92.69%

Results in bold indicate the highest scores achieved among the different classifiers.

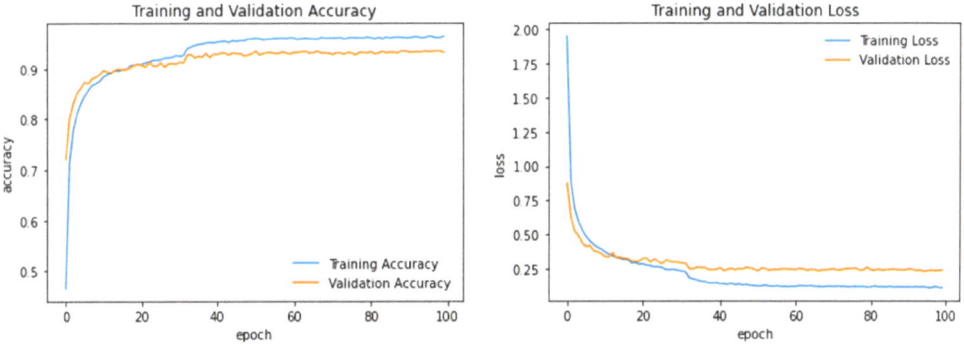

Figure 7. Learning accuracy and loss performance of Experiment 3.

Table 10. Aggregated child character recognition results and average performance of Experiments 1 to 3.

Classifier	Experiment 1		Experiment 2		Experiment 3	
	Accuracy	F1-Score	Accuracy	F1-Score	Accuracy	F1-Score
Softmax	91.78%	91.76%	78.67%	78.87%	**92.96%**	**92.92%**
SVM	91.95%	91.93%	80.17%	80.28%	**92.96%**	**92.94%**
KNN	91.50%	91.47%	79.24%	79.40%	**92.47%**	**92.42%**
RF	91.87%	91.81%	79.16%	79.15%	**92.72%**	**92.69%**
Average	91.78%	91.74%	79.31%	79.43%	**92.78%**	**92.74%**

Results in bold indicate the highest scores achieved among the different classifiers.

Experiments 4 and 5 assessed how well the suggested model could classify writers based on their handwriting into two groups: a child writer and an adult writer, with/without supplementary features. In Experiment 4, we trained and tested the CNN model without using supplementary features. Table 11 shows the model's writer-group classification performance. Moreover, Figure 8 illustrates the learning accuracy and loss performance for Experiment 4. The RF classifier received the best accuracy of 90.41%, where n_estimators was set to $C = 200$ and max_depth = 30. On the other hand, the SVM, KNN, and Softmax classifiers achieved a lower accuracy of 89.85%, 89.74%, and 88.24%, respectively, where the SVM-RBF kernel values were set to $C = 100$ and $gamma = 0.01$, and the hyperparameters of KNN were Euclidean distance and $k = 11$.

Table 11. Writer-group classification performance of Experiment 4, without supplementary features using combined Hijja and AHCD for training and testing.

Classifier	Accuracy	Precision	Recall	F1-Score
Softmax	88.24%	88.57%	91.37%	89.95%
SVM	89.85%	90.56%	91.96%	91.26%
KNN	89.74%	92.44%	89.52%	90.96%
RF	**90.41%**	**90.50%**	**89.82%**	**90.11%**
Average	89.56%	90.52%	90.67%	90.57%

Results in bold indicate the highest scores achieved among the different classifiers.

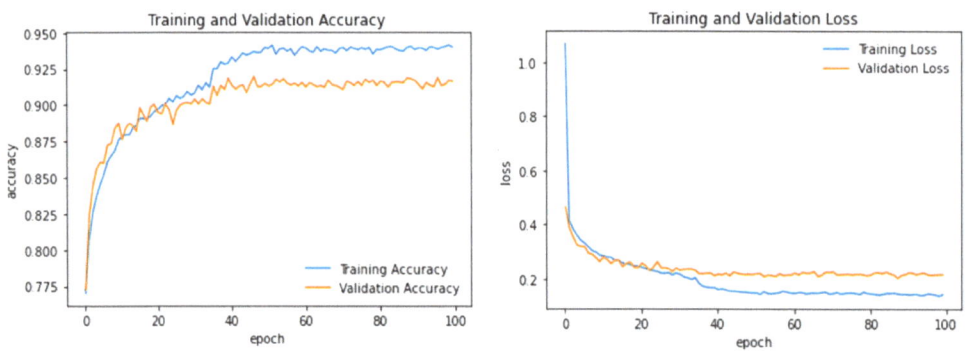

Figure 8. Learning accuracy and loss performance of Experiment 4.

The results of Experiment 4 were also analyzed and validated using the confusion matrix, as shown in Figure 9. Hence, we observed that the RF classifier, shown in Figure 9d, outperformed the Softmax, SVM, and KNN classifiers by achieving 94% accuracy for accurate adult classification and only 6% of adults were misclassified as children. Nevertheless, the best child classification accuracy was 90% using the KNN classifier, as shown in Figure 9c, whereas only 10% of the child samples were misclassified as adult ones.

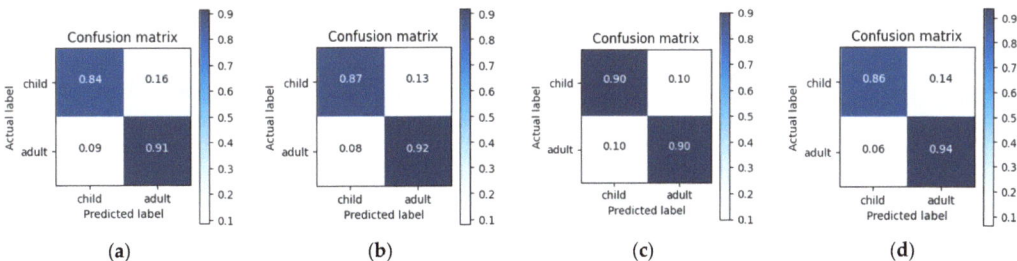

Figure 9. Confusion matrix depicting the results of Experiment 4 using three different classifiers: (**a**) Softmax classifier, (**b**) SVM classifier; (**c**) KNN classifier; (**d**) RF classifier.

For Experiment 5, we combined the CNN-extracted deep features as follows: first, with statistical-based features (SF) resulting in a 524-dimensional feature vector; second, with HOG-based features resulting in an 836-dimensional feature vector; third, with both SF and HOG features resulting in an 848-dimensional feature vector. The results reported in Table 12 show that, when all extracted features from CNN, SF, and HOG were fused, we received the highest performance for all classifiers, for all evaluation metrics, with the highest achieved accuracies of 93.98%, 92.11%, 92.06%, and 91.00% for Softmax, KNN, SVM, and RF, respectively. When CNN and HOG features were combined, they achieved the second-highest accuracy scores ranging from 90.94% to 93.88, whereas the lowest accuracies were scored by CNN and SF fusion ranging from 89.88% to 91.92%. It is worth noting that, in all three fusion cases, the Softmax classifier was superior, by all means, in writer-group classification performance over KNN, RF, and SVM. Subsequently, compared to the results of Experiment 4, by utilizing the fusion of CNN deep features with the proposed supplementary SF and HOG features, the classification performance of discriminating between adult and child handwriting was significantly improved by up to 5.74%, 5.48, 2.24%, and 3.86% for accuracy, precision, recall, and F1-score, respectively, where the average scores of these four metrics were also enhanced by up to 3.11%, 2.01%, 1.71%, and 1.84%, respectively. The hyperparameter of the SVM-RBF kernel was set to $C = 100$ and *gamma* = 0.01 for each combination of features fusion, the KNN distances were set as Euclidean with $k = 7$, Manhattan with $k = 9$, and Manhattan with $k = 11$, and the RF was set to n_estimators = 200, 300, and 200 with max_depth = 30, respectively.

Table 12. Writer-group classification performance of Experiment 5, with supplementary features using combined Hijja and AHCD for training and testing.

Classifier	CNN + SF				CNN + HOG				CNN + SF + HOG			
	A%	*P%*	*R%*	*F1%*	*A%*	*P%*	*R%*	*F1%*	*A%*	*P%*	*R%*	*F1%*
Softmax	91.92	91.85	91.57	91.70	93.88	93.91	93.54	93.71	**93.98**	**94.05**	**93.61**	**93.81**
SVM	89.88	89.75	89.48	89.61	92.03	**91.762**	91.96	91.86	**92.06**	91.760	**92.08**	**91.90**
KNN	90.00	89.66	90.01	89.81	91.75	91.42	91.87	91.60	**92.11**	**91.77**	**92.29**	**91.98**
RF	90.14	90.17	89.59	89.84	90.94	91.30	90.18	90.61	**91.00**	**91.35**	**90.23**	**90.67**
Average	90.49	90.36	90.16	90.24	92.15	92.10	91.89	91.95	**92.29**	**92.23**	**92.05**	**92.09**

Results in bold indicate the highest scores achieved among the different classifiers.

6. Discussion and Comparison

6.1. Discussion of the Results

In this work, extensive experiments were conducted to investigate how the proposed methodology can improve performance in the more challenging task of children's handwritten Arabic character recognition. The investigation was conducted by training the model on child data samples, adult data samples, and both. In addition, we evaluated the proposed approach's capabilities in classifying writers of testing handwritten character samples as adults or children and how the proposed supplementary features can help improve the classification accuracy. Based on analysis of the results of Experiments 1 to 3, it can be observed, as in Table 10, that when the (child) Hijja and (adult) AHCD datasets were merged for the training of the proposed model, as in Experiment 3, to recognize child handwritten Arabic characters, achieving 92.78% average accuracy with about 1% higher score than the 91.78% average accuracy obtained by training the proposed model on the (child) Hijja dataset only, as in Experiment 1. Such an improvement in accuracy may be a result of providing the trained model with clearer and higher-quality supportive samples of the (adult) AHCD dataset during training, which enhanced the modeling process and increased the trained model's ability to recognize more confusable children's letter samples via balanced and non-overfitted learning as possible by combining both child and adult training data. Consequently, incorporating adult data alongside children's data during the training phase improved the recognition accuracy of the (child) Hijja test datasets.

In contrast, when the same model was trained on the (adult) AHCD dataset alone and tested on the (child) Hijja dataset, as in Experiment 2, the recognition accuracy decreased noticeably, scoring a lower average accuracy of 79.31% compared to Experiment 1 and 3. The reason is that the (child) Hijja dataset is considered more complex and challenging than the (adult) AHCD dataset since it contains many letter samples that can be distorted, unclear, and low-quality, as we noticed and also stated in [16]. In other words, the model was trained using only adult samples that were fairly clear and more consistent in terms of the characteristics, shape, and size of the letter, and there were no notable distortions compared with the Hijja dataset of child samples. Due to this, the trained model could not recognize numerous confusable (child) Hijja data samples.

After analyzing the results and seeing the misclassified samples of Experiment 4, we observed that there were similarities between the handwriting of children and adults in terms of character and sizing characteristics, presence of distortions, pen stroke, and font width. Based on these similarities, some children's data samples were classified as being written by adults because they were mostly closer to the common writing style of adults. On the contrary, some adult data samples were classified as being written by children due to the presence of some adult samples with a similar style to the common children's writing style. Several factors led to such similarities between child and adult writing styles, such as growing age and educational level. The quality of a person's handwriting improves with growing age, except for exceptionally aging or sick people. Also, the higher a person's education level, the more likely it is that their handwriting will be better. However, for most of those grown people who resort to overusing technology devices rather than traditional paper-and-pen, over time, their handwriting may remain or become low-skilled, closer to children's handwriting level. In Experiment 5, combining all the proposed supplementary features (SF and HOG) or only HOG with the CNN deep features contributed to increasing discrimination accuracy between child and adult handwriting by approximately 6% using the Softmax classifier. Generally, in Experiment 5, the classification accuracy was improved in all cases compared with Experiment 4 in various ratios, as shown in Table 12.

6.2. Comparison with Existing Works

We compared our suggested methodology with different related approaches from the literature that concentrated on children's handwritten Arabic character recognition using the Hijja dataset. The comparison was made in terms of the target task, the methods used for feature extraction and classification, suggested supplementary (handcrafted) features, applied feature fusion technique, and the dataset used for training and testing the model. In these studies [4,16,17,20], researchers developed a new character recognition system using a CNN deep learning model, trained it, and tested it on children's data samples. In [18], they designed a hybrid model by combining existing CNN models as feature extractors with SVM and XGBoost machine learning models as classifiers, which were trained and evaluated using the Hijja dataset. The hybrid model has outperformed other models using the SVM classifier. However, these studies did not investigate the effect of training their suggested models on the AHCD dataset (comprising adult data samples) or on a combination of the (child) Hijja and (adult) AHCD datasets to be eventually only focused and tested on the more challenging Hijja child data samples alone.

In [19], a novel CNN model was created, trained, and evaluated using children's data. In addition, they studied how the use of data augmentation techniques affected the performance of recognition during the model training and testing when combining the two datasets, Hijja and AHCD, in various ratios. Nevertheless, they did not train their model on combined child and adult datasets in an equal proportion to exclusively be tested on the challenging child dataset. Finally, none of these studies or earlier studies addressed classifying handwritten Arabic letters into child or adult writer groups. Table 13 demonstrates different aspects of comparison between the suggested strategy and several methodologies used in earlier studies.

Table 13. Comparison between our proposed methodology and current approaches in the literature.

Ref.	Task	Feature Extraction		Feature Fusion	Classification				Dataset Used	
		CNN	Handcrafted		Softmax	SVM	KNN	RF	Training	Testing
[4]	Character Recognition	√			√				Hijja	Hijja
[16]		√			√				Hijja	Hijja
[17]		√			√				Hijja	Hijja
[20]		√			√				Hijja	Hijja
[18]		√			√	√			Hijja	Hijja
[19]		√			√				Hijja	Hijja
		√			√				A mixture of Hijja and AHCD in a different ratio	A mixture of Hijja and AHCD in a different ratio
Our study	Character Recognition	√			√	√	√	√	Hijja	Hijja
		√			√	√	√	√	AHCD	Hijja
		√			√	√	√	√	A mixture of Hijja and AHCD in an equal ratio	Hijja
	Writer-Group Classification	√	√	√	√	√	√	√	Both Hijja and AHCD	Both Hijja and AHCD

7. Conclusions

In this paper, several experiments were conducted for two tasks: handwritten character recognition and writer-group classification. First, we designed a CNN model for children's handwritten Arabic character recognition. Then, the model was used to study the impact of the training process on various handwritten Arabic character datasets belonging to children, adults, or both in particularly recognizing letter samples written by children only. We concluded that, when the model was trained on both samples of children and adult data, we achieved the best performance and obtained the highest average accuracy of 92.78%, which is rather higher than the accuracy resulting from training the model on children's data in isolation. Moreover, training the model on adult data alone, even though there are much higher-quality data compared to child data, had a negative effect on the model's performance in recognizing children's data.

The same model with necessary changes was also used to examine and assess its capability to differentiate between children's and adults' handwriting. As a result, it initially achieved an average classification accuracy of 89.28%, demonstrating after extended analysis that there could be considerably confusable similarities in writing style between adults and children. To confront such confusable similarities and improve the child handwriting discrimination performance and results, we suggested HOG-based and statistical-based supplementary features to supplement the deep features extracted from the CNN model. Amongst three proposed feature fusion approaches in Experiment 5, the approach combining CNN-based deep features with both statistical-based and HOG-based supplementary features augmented the model's performance in distinguishing between child and adult handwriting using combined Hijja and AHCD for training and testing. It yielded the highest average accuracy of 92.29%, about 2.73% higher than the result obtained using only CNN features. In addition, we trained and tested all extracted features using Softmax, SVM, KNN, and RF classifiers, where SVM with the RBF kernel gave a higher accuracy than the Softmax classifier in the character recognition task. On the other hand, in the writer-group classification task, Softmax was the superior classifier among all, according to all performance evaluation measures.

For future work, this approach can be extended and used to recognize handwritten connected Arabic letters for children and propose further useful supplementary features that may contribute to improving character recognition accuracy. Moreover, the capability of this approach using some intentional mistakes can also be investigated and analyzed. Moreover, it can also be enforced in various practical applications to discriminate between children's and adults' handwriting through texts or words.

Author Contributions: Methodology, M.S.A.; software, M.S.A.; validation, M.S.A.; formal analysis, M.S.A. and E.S.J.; investigation, M.S.A.; data curation, M.S.A. and E.S.J.; writing—original draft preparation, M.S.A.; writing—review and editing, M.S.A. and E.S.J.; visualization, M.S.A. and E.S.J.; supervision, E.S.J.; funding acquisition, E.S.J. All authors have read and agreed to the published version of the manuscript.

Funding: This research was supported and funded by KAU Scientific Endowment, King Abdulaziz University, Jeddah, Saudi Arabia, grant number 077416.

Institutional Review Board Statement: Not applicable.

Informed Consent Statement: Not applicable.

Data Availability Statement: The datasets used in this article were Hijja and AHCD. For details, please refer to [4,7].

Acknowledgments: The authors would like to thank King Abdulaziz University Scientific Endowment for funding the research reported in this paper.

Conflicts of Interest: The authors declare that they have no conflict of interest to report regarding the present study.

References

1. Albattah, W.; Albahli, S. Intelligent Arabic Handwriting Recognition Using Different Standalone and Hybrid CNN Architectures. *Appl. Sci.* **2022**, *12*, 10155. [CrossRef]
2. Alrobah, N.; Albahli, S. Arabic Handwritten Recognition Using Deep Learning: A Survey. *Arab. J. Sci. Eng.* **2022**, *47*, 9943–9963. [CrossRef]
3. Ali, A.A.A.; Suresha, M.; Ahmed, H.A.M. Survey on Arabic Handwritten Character Recognition. *SN Comput. Sci.* **2020**, *1*, 152. [CrossRef]
4. Altwaijry, N.; Al-Turaiki, I. Arabic handwriting recognition system using convolutional neural network. *Neural Comput. Appl.* **2021**, *33*, 2249–2261. [CrossRef]
5. Balaha, H.M.; Ali, H.A.; Badawy, M. Automatic recognition of handwritten Arabic characters: A comprehensive review. *Neural Comput. Appl.* **2020**, *33*, 3011–3034. [CrossRef]
6. Ghanim, T.M.; Khalil, M.I.; Abbas, H.M. Comparative study on deep convolution neural networks DCNN-based offline Arabic handwriting recognition. *IEEE Access* **2020**, *8*, 95465–95482. [CrossRef]
7. El-Sawy, A.; Loey, M.; EL-Bakry, H. Arabic handwritten characters recognition using convolutional neural network. *WSEAS Trans. Comput.* **2017**, *5*, 11–19.
8. Younis, K.S. Arabic handwritten character recognition based on deep convolutional neural networks. *Jordanian J. Comput. Inf. Technol.* **2017**, *3*, 186–200.
9. Balaha, H.M.; Ali, H.A.; Saraya, M.; Badawy, M. A new Arabic handwritten character recognition deep learning system (ahcr-dls). *Neural Comput. Appl.* **2021**, *33*, 6325–6367. [CrossRef]
10. De Sousa, I.P. Convolutional ensembles for Arabic Handwritten Character and Digit Recognition. *PeerJ Comput. Sci.* **2018**, *4*, e167. [CrossRef]

11. Boufenar, C.; Kerboua, A.; Batouche, M. Investigation on deep learning for off-line handwritten Arabic character recognition. *Cogn. Syst. Res.* **2018**, *50*, 180–195. [CrossRef]
12. Ullah, Z.; Jamjoom, M. An intelligent approach for Arabic handwritten letter recognition using convolutional neural network. *PeerJ Comput. Sci.* **2022**, *8*, e995. [CrossRef] [PubMed]
13. Alyahya, H.; Ismail, M.M.B.; Al-Salman, A. Deep ensemble neural networks for recognizing isolated Arabic handwritten characters. *ACCENTS Trans. Image Process. Comput. Vis.* **2020**, *6*, 68–79. [CrossRef]
14. AlJarrah, M.N.; Zyout, M.M.; Duwairi, R. Arabic Handwritten Characters Recognition Using Convolutional Neural Network. In Proceedings of the 2021 12th International Conference on Information and Communication Systems (ICICS), Valencia, Spain, 24–26 May 2021; pp. 182–188.
15. Ali, A.A.A.; Mallaiah, S. Intelligent handwritten recognition using hybrid CNN architectures based-SVM classifier with dropout. *J. King Saud Univ. Comput. Inf. Sci.* **2021**, *34*, 3294–3300. [CrossRef]
16. Alkhateeb, J.H. An Effective Deep Learning Approach for Improving Off-Line Arabic Handwritten Character Recognition. *Int. J. Softw. Eng. Knowl. Eng.* **2020**, *6*, 53–61.
17. Nayef, B.H.; Abdullah, S.N.H.S.; Sulaiman, R.; Alyasseri, Z.A.A. Optimized leaky relu for handwritten Arabic character recognition using convolution neural networks. *Multimed. Tools Appl.* **2021**, *81*, 2065–2094. [CrossRef]
18. Alrobah, N.; Albahli, S. A Hybrid Deep Model for Recognizing Arabic Handwritten Characters. *IEEE Access* **2021**, *9*, 87058–87069. [CrossRef]
19. Wagaa, N.; Kallel, H.; Mellouli, N. Improved Arabic Alphabet Characters Classification Using Convolutional Neural Networks (CNN). *Comput. Intell. Neurosci.* **2022**, *2022*, e9965426. [CrossRef]
20. Bouchriha, L.; Zrigui, A.; Mansouri, S.; Berchech, S.; Omrani, S. Arabic Handwritten Character Recognition Based on Convolution Neural Networks. In Proceedings of the International Conference on Computational Collective Intelligence (ICCCI 2022), Hammamet, Tunisia, 28–30 September 2022; pp. 286–293.
21. Bin Durayhim, A.; Al-Ajlan, A.; Al-Turaiki, I.; Altwaijry, N. Towards Accurate Children's Arabic Handwriting Recognition via Deep Learning. *Appl. Sci.* **2023**, *13*, 1692. [CrossRef]
22. Shin, J.; Maniruzzaman, M.; Uchida, Y.; Hasan, M.A.M.; Megumi, A.; Suzuki, A.; Yasumura, A. Important features selection and classification of adult and child from handwriting using machine learning methods. *Appl. Sci.* **2022**, *12*, 5256. [CrossRef]
23. Ahamed, P.; Kundu, S.; Khan, T.; Bhateja, V.; Sarkar, R.; Mollah, A.F. Handwritten Arabic numerals recognition using convolutional neural network. *J. Ambient Intell. Humaniz. Comput.* **2020**, *11*, 5445–5457. [CrossRef]
24. Dalal, N.; Triggs, B. Histograms of oriented gradients for human detection. In Proceedings of the 2005 IEEE Computer Society Conference on Computer Vision and Pattern Recognition (CVPR'05), San Diego, CA, USA, 20–26 June 2005; pp. 886–893.
25. Rashad, M.; Amin, K.; Hadhoud, M.; Elkilani, W. Arabic character recognition using statistical and geometric moment features. In Proceedings of the 2012 Japan-Egypt Conference on Electronics, Communications and Computers, Alexandria, Egypt, 6–9 March 2012; pp. 68–72.
26. Abandah, G.A.; Malas, T.M. Feature selection for recognizing handwritten Arabic letters. *Dirasat Eng. Sci. J.* **2010**, *37*, 242–256.
27. Elleuch, M.; Maalej, R.; Kherallah, M. A new design based-SVM of the CNN classifier architecture with dropout for offline Arabic handwritten recognition. *Procedia Comput. Sci.* **2016**, *80*, 1712–1723. [CrossRef]
28. Cervantes, J.; Garcia-Lamont, F.; Rodríguez-Mazahua, L.; Lopez, A. A comprehensive survey on support vector machine classification: Applications, challenges and trends. *Neurocomputing* **2020**, *408*, 189–215. [CrossRef]
29. Jaha, E.S. Efficient Gabor-based recognition for handwritten Arabic-Indic digits. *Int. J. Adv. Comput. Sci. Appl.* **2019**, *10*, 112–120. [CrossRef]
30. Abu Alfeilat, H.A.; Hassanat, A.B.; Lasassmeh, O.; Tarawneh, A.S.; Alhasanat, M.B.; Eyal Salman, H.S.; Prasath, V.S. Effects of distance measure choice on k-nearest neighbor classifier performance: A review. *Big Data* **2019**, *7*, 221–248. [CrossRef]
31. Hasan, M.A.M.; Nasser, M.; Ahmad, S.; Molla, K.I. Feature selection for intrusion detection using random forest. *J. Inform. Secur.* **2016**, *7*, 129–140. [CrossRef]

Disclaimer/Publisher's Note: The statements, opinions and data contained in all publications are solely those of the individual author(s) and contributor(s) and not of MDPI and/or the editor(s). MDPI and/or the editor(s) disclaim responsibility for any injury to people or property resulting from any ideas, methods, instructions or products referred to in the content.

Article

Smart Shelf System for Customer Behavior Tracking in Supermarkets

John Anthony C. Jose [1,*], Christopher John B. Bertumen [1], Marianne Therese C. Roque [1], Allan Emmanuel B. Umali [1], Jillian Clara T. Villanueva [1], Richard Josiah TanAi [2], Edwin Sybingco [1], Jayne San Juan [3] and Erwin Carlo Gonzales [4]

1. Department of Electronics and Computer Engineering, Gokongwei College of Engineering, De La Salle University, 2401 Taft Avenue, Malate, Manila 1004, Metro Manila, Philippines; christopher_bertumen@dlsu.edu.ph (C.J.B.B.); marianne_roque@dlsu.edu.ph (M.T.C.R.); allan_umali@dlsu.edu.ph (A.E.B.U.); jillian_villanueva@dlsu.edu.ph (J.C.T.V.); edwin.sybingco@dlsu.edu.ph (E.S.)
2. Department of Manufacturing Engineering and Management, Gokongwei College of Engineering, De La Salle University, 2401 Taft Avenue, Malate, Manila 1004, Metro Manila, Philippines; richard.tanai@dlsu.edu.ph
3. Department of Industrial and Systems Engineering, Gokongwei College of Engineering, De La Salle University, 2401 Taft Avenue, Malate, Manila 1004, Metro Manila, Philippines; jayne.sanjuan@dlsu.edu.ph
4. Management and Organization Department, Ramon V. del Rosario College of Business, De La Salle University, 2401 Taft Avenue, Malate, Manila 1004, Metro Manila, Philippines; erwin.carlo.gonzales@dlsu.edu.ph
* Correspondence: john.anthony.jose@dlsu.edu.ph

Citation: Jose, J.A.C.; Bertumen, C.J.B.; Roque, M.T.C.; Umali, A.E.B.; Villanueva, J.C.T.; TanAi, R.J.; Sybingco, E.; San Juan, J.; Gonzales, E.C. Smart Shelf System for Customer Behavior Tracking in Supermarkets. *Sensors* **2024**, *24*, 367. https://doi.org/10.3390/s24020367

Academic Editors: Erik Blasch and Yufeng Zheng

Received: 25 November 2023
Revised: 22 December 2023
Accepted: 4 January 2024
Published: 8 January 2024

Copyright: © 2024 by the authors. Licensee MDPI, Basel, Switzerland. This article is an open access article distributed under the terms and conditions of the Creative Commons Attribution (CC BY) license (https://creativecommons.org/licenses/by/4.0/).

Abstract: Transactional data from point-of-sales systems may not consider customer behavior before purchasing decisions are finalized. A smart shelf system would be able to provide additional data for retail analytics. In previous works, the conventional approach has involved customers standing directly in front of products on a shelf. Data from instances where customers deviated from this convention, referred to as "cross-location", were typically omitted. However, recognizing instances of cross-location is crucial when contextualizing multi-person and multi-product tracking for real-world scenarios. The monitoring of product association with customer keypoints through RANSAC modeling and particle filtering (PACK-RMPF) is a system that addresses cross-location, consisting of twelve load cell pairs for product tracking and a single camera for customer tracking. In this study, the time series vision data underwent further processing with R-CNN and StrongSORT. An NTP server enabled the synchronization of timestamps between the weight and vision subsystems. Multiple particle filtering predicted the trajectory of each customer's centroid and wrist keypoints relative to the location of each product. RANSAC modeling was implemented on the particles to associate a customer with each event. Comparing system-generated customer–product interaction history with the shopping lists given to each participant, the system had a general average recall rate of 76.33% and 79% for cross-location instances over five runs.

Keywords: smart shelves; visual analytics; retail analytics; computer vision; sensor fusion

1. Introduction

Retail analytics involves the analysis of data to provide insight into different aspects of retail operations. Through this type of analytics, trends in sales and customer behavior may be characterized [1]. Retailers are able to make data-driven decisions regarding business strategies related to pricing, product placement, and more. These strategies may help reduce costs, increase sales, and improve the overall customer experience [2]. When it comes to inventory management, neither a surplus nor deficit are beneficial. Ideally, the supply should sufficiently meet customer demand to prevent products remaining in inventory for prolonged periods of time which may lead to deterioration or eventual discarding, which would incur losses [3]. Hence, inventory management is one of the most crucial considerations when it comes to data analytics in the retail industry.

Point-of-sale (POS) systems monitor and execute transactions, making them the most common data source for retail analytics in physical establishments. Although this system effectively tracks sale volumes, more data is required to accurately provide information regarding product sales and availability relative to customer response [4]. This is also taking into consideration that additional POS system features may be kept behind a paywall that puts micro, small, and medium enterprises (MSMEs) at a disadvantage compared to their well-established competitors, who have more resources [5]. Additionally, unlike online retailers, which can monitor non-purchasing customer behavior, final purchase information from onsite retail stores is limited to data provided by POS systems [6].

Further research suggests that implementing smart shelf systems is an effective way of obtaining additional and unseen data from existing methods in physical retail stores. Several approaches have been made in designing such systems, all of which generally focus on customer interaction with products [4,7–9]. An autonomous shopping solutions provider [10] uses a scalable multi-camera approach to account for problems with occlusion. However, the processing of vision data is computationally intensive. Another approach for implementing smart shelves is making use of cameras and weight sensors [11]. However, its main limitation is the degraded system performance, due to the system not being able to detect cross-location or a situation when a person reaches for a product in different weight bins.

We take the view that the difficulty of cross-location [11] is rooted in the lack of object identifiers tracked in a video across a time period, known as track ID, from a multi-object tracker. Furthermore, a multi-product multi-person setting further leads to a noisy trajectory for track ID. Our study aims to develop a smart shelf system incorporating PACK-RMPF that solves cross-location problems while alleviating the effect of a noisy trajectory. Enhanced by this approach, our smart shelf systems evaluate customer interactions with fixed, shelved products through a combination of sensors and object recognition.

The main contributions of this study are as follows:

1. A smart shelf system that uses single camera and weight sensor arrays for a multi-person, multi-product tracking system
2. A PACK-RMPF sensor fusion algorithm that enhances the tracking of product interaction by associating weight event data and multi-object tracking data with the use of state filters and inlier estimators to address occlusion and localization problems and cross-location.
3. A simulated supermarket experiment setup that evaluates system performance based on a customer's journey of purchasing goods based on a pre-defined shopping list.

2. Related Work

Generally, inventory management seeks to maximize the collection and use of inventory-related data. In the case of retail stores, it often involves streamlining an inventory with consideration for customer demand and carrying costs, among other factors and risks. One aspect of inventory management is the tracking of goods as they move from an inventory to customers. The emergence of smart retailing, through various technologies that enable more advanced retail analytics, aims to extend the functionality of POS systems from merely tracking inventory movement to making logical inferences about how customer behavior and customer–product interactions are correlated with the turnover of products in an inventory. Smart shelves, a common approach to smart retailing, utilize a variety of ways to extract pertinent data that can be used to predict inventory with respect to customer behavior, including, but not limited to, weight sensors, RFID tags, and specialized cameras [12].

Higa and Iwamoto [13] proposed a low-cost solution involving surveillance cameras as a tool for tracking the amount of products on a shelf in 2019. The method utilized was background subtraction wherein the background of an image is removed before the foreground—which contains the products—is observed. Through the use of a CNN, the system compares fully replenished, partially stocked, and empty shelf conditions. At the end of the research, the system was able to garner an 89.6% accuracy rate. The authors proposed that this data could eventually be utilized to improve profits in retail stores by improving the shelf availability of products [13]. A similar solution was proposed back in 2015 [14]. In this solution, it was suggested that image detection and processing should alert store managers upon the need to replenish shelves, as well as the discrepancies detected such as misplaced products.

When it comes to the use of weight sensors for product tracking, physical interactions with a product have been typically tracked through changes in weight sensor data, as determined by algorithms. Ref. [8] utilized a bin system in their research wherein a single shelf row was divided into bins that had dedicated load cells for tracking a specified type of product instead of having a single platform for all available products on the row. In [15], each shelf row was equipped with load cell weight sensors arranged so that positional tracking would be possible on the shelf. This approach worked through constant weight readings from the load cells and an algorithm for detecting significant changes in the weight of a load cell in the system to be registered as a pick up. Through the use of support vector regression (SVR) and artificial neural networks (ANNs), the position of the product could also be predicted.

Unmanned retail stores, otherwise referred to as cashier-less stores, utilizing smart shelf systems typically implement human pose estimation (HPE) for customer tracking. Implementing HPE for customer tracking is advantageous—especially in solving action recognition problems—due to the plethora of pre-trained data readily available to use [16]. As an example, [17] developed a system for an unmanned retail store that makes use of visual analytics and grating sensors. With the grating sensors emitting infrared beams, the creation of an invisible curtain-like barrier was formed. Once a customer's hand passed through the curtain, the infrared beams would be interrupted and the motion would be flagged as an item being taken. One camera was integrated with a human pose estimation algorithm to classify the hand action when such an event occurred. The rest of the cameras that were installed were used to determine which product was taken. Mask R-CNN was utilized for multiple customer tracking. Under this model, the body was tracked as one silhouette. Furthermore, under this system, occlusions were solved by adding the possible losses in each part of the silhouette.

Smart shelf systems vary in terms of the extent of weight sensor and computer vision integration. For those that primarily or solely use computer vision, there is usually a focus on security, autonomous or cashier-less checkouts, or visual indicators of customer behavior. AiFi, asserting itself as a provider of autonomous shopping solutions, uses an extensive amount of cameras to be able to effectively run stores [10]. In addition to occlusion problems, there may be a question of redundancy depending on the actual system design. The in-store autonomous checkout system (ISACS) for retail proposed by [11], on the other hand, was divided into three major tracking subsystems: product tracking with weight sensors, customer tracking through cameras whose feeds underwent human pose estimation, and multi-human to multi-product tracking through sensor fusions. These systems were highly dependent on the pose estimation in associating the weight event detected with the person. However, a major limitation of these systems is the problem of collision and cross-location, especially when multiple customers are interacting with a shelf.

Hence, there are several approaches for implementing sensor fusions. An example of this approach is making use of a particle filter, which can be divided into four stages: generation, prediction, updating, and resampling. Particles are uniformly generated, and the positions of the generated particles are predicted. The sensor readings are compared with the predicted position of each particle, and the probability that relates the distance of the particles to the measured position from a sensor is calculated. The particles are then resampled so that only the nearer particles are measured. Finally, the algorithm can retrieve the trajectory of a moving object. We were able to recreate the trajectory of the wireless signals, sourcing from the moving objects, through the particle filter, and we recommend this as a solution for localization problems in sensor fusion.

3. Smart Shelf System Design

3.1. Shelf Physical Overview

The shelf tested was a double-sided gondola shelf equipped with two rows, 0.5 m apart, on each side. Each row contained three (3) weighing platforms primarily operational through a pair of straight bar load cells, hereinafter referred to as weight bins, for a total of twelve (12) weight bins. Distinct products were placed on each weight bin, corresponding to twelve (12) products. The number of units per product varied based on the dimensions of each product. The camera was located 2.43 m away from the shelf and 2.26 m from the ground, as depicted in Figure 1.

Figure 1. Physical system setup.

3.2. System Overview

PACK-RMPF utilizes the data from the weight sensors and video feed per unit time. For weight sensor data, the signals undergo a filter process as a safeguard against extreme signal spikes, movements, or errors. Extreme signal spikes and movements are referred to weight signal changes where the weight signal steps are deemed impossible. For instance, a 300 g product is deemed impossible to move higher than 1 kg of a weight signal step considering that the products are removed one at a time. Additionally, errors in retrieving weight values will send out an error string instead of float values. Thus, for extreme signal spikes, movements, or errors, forward data filling using the padding method was utilized. Afterward, the signal underwent a weight event detection algorithm. A three-period moving average and a three-period moving variance were performed. Each product contained a specific moving variance threshold based on the initial calibration results for each product. Thus, the start of an event was triggered when the moving variance was higher than the threshold two consecutive times, while the end of its event was triggered when the moving variance was lower than the threshold two consecutive times. The moving average was utilized to handle possible volatility. As such, it was utilized for determining the weight value at the start and end of an event. When a weight sensor decreased by the end of an event, it was labelled a pick up; likewise, if it increased by the end of an event, it was labelled a putback.

The data for the vision system contained the detected customers per frame attached to a unique track ID, a bounding box, and keypoint detections. Since the vision system had Re-ID features, it was able to trace a lost tracked bounding box, typically due to occlusion, back to the customer using their appearance features [18]. It was equally important that only the customer's hand were extracted from the human pose estimation (HPE), since it was the only one of significance to the system.

The integration system used a particle filter for trajectory tracking of the bounding box and the hand keypoints. These trajectories were processed via random sample consensus (RANSAC) to identify the correct customer who performed the weight event. With that, the data process of PACK-RMPF is summarized in Figure 2.

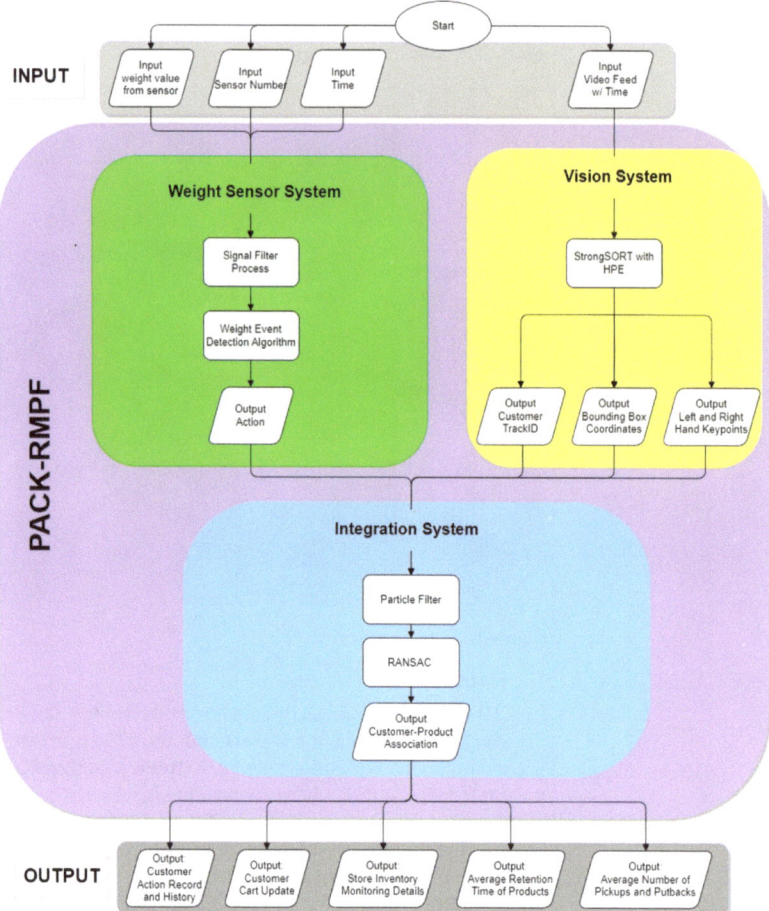

Figure 2. System overview.

At the end of the integration system where the product and the customer's behavior are associated, PACK-RMPF is able to record each customer's action. Likewise, the system is able to process, derive, and obtain the history of each customer's action, the current products in their cart, the current inventory status of the shelf, the average retention time of each product, and the number of specific actions the customer performed.

4. Weight Change Event Detection System

4.1. Weight Bin

Each weight bin consisted of a pair of straight bar load cells. An acrylic sheet was mounted on top of the load cells to aid in weight distribution. The load cells were also screwed onto the shelf. Each load cell had a capacity of 3 kg for a total capacity of 6 kg per weight bin. The dimensions of each weight bin are listed on Figure 3.

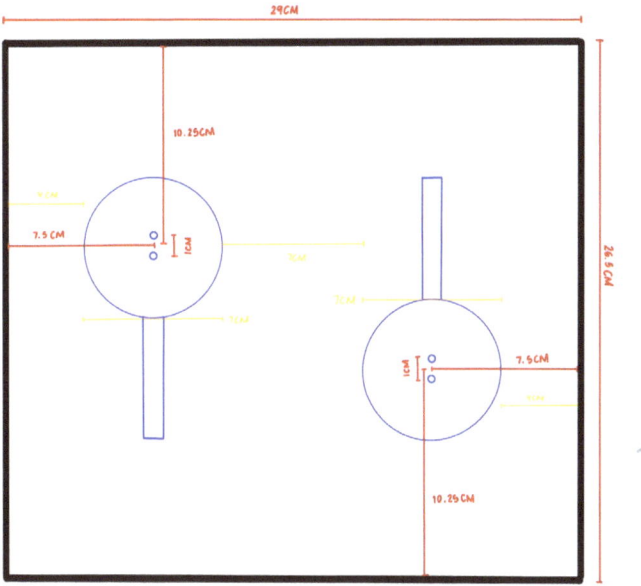

Figure 3. Dimension of each weight bin.

As illustrated in Figure 4, the wires of each load cell were connected in parallel when interfaced with the inputs of an HX711 amplifier module. By default, the module had a gain of 124 and a corresponding sampling rate of 80 Hz. Each module was connected to a Raspberry Pi 4B that would then process the readings from each weight bin concurrently. The ground, supply, and clock pins for each HX711 module were the same, while each digital output was connected to a dedicated GPIO pin.

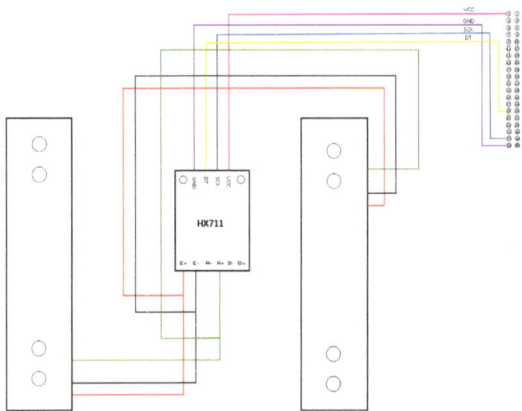

Figure 4. Circuit connection of a weight bin.

4.2. Weight Sensor Array Hardware

The weight sensor hardware system is shown in Figure 5. Every weight bin has a unique assignment on the physical location along the double-sided shelf. Furthermore, each weight bin is assigned a unique product item. Each row on one side of the shelf is partitioned into three slots, leading to three weight bins per row. On each side of the shelf, there are two rows, a top row and a bottom row. A total of 18 weight bins are connected to Raspberry Pi 4B, which acts as the controller.

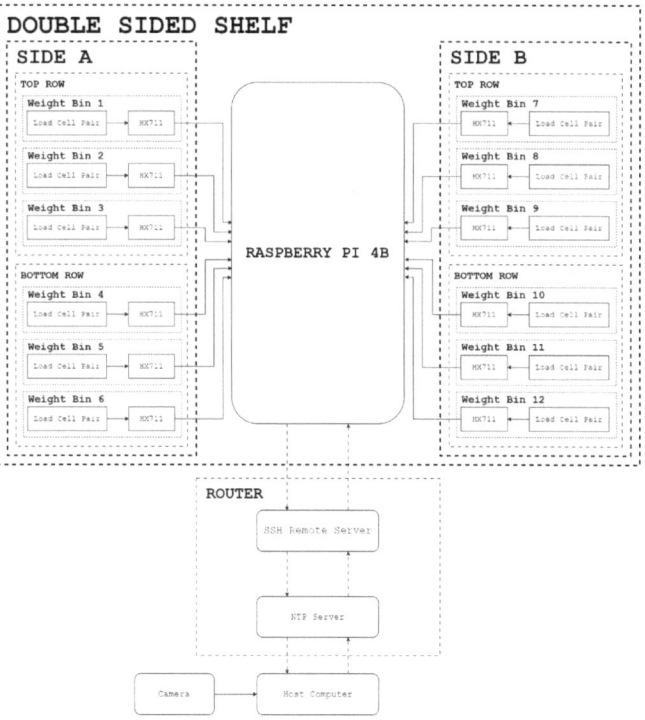

Figure 5. Hardware diagram of the weight sensor array.

The controller was interfaced with a host computer using secure shell (SSH) protocol. The host computer also served as the network time protocol (NTP) server to set the weight reading timestamps. Through the network, a server would be initialized to enable the RPi to continuously send timestamps and weight readings for each weight bin to a client device.

4.3. Weight Change Event Detection System

An overview of the weight sensor data processing, with the main goal of identifying and classifying weight change events, is found in Figure 6. The weight sensor data acquired from the controller are grouped by specific weight bin location λ. Forward filling is also implemented so that invalid readings from a weight bin assume the value of the last valid reading from the same weight bin. Then, processing is conducted based on three-period moving variance values.

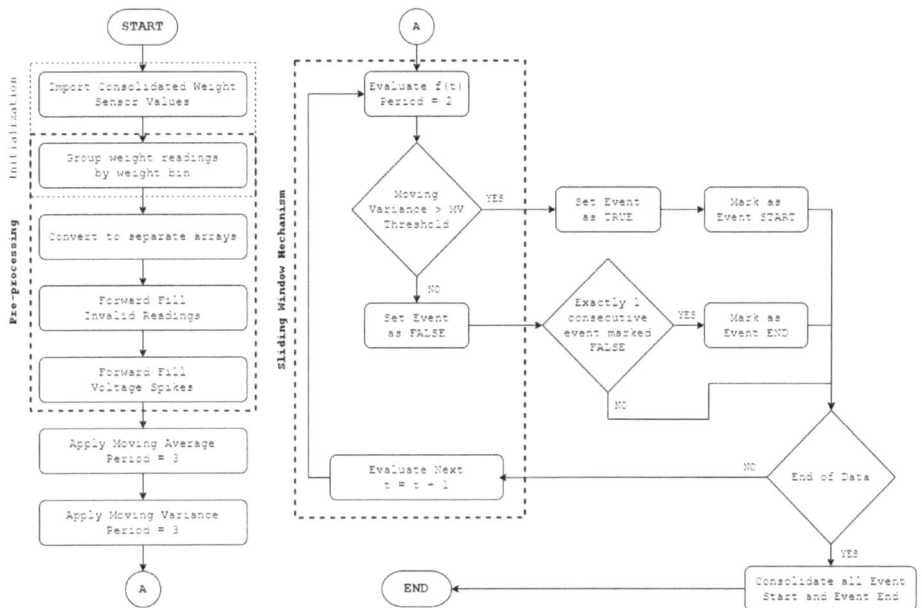

Figure 6. Weight event detection system.

The moving mean μ_t and moving variance v_t calculate the mean and variance along a specific windowing period, respectively. N is the window length and w_t is the weight values at time t. We define the ff as follows:

$$\mu_t = \frac{1}{N} \sum_{i=t-N}^{t} w_i \tag{1}$$

$$v_t = \frac{1}{N} \sum_{i=t-N}^{t} (w_i - \mu_t)^2 \tag{2}$$

Each weight data stream is continuously processed via its moving variance, v_t^λ. The weight change event detection, Δv_t^λ, is shown in Equation (3). It is considered to be detected once a weight data stream, v_t^λ, goes above a pre-defined product threshold, T_p^λ.

$$\Delta v_t^\lambda = \begin{cases} 1 & , v_t^\lambda \geq T_p^\lambda \\ 0 & , v_t^\lambda < T_p^\lambda \end{cases} \tag{3}$$

The product threshold T_p^λ is calibrated based on the product weight per unit and empirical repeated testing conducted with the corresponding weight bin. Table 1 summarizes the thresholds used for each weight bin.

Once a weight change event is detected, it must be further classified as either a customer picking up a product (pick up) or a customer putting back a product (put back).

When the moving variance threshold is surpassed, the algorithm considers the specific weight event involved as a significant weight change that must be classified as a pick up or a put back. With this, a sliding window mechanism is implemented to determine two consecutive events. The start of an event for each product is determined when the moving variance exceeds the moving variance threshold two consecutive times. Likewise, the end of an event for each product is determined when it is below the moving variance threshold two consecutive times. The average readings of the start and the end of an event are then compared. A weight change event would be flagged as a pick up if the reading at the end of the event was less than that at the start of the weight event. On the other hand, a weight event would be flagged as a put back if the reading at the end of the event was greater than that at the start of the weight event.

Table 1. Moving variance threshold for each weight sensor.

Product	Weight Bin	Weight (g)	Threshold, T_p^λ
Piattos	HX1	40	100
Cream-O	HX2	33	100
Whattatops	HX3	35	100
Loaded	HX4	32	100
Bingo	HX5	28	50
Cheesecake	HX6	42	7000
Water	HX7	360	1000
Zesto (Orange)	HX8	200	200
Lucky Me (Beef)	HX9	55	200
Mogu Mogu	HX10	330	8000
Zesto (Apple)	HX11	200	200
Lucky Me (Calamansi)	HX12	80	200

5. Vision System

5.1. Physical Camera

In this study, a webcam was considered. The A4Tech PK910P has a resolution of 720p and served as the primary camera during the testing of the complete, integrated system. Moreover, the live processing of vision data was simplified with this camera. To facilitate this processing, a computer equipped with an Intel Core i7 CPU and a dedicated NVIDIA GeForce RTX 3060 GPU was used.

5.2. Extraction of Multi-Object Trajectory

Primarily, the vision system aims to track and assign an identifier to customers interacting with the shelf. We present the vision system in Figure 7. This process is handled by Keypoint R-CNN [19] and StrongSORT [18].

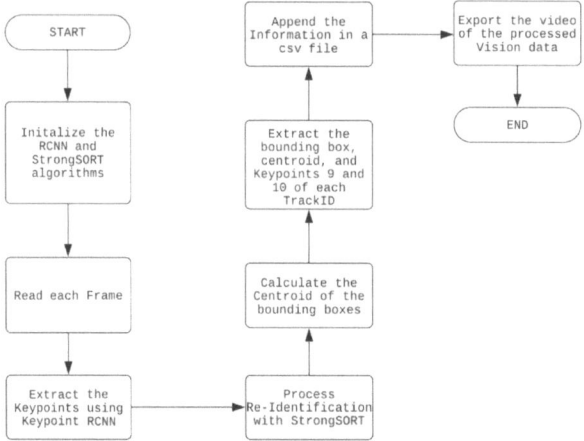

Figure 7. Summary of vision system.

The Keypoint R-CNN used was the built-in PyTorch model (accessed on 7 November 2023 https://pytorch.org/vision/main/models/keypoint_rcnn.html) with ResNet-50 [20] and feature pyramid networks [21] as the backbone. Our Keypoint R-CNN used a pre-trained model from the COCO dataset [22]. Each person identified via the model was assigned to have 17 keypoints. Alongside with these keypoints, the corresponding bounding box was also detected via the model. Only those detections having a confidence score of 80% or higher were considered—this rule was put in place to filter out false positive detections. Each bounding box was assigned an ID through StrongSORT, which is an MOT baseline that utilizes a tracking-by-detection paradigm approach. This tracked and associated objects in a scene through the appearance and velocity of the objects. The most pertinent keypoints from Keypoint RCNN were keypoints 9 and 10 which tracked the left and right hands, respectively. After performing both vision processes, the bounding box information and ID, keypoints 9 and 10, and the timestamps for each frame were exported to a comma separated- values (CSV) file.

Furthermore, the vision system utilized a pre-trained StrongSORT model for assigning each customer a track ID and for Re-ID purposes. Its detector was trained on the CrowdHuman dataset [20,21] and the MOT17 half-training set [20]. The training data were generated by cutting annotated trajectories, not tracklets, with random spatiotemporal noise at a 1:3 ratio of positive and negative samples [20]. Adam was utilized as the optimizer, and cross-entropy loss was utilized as the objective function and was trained for 20 epochs with a cosine annealing learning rate schedule [20]. For the appearance branch, the model was pretrained on the DukeMTMC-reID dataset [20]. In the study, strongSORT Re-ID weights utilized osnet_x_25_msmt17. Additionally, strongSORT was equipped with human pose estimation using a pre-trained Keypoint RCNN, Resnet 50 fpn model. Its purpose was to detect the hands of the customers, which would later be utilized in the integration system to track the trajectories of a customer's hand.

6. PACK-RMPF

6.1. PACK-RMPF Initialization

The fusion of the weight detection system data with the vision data system is performed via the product association with customer keypoints (PACK) through RANSAC modelling and particle filtering (RMPF). This includes the implementation of a particle filter and random sample consensus (RANSAC). For the particle filter, objects are tracked with respect to a landmark. Specifically, the landmarks were placed at the center of each weight bin based on its location in a frame. A sample of the points of each landmark is provided in Figure 8.

Figure 8. Sample of weight bin landmarks.

A pre-defined rectangular area is created for each weight bin, as shown in Figure 9, for which a RANSAC model of all possible locations of the keypoints bounded by the box is created.

Figure 9. Sample of weight bin pre-defined rectangular area.

6.2. PACK-RMPF System and Cross-Location

The PACK-RMPF algorithm associates the customers with the interacted product using the multiple particle filter and RANSAC modeling. The timestamps from the weight and vision data are matched for the purpose of synchronization. Each detected customer's keypoints are filtered in the multiple particle filter where the particles are compared to the RANSAC model. Each detected customer outputs an inlier score. The customer with the highest inlier score is associated with the interacted item. Each block of the PACK-RMPF algorithm shall be discussed in detail in the following sections. Furthermore, a summary of all the blocks is provided in Figure 10.

Figure 10. Summary of the PACK-RMPF algorithm.

6.2.1. Timestamp Matching

As shown in Algorithm 1, the timestamp matching algorithm aims to match a weight event action w_t, through its weight event timestamp t, to vision attributes V_τ and its event timestamp τ. The vision attributes include the track ID information with its corresponding bounding box and left and right hand keypoint coordinates. The algorithm iterates based on the length of the weight event actions w_t. The timestamp matching algorithm begins by calculating the $\Delta time$ of w_t. The following is the formula for $\Delta time$:

$$\Delta time = t - 3.5 \text{ seconds} \qquad (4)$$

Then, the algorithm tries to find where the timestamps of t and τ are equal—or if there are no equal timestamps—within the 500 ms latency. Then, from Equation (1), all vision attributes associated with a vision timestamp V_t within the range of the $\Delta time$ are selected. The selected V_t are stored alongside τ, t, and w_t for processing in the multiple particle filter block.

Algorithm 1: Timestamp Matching Algorithm

1: **Input:** weight event timestamp (t), weight event action (w_t), vision event timestamp (τ), vision Attributes (V_τ)
2: **Output:** matched weight event timestamp with vision event timestamp (M_τ)
3: **For Each** iterations conducted
4:     Calculations are made as follows:
5:     Find where $t = \tau$ or τ is within 500 ms of w_t
6:     Store w_t and selected vision attributes in one M_τ array
7:     Store t, w_t, and selected vision attributes in one M_τ array
8: **End**

6.2.2. Multiple Particle Filter

The purpose of the multiple particle filter is to be able to track the trajectory of the bounding box and hand keypoints. The knowledge of the trajectory of the system is what enables PACK-RMPF to be able to determine events where cross-location occurred. Algorithm 2 describes the system flow of the particle filter.

Algorithm 2: Multiple Particle Filter Algorithm

1: **Input:** (M_τ) containing matched vision attributes (V_τ) : customer track ID ($V_{\tau T\_ID}$), bounding box coordinates ($V_{\tau bbox}$), Keypoint 9 coordinates ($V_{\tau K9}$), Keypoint 10 coordinates ($V_{\tau K10}$)
2: **Output:** Bounding box particles (P_{bbox}), Keypoint 9 particles (P_{K9}), Keypoint 10 particles (P_{K10}) of each track ID ($V_{\tau T\_ID}$)
3: **For** len(M_τ) iterations, follow the following procedure:
4:     Extract the vision attributes inside M_τ
5:     **If** V_τ^1, **then**
6:         Initialize random particles P with coordinates (P_x, P_y) based on the initial coordinates of V_τ^1
7:         Compute the distance between the landmark and initial coordinates V_τ^1
8:         Move the particles diagonally by adding a diagonal distance to the particles
9:         Calculate for the weights of each particle
10:        Apply systemic resampling
11:        Store particles (P) to global particles repository
12:     **Otherwise**
13:        Extract global particles (P) containing current P_{bbox}, P_{K9}, P_{K10}
14:        Compute the distance between the landmark and the current coordinates V_τ^n
15:        Move the particles diagonally by adding a diagonal distance to the particles
16:        Calculate the weights of each particle
17:        Apply systemic resampling
18:        Store particles (P) to global particles repository
19: **End**
20: Collect final particles (P) containing final P_{bbox}, P_{K9}, P_{K10}

The particle filter process can be summarized into three steps. These steps are predict, update and resample. The predict step is used to predict the trajectory of the keypoints [22]. For the context of PACK-RMPF, the predict step begins by generating uniform random particles based on the initial position of the vision attributes, V_τ, consisting of the coordinates of the position of the bounding box and the hands. Then, each particle is moved, with one unit moved to the x axis and one unit moved to the y axis.

Then, the particles are updated based on their position relative to the weight bin landmarks. The following equation was utilized to calculate the position of the particles with respect to the landmarks [22]:

$$\text{distance} = \sqrt{(V_{Ax} - x_l)^2 + (V_{Ay} - y_l)^2} \qquad (5)$$

Figure 11. Weight event detection with quick pick ups and putbacks.

For each weight bin, there were minimal false or error weight readings. Weight Bin 7, corresponding to the platform with water, exhibited the most errors at 0.10% with 10 false readings among 9,753 data points. On average, 0.03% of the readings per weight bin returned were false. Evaluating each run, as low as 0.005866% and not more than 0.01725% of readings returned false. The overall performance report of each weight bin is found in Table 2.

Table 2. Overall performance report of each weight bin.

Weight Bin	Ratio	False Readings	Total Count
HX1	0.00%	0	9836
HX2	0.04%	4	9875
HX3	0.01%	1	9881
HX4	0.02%	2	9797
HX5	0.02%	2	9828
HX6	0.04%	4	9836
HX7	0.10%	10	9753
HX8	0.03%	3	9759
HX9	0.03%	3	9839
HX10	0.03%	3	9846
HX11	0.01%	1	9835
HX12	0.01%	1	9855
Average	0.03%	2.833333333	9828.333333

7.3. Vision System Results

The vision system plays an important role in identifying the customers interacting with the shelf. The vision system assigns each customer a track ID. The track ID serves as the identifier and is used to associate a weight event with a customer.

Figure 12 illustrates that the vision system could reliably track up to three customers at a time roaming around the double-sided shelf. It was possible for the system to detect and track upward of five people at a time, even without pre-training the StrongSORT model. However, adjustments to StrongSORT parameters, a different camera angle, or multiple cameras may still be needed to improve reliability.

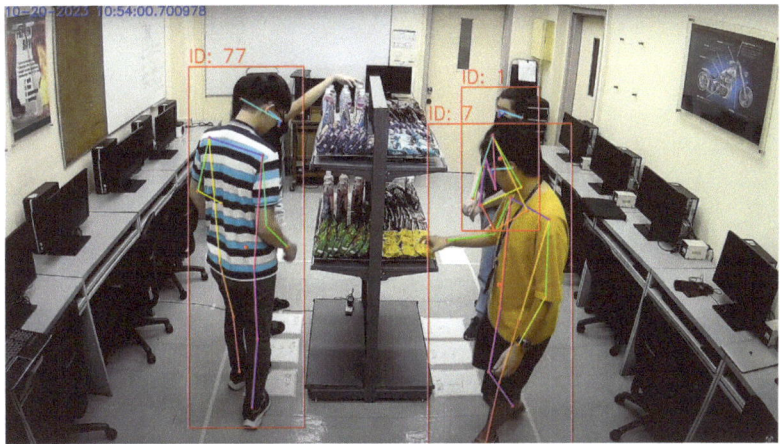

Figure 12. Three people tracked while interacting with the shelf.

Based on these results, the developed smart shelf system reliably accommodates up to two customers at a time. It must be noted that the reliability was greater in instances where only one customer at a time was present on each side of the shelf. In cases where adjacent occlusion was minimal or did not occur around the time that weight events occurred, up to three customers could be detected, tracked, and associated with some reliability.

7.4. PACK-RMPF Simulated Supermarket Setup Results

The simulated supermarket setup aims to check the system performance of PACK-RMPF. PACK-RMPF tracks the bounding box and the hands of each customer. These keypoints play a big role in associating product movement with the customers in the frame. Furthermore, tracking these keypoints is what enables PACK-RMPF to associate a product with a customer during scenarios with cross-location. Hence, the proponents performed experimentation and tried gauging the performance of PACK-RMPF if PACK-RMPF only tracked (1) the bounding box, (2) the hands, and (3) both the bounding box and the hands of each customers.

The previous literature has added emphasis on tracking head keypoint [11]. However, PACK-RMPF gives weight on tracking the centroid of the bounding box. Table 3 illustrates the average percent of associations per run when PACK-RMPF is calibrated to track only the bounding box of each customer.

Table 3. Average percent of associations per run of the bounding box approach of PACK-RMPF.

Run	Duration	Average Percent of Correct Association
1	30 min	74.76%
2	30 min	72.79%
3	30 min	75.77%
4	30 min	91.67%
5	60 min	78.36%
Overall Average Percent of Association		78.67%

The iteration of each run consisted of scenarios where (1) there were multiple participants interacting with the shelf and (2) there was a single participant interacting with the shelf. Overall, PACK-RMPF, when tracking only the bounding box of the customer, yields an overall average percent association of 78.67%.

In the previous literature, tracking only the head has worked well when there was only one customer interacting with the shelf. The accuracy could reach as high as 100% [11].

PACK-RMPF when tracking the bounding box also reaches this same accuracy in single customer scenarios. However, incorrect associations increase as the number of participants interacting with the shelf increases. In the case of PACK-RMPF, the overall percent of association—only considering the scenarios with multiple customers—yields an average percent of association of around 57%.

The density of people interacting with the shelf affects the system performance of smart shelves due to occlusions [11]. Similarly, this was the case with PACK-RMPF. Correct associations considered whether the expected weight events triggered by a customer were associated with that customer. Upon analyzing the integrated data and reviewing the camera footage, false associations were mainly attributed to instances of adjacent occlusions—where customers were occluded from the camera's point of view while another customer is on the same side of the shelf—due to customer stalling or customers entering in groups.

In the example shown in Figure 13, the keypoints of the occluded customer were able to be determined by the system. However, the tracked bounding box was not assigned since the confidence score threshold was not met. Although there were instances where the customer was not occluded, factors such as the 3.5 s interval considered by the integration system typically led to a higher RANSAC inlier score for the non-occluded customer standing closer to the camera.

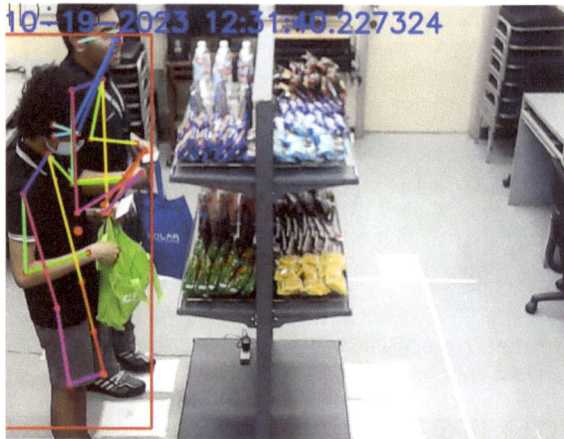

Figure 13. Example of adjacent occlusion.

Similarly, the same trend, as illustrated in Table 4, can also be seen with PACK-RMPF that only tracks the hands of the participants.

Table 4. Average percent of associations per run of the hands approach of PACK-RMPF.

Run	Duration	Average Percent of Correct Association
1	30 min	78.36%
2	30 min	81.97%
3	30 min	66.92%
4	30 min	87.77%
5	60 min	75.25%
Overall Average of Percent Association		78.05%

The hands approach of tracking in PACK-RMPF garnered an almost similar 78% overall average percent of association. But, what sets the hands approach of PACK-RMPF apart is the fact that it was able to garner an average percent of association of 58% in the multiple people tracking scenario. Even if the hands were occluded from the shelf, the system was still able to garner a higher score compared to previous approaches that only garnered 40% accuracy [11]. Similarly, the probable source of inaccuracies in associating the customer with the product is adjacent occlusion which was discussed in the previous paragraph.

Finally, Table 5 illustrates the results of using PACK-RMPF as a whole system, which combines both hand tracking and bounding box tracking, achieved an average of 76.33% correct associations of customers with weight events per run.

Table 5. Average percent of associations per run of PACK-RMPF.

Run	Duration	Average Percent of Correct Association
1	30 min	74.22%
2	30 min	77.55%
3	30 min	68.17%
4	30 min	86.11%
5	60 min	75.62%
Overall Average Percent of Association		76.33%

7.5. PACK-RMPF with Cross-Location Results

A persistent limitation of threshold-based approaches in related studies that sought to associate customers with weight sensor data is the instance wherein customers interact with products that they were not standing directly in front of [11]. This includes the picking up or putting back of products on the same side of the shelf. Figure 14 provides a visual example of an instance of a cross-location pick up.

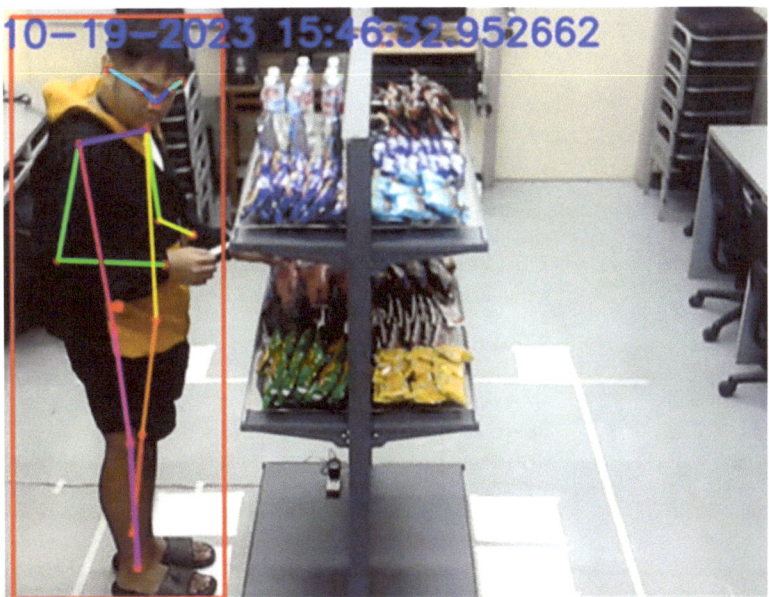

Figure 14. Example of cross-location pick up.

In this study, these instances are referred to as cross-location events. By utilizing a particle filter to keep track of the trajectory of the left and right wrists of each customer, along with the assignment of the RANSAC inlier scores for the tracked keypoints, this problem was addressed. Table 6 illustrates the performance of this system in terms of associating customers and products with cross-location.

Table 6. PACK-RMPF cross-location results.

Run	Duration	Number of Items with Cross-Location	Number of Items Correctly Identified	Percent of Correct Association with Cross-Location
1	30 min	5	5	100%
2	30 min	2	2	100%
3	30 min	5	3	60%
4	30 min	4	3	75%
5	60 min	5	3	60%
	Overall Average Percent of Association			79%

Overall, the system was able to achieve an average of 79% correct associations across all runs. Similarly, the deviations in the percent of correct association could be associated with adjacent occlusion. Adjacent occlusion is the occlusion of vision data needed to process and gather pieces of information, such as the location of the keypoints and the bounding box with respect to the shelf. These pieces of information are crucial in the processing of particle filters and RANSAC. Despite this, the percent of correct association per run was at least 60%.

8. Real-World Business Use for Case-Enhancing Retail Experience and Optimization

The smart shelf system offers enormous potential for improving customer experience and optimizing operations in brick-and-mortar retail stores.

8.1. Inventory Management and Product Placement

The real-time tracking of inventory levels, facilitated by the weight sensors on shelves, detect stock levels and can assist in preventing inventory shortage. This solution reduces the costs associated with manual stock checking labor. Moreover, strategic product placement is enabled by the analysis of customer interaction data, identifying high-traffic areas and providing insights into product appeal. For example, high velocity impulse purchase items can be strategically placed near checkout counters to maximize sales, or slow-moving items could be placed in locations that would improve their velocity.

For the proposed system, the expected replacement policy is that a member of the supermarket staff is responsible for handling item replacements requested by the customer.

Furthermore, to optimize product placement, the ability of the system to detect pickups and putbacks at specific locations of the smart shelf enables the identification of hotspots where customers frequently interact with the shelf. These hotspots can be helpful in formulating data-driven product selection and placement strategies leading to product purchases.

8.2. Operational Efficiency

The integration of the smart shelf system with existing retail management systems, such as point-of-sale and inventory management, could facilitate the improvement of data-driven decision making across various business operations [24]. Sale trends and interaction data could be used to optimize purchase decisions and supply chain activities, thereby contributing to more efficient operational processes.

Hence, the ability of the smart shelf to track the frequency of pickups and putbacks gives retailers an insight into customer patterns. For example, the data on the frequency of pickups and putbacks can give the retailer an insight into which products are frequently picked up and returned. Then, from these patterns, they can deduce plans or strategies that could help facilitate a higher probability of purchase for products.

Moreover, the smart shelf system enables retailers to undertake a data-driven approach in optimizing their product placement and strategy. By continuously monitoring pickups and putbacks, the retailers are equipped with a closed-loop cycle of formulated strategies and can execute a new strategy, evaluate against conventional placements, and re-formulate strategies.

8.3. Addressing Challenges and Considerations

While the potential of such smart retail technologies is promising, careful consideration must be given to customer privacy and the perception of surveillance [25]. Establishing clear policies and maintaining transparency with customers about data collection and usage is critical. In addition, current brick-and-mortar retail businesses could face technical challenges in the adoption of these AI solutions.

9. Conclusions

Weight sensor and computer vision systems were successfully integrated as a smart shelf system. Through data processing and sensor fusion, weight change events detected through the weight system could be associated with a specific customer to help characterize touch-based customer behavior and generate a log of customer–product interactions. The use of a network time protocol server allowed timestamps of both systems to be matched, which then enabled the integration of their data. The weight system performed well with an average of 98% correct event detections and a maximum of 0.01725% invalid readings in a single run. It was also able to detect events that happened in quick succession of each other. The vision system had particularly good detection rates when handling up to three customers at a time. Customers were reidentified by the system after short periods of occlusion. The particle filter and RANSAC implemented in the integration system were able to associate the weight change events with each customer, with a recall rate of 68% or higher. Furthermore, the problem of cross-location was resolved, which can be exemplified by the 79% overall recall rate of the system with cross-location.

Despite the overall success, certain challenges were noted. These include customer stalling and larger groups on the vision data leading to relatively low detection percentages for specific customers. Recommendations for improvement include considering alternative materials and components for shelf and prototype construction, optimizing load cell configurations for weight distribution, and refining wiring through proper PCB implementation. Prospects involve scaling up the system to accommodate additional shelves, exploring the use of multiple cameras in terms of reliability and scalability, and investigating real-time data processing for enhanced system responsiveness. Furthermore, future studies could take the opportunity to fine tune the PACK-RMPF system so that it can take advantage of tracking both the bounding box and the hands, especially in scenarios where multiple people interact with the shelf. Moreover, future experimentation involving multiple product pickups and putbacks is also recommended. Use cases of the smart shelf system could be deployed in small-scale grocery stores.

Author Contributions: Vision system conceptualization, C.J.B.B., A.E.B.U. and J.C.T.V.; sensor system conceptualization, C.J.B.B., M.T.C.R. and J.C.T.V.; integration system conceptualization, C.J.B.B. and J.A.C.J.; vision programming, C.J.B.B. and A.E.B.U.; hardware programming, R.J.T., M.T.C.R. and A.E.B.U.; hardware prototyping, R.J.T., M.T.C.R. and J.C.T.V.; hardware validation, M.T.C.R. and J.C.T.V.; equipment provision, R.J.T.; data acquisition, C.J.B.B., M.T.C.R., A.E.B.U. and J.C.T.V.; writing—original draft preparation, C.J.B.B., M.T.C.R., A.E.B.U. and J.C.T.V.; writing—review and editing, M.T.C.R., J.C.T.V., J.A.C.J., R.J.T., E.C.G. and J.S.J.; supervision, J.A.C.J., R.J.T., E.S. and E.C.G. All authors have read and agreed to the published version of the manuscript.

Funding: This research was funded by the Research and Grants Management Office Project No. 21 IR S 3TAY20-3TAY21 of De La Salle University.

Institutional Review Board Statement: The study was conducted in accordance with the Declaration of Helsinki, and approved by the ethics review committee of Velez College with accreditation number PHREB L2-2019-032-01 approved on 13 October 2021.

Informed Consent Statement: Informed consent was obtained from all subjects involved in the study.

Data Availability Statement: Data are contained within the article.

Conflicts of Interest: The funders had no role in the design of the study; in the collection, analyses, or interpretation of data; in the writing of the manuscript; or in the decision to publish the results.

References

1. Chandramana, S. Retail Analytics: Driving Success in Retail Industry with Business Analytics. *Res. J. Soc. Sci. Manag.* **2017**, *7*, 159–166. [CrossRef]
2. Hickins, M. What Is Retail Analytics? The Ultimate Guide. Oracle. Available online: https://www.oracle.com/in/retail/what-is-retail-analytics/ (accessed on 17 September 2023).
3. Kolias, G.D.; Dimelis, S.P.; Filios, V.P. An empirical analysis of inventory turnover behaviour in Greek retail sector: 2000–2005. *Int. J. Prod. Econ.* **2011**, *133*, 143–153. [CrossRef]
4. Pascucci, F.; Nardi, L.; Marinelli, L.; Paolanti, M.; Frontoni, E.; Gregori, G.L. Combining sell-out data with shopper behaviour data for category performance measurement: The role of category conversion power. *J. Retail. Consum. Serv.* **2022**, *65*, 102880. [CrossRef]
5. Anoos, J.M.M.; Ferrater-Gimena, J.A.O.; Etcuban, J.O.; Dinauanao, A.M.; Macugay, P.J.D.R.; Velita, L.V. Financial Management of Micro, Small, and Medium Enterprises in Cebu, Philippines. *Int. J. Small Bus. Entrep. Res.* **2020**, *8*, 3–76. [CrossRef]
6. Venkatesh, V.; Speier-Pero, C.; Schuetz, S. Why do people shop online? A comprehensive framework of consumers' online shopping intentions and behaviors. *Inf. Technol. People* **2022**, *35*, 1590–1620. [CrossRef]
7. Zhang, Y.; Xiao, Z.; Yang, J.; Wraith, K.; Mosca, P. A Hybrid Solution for Smart Supermarkets Based on Actuator Networks. In Proceedings of the 2019 7th International Conference on Information, Communication and Networks (ICICN), Macao, China, 24–26 April 2019; pp. 82–86. [CrossRef]
8. Ruiz, C.; Falcao, J.; Pan, S.; Noh, H.Y.; Zhang, P. AIM3S: Autonomous Inventory Monitoring through Multi-Modal Sensing for Cashier-Less Convenience Stores. In Proceedings of the 6th ACM International Conference on Systems for Energy-Efficient Buildings, Cities, and Transportation, New York, NY, USA, 13–14 November 2019; pp. 135–144. [CrossRef]
9. Xu, R.; Nikouei, S.Y.; Chen, Y.; Polunchenko, A.; Song, S.; Deng, C.; Faughnan, T.R. Real-Time Human Objects Tracking for Smart Surveillance at the Edge. In Proceedings of the 2018 IEEE International Conference on Communications (ICC), Kansas City, MO, USA, 20–24 May 2018; pp. 1–6. [CrossRef]
10. "AiFi". AiFi. Available online: https://aifi.com/platform/ (accessed on 18 November 2023).
11. Falcao, J.; Ruiz, C.; Bannis, A.; Noh, H.; Zhang, P. ISACS: In-Store Autonomous Checkout System for Retail. *Proc. ACM Interact. Mob. Wearable Ubiquitous Technol.* **2021**, *5*, 1–26. [CrossRef]
12. Völz, A.; Hafner, P.; Strauss, C. Expert Opinions on Smart Retailing Technologies and Their Impacts. *J. Data Intell.* **2022**, *3*, 278–296. [CrossRef]
13. Higa, K.; Iwamoto, K. Robust Shelf Monitoring Using Supervised Learning for Improving On-Shelf Availability in Retail Stores. *Sensors* **2019**, *19*, 2722. [CrossRef] [PubMed]
14. Moorthy, R.; Behera, S.; Verma, S. On-Shelf Availability in Retailing. *Int. J. Comput. Appl.* **2015**, *115*, 47–51. [CrossRef]
15. Lin, M.-H.; Sarwar, M.A.; Daraghmi, Y.-A.; İk, T.-U. On-Shelf Load Cell Calibration for Positioning and Weighing Assisted by Activity Detection: Smart Store Scenario. *IEEE Sens. J.* **2022**, *22*, 3455–3463. [CrossRef]
16. Martino, L.; Elvira, V. Compressed Monte Carlo with application in particle filtering. *Inf. Sci.* **2021**, *553*, 331–352. [CrossRef]
17. Ullah, I.; Shen, Y.; Su, X.; Esposito, C.; Choi, C. A Localization Based on Unscented Kalman Filter and Particle Filter Localization Algorithms. *IEEE Access* **2020**, *8*, 2233–2246. [CrossRef]
18. Du, Y.; Zhao, Z.; Song, Y.; Zhao, Y.; Su, F.; Gong, T.; Meng, H. StrongSORT: Make DeepSORT great again. *IEEE Trans. Multimed.* **2023**, *25*, 8725–8737. [CrossRef]
19. He, K.; Gkioxari, G.; Dollar, P.; Girshick, R. Mask R-CNN. *IEEE Trans. Pattern Anal. Mach. Intell.* **2020**, *42*, 386–397. [CrossRef]
20. He, K.; Zhang, X.; Ren, S.; Sun, J. Deep Residual Learning for Image Recognition. In Proceedings of the 2016 IEEE Conference on Computer Vision and Pattern Recognition (CVPR), Las Vegas, NV, USA, 27–30 June 2016; pp. 770–778. [CrossRef]
21. Lin, T.-Y.; Dollár, P.; Girshick, R.; He, K.; Hariharan, B.; Belongie, S. Feature Pyramid Networks for Object Detection. In Proceedings of the 2017 IEEE Conference on Computer Vision and Pattern Recognition (CVPR), Honolulu, HI, USA, 21–26 July 2017; pp. 936–944. [CrossRef]
22. Lin, T.-Y.; Maire, M.; Belongie, S.; Hays, J.; Perona, P.; Ramanan, D.; Dollár, P. Microsoft COCO: Common Objects in Context. In Proceedings of the Computer Vision—ECCV 2014: 13th European Conference, Zurich, Switzerland, 6–12 September 2014; Springer: Cham, Swizerland, September 2014; pp. 740–755.
23. Khair, U.; Fahmi, H.; Hakim, S.A.; Rahim, R. Forecasting Error Calculation with Mean Absolute Deviation and Mean Absolute Percentage Error. *J. Phys. Conf. Ser.* **2017**, *930*, 012002. [CrossRef]
24. Sharma, P.; Shah, J.; Patel, R. Artificial Intelligence Framework for MSME Sectors with Focus on Design and Manufacturing Industries. *Mater. Today Proc.* **2022**, *62*, 6962–6966. [CrossRef]
25. Sariyer, G.; Kumar Mangla, S.; Kazancoglu, Y.; Xu, L.; Ocal Tasar, C. Predicting Cost of Defects for Segmented Products and Customers Using Ensemble Learning. *Comput. Ind. Eng.* **2022**, *171*, 108502. [CrossRef]

Disclaimer/Publisher's Note: The statements, opinions and data contained in all publications are solely those of the individual author(s) and contributor(s) and not of MDPI and/or the editor(s). MDPI and/or the editor(s) disclaim responsibility for any injury to people or property resulting from any ideas, methods, instructions or products referred to in the content.

Article

Research on an Algorithm of Express Parcel Sorting Based on Deeper Learning and Multi-Information Recognition

Xing Xu [1], Zhenpeng Xue [1] and Yun Zhao [2,*]

[1] School of Mechanical and Energy Engineering, Zhejiang University of Science and Technology, Hangzhou 310023, China
[2] School of Information and Electronic Engineering, Zhejiang University of Science and Technology, Hangzhou 310023, China
* Correspondence: zy_super0201@163.com

Abstract: With the development of smart logistics, current small distribution centers have begun to use intelligent equipment to indirectly read bar code information on courier sheets to carry out express sorting. However, limited by the cost, most of them choose relatively low-end sorting equipment in a warehouse environment that is complex. This single information identification method leads to a decline in the identification rate of sorting, affecting efficiency of the entire express sorting. Aimed at the above problems, an express recognition method based on deeper learning and multi-information fusion is proposed. The method is mainly aimed at bar code information and three segments of code information on the courier sheet, which is divided into two parts: target information detection and recognition. For the detection of target information, we used a method of deeper learning to detect the target, and to improve speed and precision we designed a target detection network based on the existing YOLOv4 network, Experiments show that the detection accuracy and speed of the redesigned target detection network were much improved. Next for recognition of two kinds of target information we first intercepted the image after positioning and used a ZBAR algorithm to decode the barcode image after interception. The we used Tesseract-OCR technology to identify the intercepted three segments code picture information, and finally output the information in the form of strings. This deeper learning-based multi-information identification method can help logistics centers to accurately obtain express sorting information from the database. The experimental results show that the time to detect a picture was 0.31 s, and the recognition accuracy was 98.5%, which has better robustness and accuracy than single barcode information positioning and recognition alone.

Keywords: deeper learning; multiple information fusion; YOLOv4; express sorting; information to identify

Citation: Xu, X.; Xue, Z.; Zhao, Y. Research on an Algorithm of Express Parcel Sorting Based on Deeper Learning and Multi-Information Recognition. *Sensors* **2022**, *22*, 6705. https://doi.org/10.3390/s22176705

Academic Editors: Yufeng Zheng and Erik Blasch

Received: 9 August 2022
Accepted: 31 August 2022
Published: 5 September 2022

Publisher's Note: MDPI stays neutral with regard to jurisdictional claims in published maps and institutional affiliations.

Copyright: © 2022 by the authors. Licensee MDPI, Basel, Switzerland. This article is an open access article distributed under the terms and conditions of the Creative Commons Attribution (CC BY) license (https://creativecommons.org/licenses/by/4.0/).

1. Introduction

With the rapid development of the world economy, people's living standards are improving day by day; At the same time, the rapid development of the internet enables more and more consumers to choose convenient online shopping. According to the statistics of the China Post Bureau, the business volume of express service enterprises in China has reached 108.30 billion, up 29.9% year on year, and business revenue has reached 1033.23 billion yuan, up 17.5% year on year. In 2021, because of the increase of express business, this has put existing logistics systems to a huge test. At present, the sorting of express delivery is mainly by a courier sheet that can be divided into automatic sorting, semi-automatic sorting and manual sorting. Automatic sorting is by use of infrared bar code detection based on radio frequency identification (rfid) technology for delivery information [1]. This method is costly, difficult to popularize, and mainly used in large-scale logistics express sorting centers [2]. Semi-automatic sorting involves a semi-automatic sorting machine based on machine vision for sorting. The staff put the courier sheet upward and then put it on a conveyor belt. A camera above the conveyor belt acquires pictures of the courier sheet, and then the central processor identifies the barcode information to generate an electrical signal and control the conveyor belt to send the express to different areas. Manual sorting done directly by a labor force, which is inefficient. With the continuous development of smart logistics and the limitations of cost, most small logistics sorting centers now adopt a semi-automatic sorting method. However, due to the limitations of the sorting environment and equipment, barcodes cannot normally be recognized in the identification process, which requires manual intervention to increase sorting efficiency.

The courier sheet mainly contains a one-dimensional barcode, three segments of express code and the user's personal information. The one-dimensional bar code is economical, time-dependent and has abundant logistics information. When this semi-automatic sorting method scans the one-dimensional bar code for sorting, the bar code can be disturbed obscured by stains and distortion after layer-by-layer sorting, which seriously affects express sorting. The light in the warehouse, creases and other factors affect detection of the barcode. An express three-segment code is composed of characters with larger fonts, and each segment represents different information, including cities, outlets and salesmen. However, the three-segment code is also easily affected by the characteristics of its surface, resulting in low accuracy in detecting the three-segment code area. Therefore, in the process of express sorting, the positioning of target information is key to sorting an determines its efficiency. Existing sorting methods based on single information on the courier sheet can be divided into two categories. One is a method based on traditional digital image processing. For example, Huang et al. [3] used Halcon visual recognition technology to extract single three-segment code information from the courier sheet with good accuracy but high cost. Weihao et al. [4] used a Hough transform to detect regions containing barcodes. Katona M [5] used algorithms based on morphological operations to detect barcodes, and an improved version [6] used Euclidean distance maps to match barcode candidates. However, such methods are subject to environmental factors and depend on digital image processing, so it is difficult to detect barcodes accurately and efficiently in complex environments. Another method is based on deeper learning [7]. With the continuous development of convolutional neural networks in object detection, these have been widely used in sorting. Zamberletti A et al. [8] first used a deep neural network for barcode detection, but it was not very effective in practice because it was developed using experimental images. Kolekar A and Ren Y [9,10] used the deeper learning detector of SSD for barcode detection, and achieved good performance under a complex background. Li J et al. [11] used the Faster R-CNN network for barcode detection and achieved better detection results with higher accuracy and strong robustness. Z. Pan [12] used a YOLO algorithm for detection of packages when dealing with the problem of express stacking. R. Shashidhar [13] used a YOLOv3 model and OCR to detect and recognize license plates and achieved good results. Methods based on traditional digital processing are easily affected by environmental factors when detecting the target area, which leads to inaccurate detected areas and may cause recognition errors. The use of deeper learning methods to detect the target area has advantages. In image recognition, a convolutional neural network is used to learn various features of the target, which is more robust in target detection. In this regard, considering the complexity of the express sorting environment, we decided to use a deeper learning method to detect the target information. At the same time, we considered that it would be too simple to select only certain information as the sorting information in the real-time sorting process, and the robustness would be poor. In a complex logistics environment, false detection will occur, which will affect sorting efficiency. Therefore, we chose the multi-information method t better guarantee the accuracy of identification.

In summary, we propose a deeper learning-based multi-information fusion method for courier sheet recognition. The method is mainly divided into two stages, one being the positioning of the one-dimensional barcode and the three-segment code on the courier sheet, and the other the decoding of the barcode and the recognition of the three-segment code. As shown in Figure 1, the images are first input into the target detection network for positioning of the two kinds of information. In enable accurate detection in complex environments, we redesigned the key positioning network, which was optimized based on YOLOv4, and ensured the speed and accuracy of the optimized network. The network includes the backbone feature extraction network of YOLOv4, the spatial pooling layer after adding the cross-stage module, the attention module SE, and the use of FPN structure. As a backbone feature extraction network, CSPDarKet53 can ensure accuracy and greatly reduce the number of parameters. A spatial pooling layer with inter-phase modules was used instead of the original spatial pooling layer, which helped to ensure the accuracy and reduce the number of parameters. The attention module SE was added to enhance features that improved the accuracy of detection. Replacing the original structure with an FPN structure effectively reduces the network complexity and parameter number, and the optimized positioning network is much better than the YOLOv4 network. The next step is the recognition of target information. First, the rectangular boxes containing barcode information and three-segment code information are captured, and the pictures containing barcode information are decoded and output by the ZBAR algorithm. The characters of the three-segment code information box are recognized by Tesseract-OCR text recognition. The string information of the two is written into the text so that the express sorting information can be accurately obtained from the database. Our main contributions are as follows.

- We propose a multi-information fusion courier sheet recognition method instead of single information target recognition in the process of express sorting to improve the recognition rate of sorting.
- The YOLOv4 target detection model was optimized for target information positioning. Compared with other detection networks, the performance of courier sheet detection is more powerful.

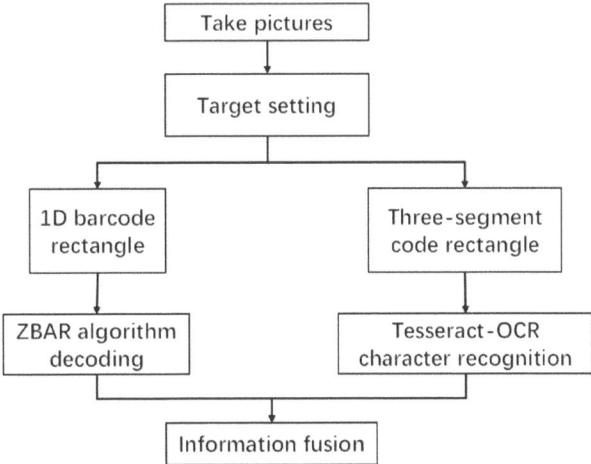

Figure 1. Method structure diagram.

The rest of this article is as follows. Section 2 describes the target detection network used by the method. Section 3 mainly describes the recognition method of target information. In Section 4, experiments are described to verify the reliability of the detection model and the recognition algorithm, and the whole method is evaluated to verify the feasibility of the proposed method. Section 5 provides the conclusion.

2. Target Detection Network

Widely used target detection algorithms in current target detection tasks are all based on deep convolutional neural networks [14] that can learn features from a large amount of data. At present, detection is mainly divided into two-stage with detectors such as R-CNN,

Fast R-CNN, Faster R-CNN [15–17] and single-stage detectors such as YOLO [18] series and SSD [19]. The output of the single-stage detector only needs a CNN operation to obtain the result directly. The two-stage detector needs to be divided into two steps. The first step is to perform a simple CNN operation, and the second step is to score the results obtained in the first step. Then, the candidate regions with high scores are input into CNN for final prediction. Because of the existence of candidate regions, the two-stage detector has high accuracy but is not as fast as the single-stage detector. Therefore, for fast real-time target detection, a single-stage detector is preferred. Whether target detectors are efficient free (e.g., CenterNet [20]) or anchor based (e.g., EfficientDet and YOLOv4. [21,22]) divides them into two types based on anchor points. The biggest advantage of the former is that the speed of the detector is very fast, and there is no need for preset anchor and direct regression, which greatly reduces time consumption and computational power. The latter has higher accuracy and can extract richer features, but it takes more time and computational power. Therefore, our research considers this selection.

YOLOv4 was improved on the basis of YOLOv3 [23]. As an efficient and powerful target detection model, it takes into account both speed and accuracy. It is mainly composed of three parts: a feature extraction network, Backbone; a Neck for feature fusion, and a detection Head, Yolo Head, for classification and regression operation.

As shown in Figure 2, a picture of the courier sheet captured is input into the YOLOv4 network. The network first adjusts the picture to the size of [3, 416, 416], and then trunk feature extraction network CSPDarknet53 extracts target features. A shallow feature map, deep feature number, and a deep feature map are introduced into the Neck part. After using the SPP structure to enhance the receptive field on the deep feature map, the three feature maps are put into the path aggregation network PANet [24] to extract features repeatedly, and finally into the Yolo Head. The image can be divided into [52, 52, N], and [13, 13, N] feature maps of different sizes for detection of large targets, medium targets and small targets, where $N = 3 \times (5 + C)$, which depends on the model category.

The loss function of YOLOv4 can be divided into three parts: confidence loss L_{conf}, classification loss l_{class}, and regression frame loss l_{CoU}. L_{conf} and l_{class}. These are expressed in Equations (1) and (2).

$$L_{conf} = \sum_{i=0}^{s^2}\sum_{j=0}^{B} I_{ij}^{obj}(C_i - \hat{C}_i)^2 + \lambda_{noobj}\sum_{i=0}^{s^2}\sum_{j=0}^{B} I_{ij}^{noobj}(C_i - \hat{C}_i)^2 \qquad (1)$$

$$l_{class} = \sum_{i=0}^{s^2} I_i^{obj} \sum_{c \in classes} (p_i(c) - \hat{p}_i(c))^2 \qquad (2)$$

Regression box loss represents the error between the prediction box and the real box. To ensure more accurate calculation results, several aspects are considered, including the overlapping area of the detection frame, the distance of the center point, and the length-width ratio. Regression box loss l_{CoU} formula is shown in Equation (5).

$$v = \frac{4}{\pi^2}\left(\arctan\frac{w^{gt}}{h^{gt}} - \arctan\frac{w}{h}\right)^2 \qquad (3)$$

$$\alpha = \frac{v}{(1 - \text{IoU}) + V} \qquad (4)$$

$$l_{CoU} = 1 - \text{IoU} + \frac{d^2}{c^2} + \alpha v \qquad (5)$$

Note that α and v are penalty terms for the aspect ratio, w^{gt} and h^{gt} are the width and height of the real box, w and h are the width and height of the predicted box, d is the Euclidean distance between the two center points, and c is the diagonal distance of the closure.

The loss function of YOLOv4 is expressed in Equation (6).

$$loss(object) = l_{CoU} - l_{conf} - l_{class} \qquad (6)$$

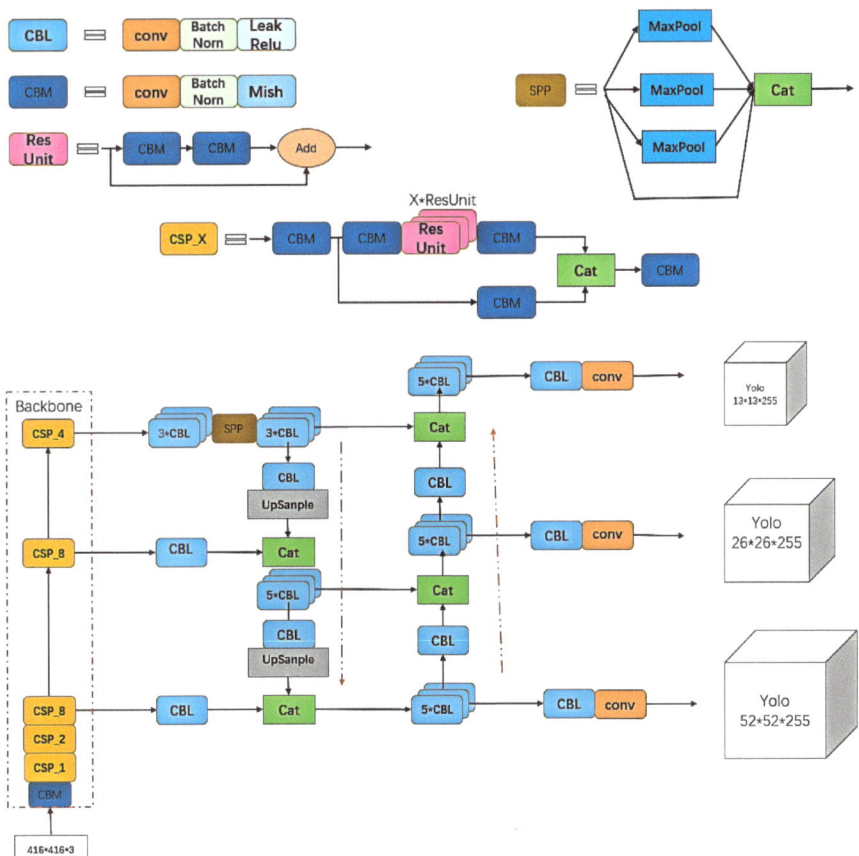

Figure 2. YOLOv4 network.

2.1. SPP Module of Csp Modularization

In deeper learning, the high-level network layer has a large receptive field, so it has a strong ability to represent semantic information. However, the feature map has low resolution and poor ability to represent spatial information. The receptive field ratio of low layer network layer is small, in contrast to that of high layer network layer. Therefore, spatial pyramid pooling SPP [25] was proposed to deal with these problems [26]. This structure is mainly about the maximum pooling of 5×5, 9×9 and 13×13 with different sizes after convolution, batch normalization and activation function. The maximum pooling of the characteristic graph is joined together to change the channel to 2048 with the original size unchanged. Such operation by integrating different receptive fields can enrich the semantic information of feature maps and effectively improve model performance [27]. At the same time, we know that CSPDarknet53, the backbone feature extraction network of YOLOv4, is the key factor in obtaining good results with this network. The cross-stage part network (CSPNet [28]) is a structure proposed from the perspective of network architecture, as shown in Figure 2, CSP_X. This structure divides the input part into two parts, and the backbone part continues the residual The stacking of the other part is directly connected to the end to achieve channel splicing with the backbone part, which is equivalent to a large residual edge. Splitting first and then overlapping greatly reduces the number of parameters and computation, and meanwhile strengthens the CNN's learning ability and eliminates a computing bottleneck [29]. A K layer CNN with B basic layer channels is shown in Table 1 below.

Table 1. Add CSP structured Dark Layer FLOPs.

Model	Original	To CSP
Dark layer	5whkb2	whb2(3/4 + 5k/2)

In addition to the CSP structure of the trunk network, we considered combining the SPP structure mentioned above with the CSP module and optimizing it in the network. This KIND of CSP modular SPP structure reduces the amount of calculation resulting from increasing the SPP module and improves accuracy, achieving the purpose of reducing parameters but ensuring accuracy [28]. The improved CSP-SPP module is shown in Figure 3.

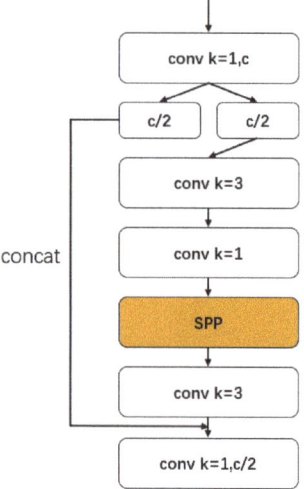

Figure 3. Csp-SPP module.

2.2. Attention Module SE

The attention model was originally used in machine translation and has become an important part of neural networks. The attentional mechanism module can pick out helpful features by attaching weights to different concerns within the network. Among many attention modules, the SE module is the classic. This focuses on the relationships between channels so that the model learns only useful channel characteristics. It first reduces the dimension of spatial features to 1×1 by global average pooling based on the width and height of feature graphs, as shown in Equation (7). Then, two fully connected layers and nonlinear activation functions are used to establish connections between channels, as shown in Equation (8).

$$z_C = F_{sq}(x_c) = \frac{1}{1 - 1 \times w} \sum_{i=1}^{H} \sum_{j=1}^{W} x_c(i,j) \qquad (7)$$

$$\hat{z} = T_2(ReLU(T_1(z))) \qquad (8)$$

The normalized weight is obtained by a Sigmoid activation function, and weighted to each channel of the original feature map by multiplication to complete the re-calibration of the original feature by channel attention, as shown in Equation (9) below.

$$\hat{x} = x \cdot \sigma(\hat{z}) \tag{9}$$

After global average pooling, the global receptive field can be obtained. During the first full connection, the parameters and calculation amount are greatly reduced by reducing the dimension of the feature graph. Following the nonlinear activation function, the correlation between channels is completed by restoring the original channel number through a full connection. See Figure 4.

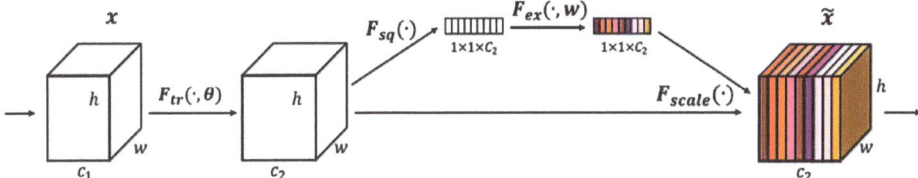

Figure 4. Attention module [30].

2.3. Use of the Feature Pyramid Structure

We used a feature pyramid structure, FPN, to replace the PANet path aggregation structure. PANet is an improved version of FPN, which adds a top-down path after a top-down path to achieve feature fusion. Such a structure can be more beneficial to classification and positioning, but at the same time greatly increases the cost of computing. The object features to be detected in our study are not complex, and the difference between the two structures is not obvious. However, it was hoped that the computation and complexity of the network would be reduced, so the FPN structure was used for feature fusion.

2.4. Improved YOLOv4 Algorithm

The structure of the detection network is shown in Figure 5. We continued to use the backbone feature extraction network of YOLOv4, and added the SE module after three output layers and after up-sampling to improve positioning accuracy. After that, the backbone part was used with the above-mentioned Csp-spp module to reduce more parameters while improving the receptive field, and finally we used the FPN structure to fuse the features and then output the targe.

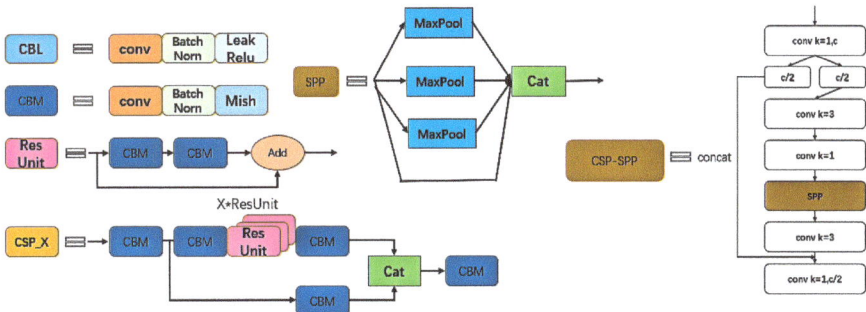

Figure 5. Cont.

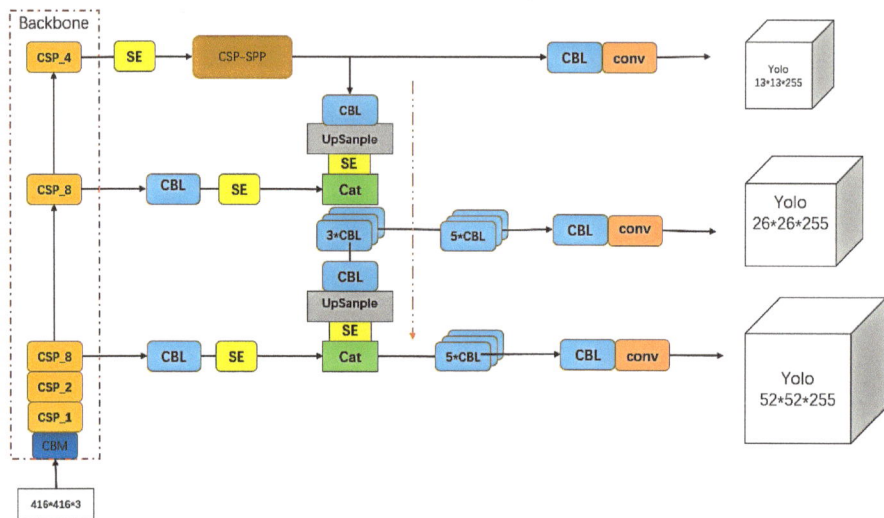

Figure 5. Improved YOLOv4 network.

3. Identification of Target Information

3.1. Barcode Decoding

To decode a barcode on the courier sheet, we chose the Zbar algorithm for the decoding operation. The Zbar algorithm is an open-source barcode detection algorithm online. The algorithm can not only read a variety of sources of barcode, such as image files, and videos, but also supports a variety of barcode types, including EAN-13/UPC-A, UPC-E, EAN-8, Code128, Code38, and QR. Our form of bar code was mainly code128. This is shown in Figure 6. Code128 consists of a series of parallel bars and blanks divided from left to right into left margin, start bit, data, validator, end bit, and right margin.

Figure 6. Diagram of bar code.

(1) Band Code. Four values of 1, 2, 3 and 4 are assigned according to the thickness, thickness and width of the bar, and the blank respectively. The Band Code of the barcode can be obtained successively.

(2) Left and right-side blank area. A blank space should be left on both sides of the bar code and the width should be 10 times the unit width (note: the unit width is the stripe width of width (1), allowing the bar code reader to enter the readability stage.

(3) Starting bit. The bar and blank detected in the first area of the barcode, which is the beginning of the visible part of the barcode, is composed of six interwoven bars and blanks of different thickness, with a total of 11-unit widths. In Code128, the starting bits of code A, B, and C are 211412, 211214, and 211232 respectively. The type of Code128 is determined by the start bit.

(4) Data. The data area expresses the coding information of the barcode, which is composed of multiple characters. Each character also consists of six bars and blanks.

(5) Validator. This is used to verify the validity of the barcode. The method of checksum module 103 was adopted, and the calculation method [31] is shown in Equation (10).

$$C = I \cdot N \bmod 103 \tag{10}$$

N is the value of the bit data.

(6) End character. This indicates the end-state of the barcode, which is fixed, and the corresponding Band Code is 2331112;

After the image is put into the detection network to detect the area of the bar code, the rectangular box containing the bar code must be captured. After the captured image is put into the ZBar algorithm, the algorithm analyzes and scans the image, and determines the Band Code of the bar code by the width of the bar and the empty, to extract the character information contained in the bar code.

As shown in Figure 6, the string "ST089030003" was identified by this algorithm, so that the sorting information of the express could be retrieved from the database.

3.2. Recognition of Three Segments of Code

Three-segment code characters are mainly printed bodies combining digits, hyphens and English letters. After obtaining pictures containing three-segment code characters, OCR (Optical Character Recognition) Character Recognition is required. Only when the string information is obtained can the sorting information corresponding to the character be obtained through the database. From the collected data, it was seen that the character distortion of the package would inevitably occur during the transportation process, and the recognition environment could be complex, which would affect the accuracy of recognition. Therefore, Tesseract was used for recognition. Tesseract is an open-source OCR engine. The fourth-generation version can support deep learning OCR, can recognize multiple formats of image files, and convert these to text. Figure 7 shows the single three-segment code style on the express side.

Figure 7. Courier sheet three-segment code style.

After obtaining the rectangular box containing three sections of code information, we used OpenCV and Tesseract together to obtain text recognition. As shown in Figure 8, after the image is first input, we use OpenCV's EAST text detector to detect the text in the image. The EAST text detector provides the bounding box coordinates of the text ROI. We extract each text ROI and input these into the LSTM deep learning text recognition algorithm of Tesseract V4. Finally, the output of the LSTM provides the actual OCR result, which is a string. After obtaining the string, we find the sorting information represented by the corresponding number through the database.

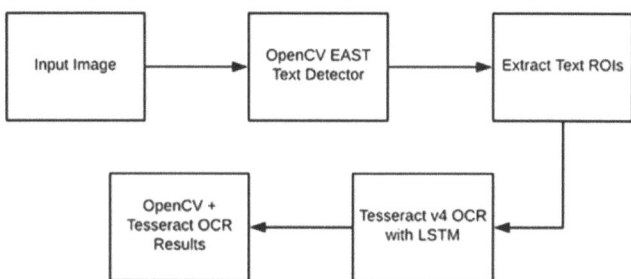

Figure 8. OpenCV OCR flow chart.

4. Experimental Design

4.1. Dataset

A dataset was created to simulate the environment of logistics and contained a total of 1680 images, mainly captured by cameras. The pictures of express delivery sheets included multiple express companies and different materials and different sizes, and were sampled under different lighting conditions and different angles. After obtaining the dataset, we used the open-source labeling tool labelimg to label in the VOC dataset format. Before image training, we also augmented the data set, and improved the generalization performance of the model by adjusting the image rotation angle, hue, saturation and other operations. Finally, the data set was divided into a training set and a test set with a ratio of 9:1. Each sample corresponded to two files, namely (1) a JPG file with the image of the package containing the express receipt, and (2) an xml file that stores image information, labels and coordinates corresponding to the region of interest in the image.

4.2. Experimental Environment and Training Process

This experiment used the operating system win 10 64 and the neural network framework pytorch. The hardware configuration included a CPU with Intel(R) Core (TM) i9-10900K CPU @ 3.70 GHz 3.70 GHz; RAM is 64 GB; GPU is NVIDIA GeForce RTX 2080 Ti.

In the object detection network experiment, the size of the input image was 416 × 416, the batch size was 16, the maximum number of iterations was 100, the initial learning rate was 0.001, and the attenuation coefficient was 0.0005. The ratio of training set to test set was 9:1.

At the same time, using the pre-trained model in the detection network, an accurate model could be obtained in a short time by transfer learning.

4.3. Analysis of Detection Experiment Results

4.3.1. Evaluation Index of Experimental Results

The target detection model was applied to the distribution center for real-time detection, so the detection speed and accuracy were more important evaluation criteria. The experiment used frame rate per second (FPS) as the speed evaluation index. The FPS value reflects the number of pictures that can be processed per second. The higher the FPS, the faster the detection speed. After that, the FPS data was obtained in the above configuration. Finally, it was decided to use the average precision (AP), precision rate (P), recall rate (R), F1-measure (F 1) value, model size and FPS in the detection network to evaluate the network performance. The calculation formulas of P, AP and $F1$ are expressed as Equations (10)–(12):

$$P = \frac{TP}{TP + FP} \quad (11)$$

$$AP = \int_0^1 P(R)d_R \quad (12)$$

$$F1 = \frac{2TP}{2TP + FP + FN} \quad (13)$$

Among them, TP represents positive samples predicted to be positive, FP represents negative samples predicted to be positive, and FN represents positive samples predicted to be negative.

4.3.2. Improved YOLOv4 Model Evaluation

The results of the improved YOLOv4 model are shown in Figure 9. It can be seen from various indicators that the experimental results of this positioning network model were good.

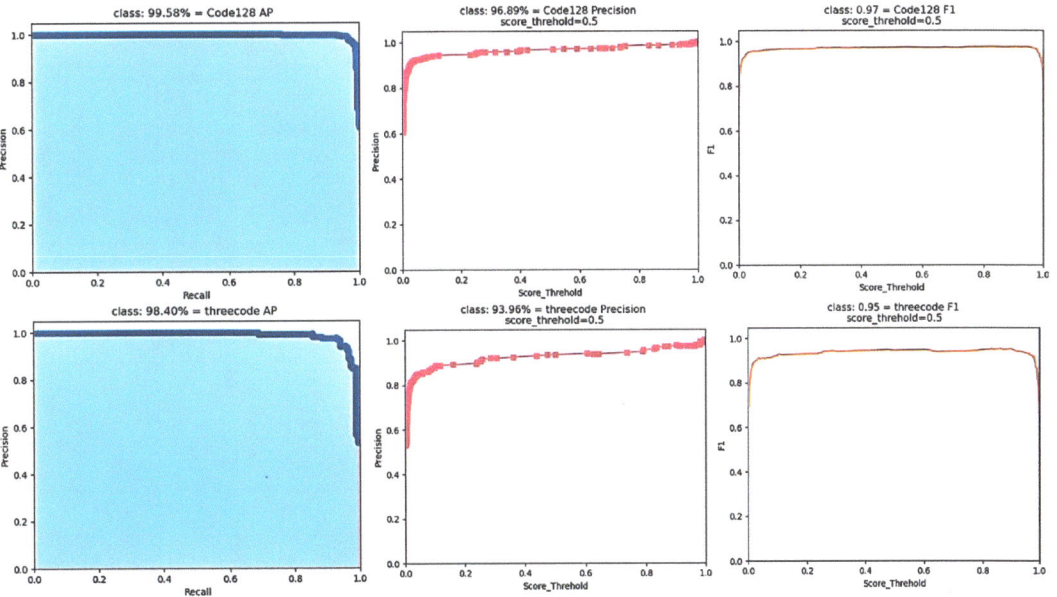

Figure 9. Model result diagram.

Ablation Experiments

In this section, the SPP structure combined with CSP structure is denoted as CS-YOLOv4, and the SE module is denoted as SCS-YOLOv4. Through experimental testing, we found that the performance of our model was improved in various aspects.

It can be seen from Tables 2 and 3 that the optimized YOLOv4 network SCS-YOLOv4 has different degrees of improvement in AP, P, FPS and size compared with the YOLOv4 network. In particular, in the detection of three-segment codes, the AP value increased by 1.7 percentage points, and the p value increased by 3.5 percentage points.

Table 2. Comparison of model performance after optimization (bar code).

Model	P	FPS	Size (MB)
YOLOv4	96.6%	21	245
CS-YOLOv4	96.3%	22	174
SCS-YOLOv4	96.9%	26	174

Table 3. Comparison of model performance after optimization (three-segment code).

Model	P	FPS	Size (MB)
YOLOv4	73.6%	21	245
CS-YOLOv4	92.4%	22	174
SCS-YOLOv4	94.9%	26	174

Comparative Experiments of Different Models

In our study, the common positioning model and SCS-YOLOv4 model were selected to compare their performance. Tables 4 and 5 show model comparisons with respect to five aspects of AP, F1, P, FPS and size. All the results were obtained from the same data set.

Table 4. Performance comparison of different models (bar code).

Model	AP	F1	P	FPS	Size
SSD	96.1%	0.86	96.4%	70	110
YOLOv3	85.3%	0.78	87.9%	34	236
YOLOv4	96.9%	0.97	91.44%	21	245
Faster R-CNN	99.8%	0.99	99.9%	12	522
Ours	99.6%	0.97	96.9%	26	174

Table 5. Performance comparison of different models (three-segment code).

Model	AP	F1	P	FPS	Size
SSD	86.5%	0.66	94.4%	70	110
YOLOv3	85.3%	0.78	87.8%	34	236
YOLOv4	96.7%	0.93	91.4%	21	245
Faster R-CNN	99.0%	0.98	95.7%	12	522
Ours	98.40%	0.95	93.9%	26	174

The two tables above clearly show the differences between the models. The two-stage target detection network Faster R-CNN model has a significant advantage in accuracy, but the model detection speed is too slow and the model is too large. The SSD300 test model has faster speed and more suitable size, but the accuracy is slightly different from other models. YOLOv4 network takes into account both speed and accuracy, and performs well as a whole. Considering that speed and accuracy are important indicators for sorting of express deliveries in a logistics center, and our network is more powerful than the YOLOv4 network and has been improved in various aspects, we used our network to achieve the positioning of target information.

4.4. Experimental Results Analysis

After the SCS-YOLOV4 algorithm was used to complete positioning, information was identified. During recognition, we found that it was difficult to accurately and quickly identify a picture with a large deflection angle, so it was necessary to correct this. We intercepted the extent of the bounding box, then the edge detection algorithm in OpenCV is used to process the captured image, and the minAreaRect () method was used to obtain the deflection angle of the image. Finally, affine transformation was used to correct the deflection image. After the corrected picture was obtained, we used the ZBar algorithm to identify the bar code, and the Tesseract to identify the three-segment code.

4.4.1. Evaluation of Experimental Results

In the logistics environment, if you want to carry out real-time recognition, recognition accuracy P and recognition speed S are important indicators. P is defined in Equation (14).

$$p = \frac{N_1}{N} \times 100\% \qquad (14)$$

where N is the number of samples, and N_1 is the number of correctly identified samples.

Barcode Decoding Test

We selected 200 bar code pictures as samples for the bar code recognition experiment. The experimental results are shown in Table 6.

Table 6. Barcode decoding test results.

The Number of Samples	Correct Number of Decoders	Decoding Success Rate
200	194	97%

Three-Segment Code Identification Test

We selected 200 images of three sections of code as samples for the barcode recognition experiment. We found that the image had a lot of interference information which affected recognition accuracy. Therefore, after rotation correction, three code regions were positioned to reduce the interference information and increase the recognition accuracy. The experimental results are shown in Table 7.

Table 7. Barcode decoding test results.

The Number of Samples	Correct Number of Decoders	Decoding Success Rate
200	192	96%

Multi-Information Target Recognition Test

We analyzed the recognition results of 200 pictures. From the perspective of the overall method, the success rate of express sorting recognition was 98.5%, whether it was single information recognition or multiple information recognition. The results are shown in Table 8.

Table 8. Method identification success rate.

The Number of Samples	Number of Correct Output Messages	Decoding Success Rate
200	197	98.5%

4.5. Time Performance

In our research, the above three algorithm modules were tested separately using 16GB RAM on a 64-bit Windows operating system, with an Intel(R) Core (TM) I7-10875H CPU @ 2.30 GHz, and a main frequency of 2.30 GHz. The running time is shown in Table 7 in seconds (s), and the average running time was 0.31 s when processing a 416 × 416 size package image, as shown in Table 9.

Table 9. Method time performance.

Object Detection	Decoding of Bar Code	Three-Segment Code Identification
0.04	0.08	0.19

4.6. Comparison of Different Express Sorting Methods

We compared several express sorting methods as shown in Table 10.

Table 10. Different Identification methods of the courier sheet.

Method	Target Detection Time	Total Time	Information Recognition Accuracy
Katona M [5]	0.05	0.15	91.6%
Liu W [8]	0.02	0.11	93.6%
Polat E [32]	0.48	0.57	97.4%
Ours	0.04	0.31	98.5%

It can be seen from Table 9 that the method of Liu W et al. is better in time performance, and can reach 0.11 s, but the recognition accuracy is low. The time performance of our method is not outstanding, only 0.31 s, but our method can attain 98.5% accuracy, which is more robust than other methods.

5. Conclusions

Aiming at the problem of the low recognition rate of single information sorting methods in small semi-automatic sorting centers, our research proposes a fast recognition method for courier sheet analysis based on deeper learning using multi-information fusion of a one-dimensional barcode and three-segment code. The experimental results show that the method can obtain the information on the courier sheet accurately. At the same time, considering that the overall recognition time is slower than that of single information, we can take the barcode information as the main information and the three-segment code information as the auxiliary information. Only when the barcode cannot be identified is the three-segment code information identified to reduce the sorting time. In general, although this method is slower than the single information recognition method, the multi-information recognition method ensures the accuracy of recognition and has good robustness.

Author Contributions: Conceptualization, Z.X. and X.X.; methodology, Z.X.; software, Z.X.; validation, Z.X., X.X. and Y.Z.; formal analysis, X.X.; investigation, Z.X.; re-sources, Y.Z.; data curation, Z.X.; writing—original draft preparation, Z.X.; writing—review and editing, X.X. and Y.Z.; supervision, X.X.; project administration, Z.X.; funding acquisition, X.X. All authors have read and agreed to the published version of the manuscript.

Funding: This work was supported by the National Key Research and Development Program of China (No. 2019YFE0126100), the Key Research and Development Program in Zhejiang Province of China (No. 2019C54005).

Institutional Review Board Statement: Not applicable.

Informed Consent Statement: Not applicable.

Data Availability Statement: Data available on request due to restrictions e.g., privacy or ethical. The data presented in this study are available on request from the corresponding author. The data are not publicly available due to our dataset is about courier package, there is a lot of personal information on it.

Conflicts of Interest: The authors declare no conflict of interest.

References

1. Reyes, P.M.; Li, S.; Visich, J.K. Determinants of RFID adoption stage and perceived benefits. *Eur. J. Oper. Res.* **2016**, *254*, 801–812. [CrossRef]
2. Feng, W.; Wu, X.; Liu, D. High-speed barcode recognition system. *J. Comput. Syst. Appl.* **2015**, *24*, 38–43.
3. Huang, M.; Li, Y. Express Sorting System Based on Two-Dimensional Code Recognition. In Proceedings of the 2018 International Conference on Sensor Networks and Signal Processing (SNSP), Xi'an, China, 28–31 October 2018; pp. 356–360.

4. Liu, W.; Chen, J.; Wang, N.; Shen, J.; Li, W.; Jiang, L.; Chen, X. Fast segmentation identification of express parcel barcode based on MSRCR enhanced high noise environment. In Proceedings of the 2019 2nd International Conference on Safety Produce Informatization (IICSPI), Chongqing, China, 28–30 November 2019; pp. 85–88.
5. Katona, M.; Nyúl, L.G. A novel method for accurate and efficient barcode detection with morphological operations. In Proceedings of the 2012 Eighth International Conference on Signal Image Technology and Internet Based Systems, Sorrento, Italy, 25–29 November 2012; pp. 307–314.
6. Katona, M.; Nyúl, L.G. Efficient 1D and 2D barcode detection using mathematical morphology. In Proceedings of the International Symposium on Mathematical Morphology and Its Applications to Signal and Image Processing, Uppsala, Sweden, 27–29 May 2013; Springer: Berlin/Heidelberg, Germany, 2013; pp. 464–475.
7. Zhang, W.J.; Yang, G.; Lin, Y.; Ji, C.; Gupta, M.M. On definition of deep learning. In Proceedings of the 2018 World Automation Congress (WAC), Stevenson, WA, USA, 3–6 June 2018; pp. 1–5.
8. Zamberletti, A.; Gallo, I.; Carullo, M.; Binaghi, E. Neural Image Restoration for Decoding 1-D Barcodes using Common Camera Phones. *VISAPP* **2010**, *1*, 5–11.
9. Kolekar, A.; Dalal, V. Barcode detection and classification using SSD (single shot multibox detector) deep learning algorithm. In Proceedings of the 3rd International Conference on Advances in Science & Technology (ICAST), Madrid, Spain, 28–30 October 2020.
10. Ren, Y.; Liu, Z. Barcode detection and decoding method based on deep learning. In Proceedings of the 2nd International Conference on Information Systems and Computer Aided Education (ICISCAE), Dalian, China, 28–30 September 2019; pp. 393–396.
11. Li, J.; Zhao, Q.; Tan, X.; Luo, Z.; Tang, Z. Using deep ConvNet for robust 1D barcode detection. In Proceedings of the International Conference on Intelligent and Interactive Systems and Applications, Beijing, China, 28–30 December 2017; pp. 261–267.
12. Pan, Z.; Jia, Z.; Jing, K.; Ding, Y.; Liang, Q. Manipulator Package Sorting and Placing System Based on Computer Vision. In Proceedings of the 2020 Chinese Control and Decision Conference (CCDC), Hefei, China, 22–24 August 2020; pp. 409–414. [CrossRef]
13. Shashidhar, R.; Manjunath, A.S.; Kumar, R.S.; Roopa, M.; Puneeth, S.B. Vehicle Number Plate Detection and Recognition using YOLO-V3 and OCR Method. In Proceedings of the 2021 IEEE International Conference on Mobile Networks and Wireless Communications (ICMNWC), Tumkur, India, 3–4 December 2021; pp. 1–5. [CrossRef]
14. Carion, N.; Massa, F.; Synnaeve, G.; Usunier, N.; Kirillov, A.; Zagoruyko, S. End-to-end object detection with transformers. In Proceedings of the European Conference on Computer Vision, Glasgow, UK, 23–28 August 2020; pp. 213–229.
15. Girshick, R.; Donahue, J.; Darrell, T.; Malik, J. Rich feature hierarchies for accurate object detection and semantic segmentation. In Proceedings of the 27th IEEE Conference on Computer Vision and Pattern Recognition (CVPR), Columbus, OH, USA, 23–28 June 2014; pp. 2373–2384.
16. Girshick, R. Fast r-cnn. In Proceedings of the IEEE International Conference on Computer Vision, Santiago, Chile, 7–13 December 2015; pp. 1440–1448.
17. Ren, S.; He, K.; Girshick, R.; Sun, J. Faster r-cnn: Towards real-time object detection with region proposal networks. In *Advances in Neural Information Processing Systems 28 (NIPS 2015)*; Curran Associates, Inc.: Red Hook, NY, USA, 2015.
18. Redmon, J.; Divvala, S.; Girshick, R.; Farhadi, A. You only look once: Unified, real-time object detection. In Proceedings of the IEEE Conference on Computer Vision and Pattern Recognition, Las Vegas, NV, USA, 27–30 June 2016; pp. 779–788.
19. Liu, W.; Anguelov, D.; Erhan, D.; Szegedy, C.; Reed, S.; Fu, C.Y.; Berg, A.C. Ssd: Single shot multibox detector. In Proceedings of the European Conference on Computer Vision, Amsterdam, The Netherlands, 11–14 October 2016; pp. 21–37.
20. Zhou, X.; Wang, D.; Krähenbühl, P. Objects as points. *arXiv* **2019**, arXiv:1904.07850.
21. Tan, M.; Pang, R.; Le, Q.V. Efficientdet: Scalable and efficient object detection. In Proceedings of the IEEE/CVF Conference on Computer Vision and Pattern Recognition, Seattle, WA, USA, 14–19 June 2020; pp. 10781–10790.
22. Bochkovskiy, A.; Wang, C.Y.; Liao, H.Y.M. Yolov4: Optimal speed and accuracy of object detection. *arXiv* **2020**, arXiv:2004.10934.
23. Redmon, J.; Farhadi, A. Yolov3: An incremental improvement. *arXiv* **2018**, arXiv:1804.02767.
24. Liu, S.; Qi, L.; Qin, H.; Shi, J.; Jia, J. Path aggregation network for instance segmentation. In Proceedings of the IEEE Conference on Computer Vision and Pattern Recognition, Salt Lake City, UT, USA, 18–22 June 2018; pp. 8759–8768.
25. He, K.; Zhang, X.; Ren, S.; Sun, J. Deep residual learning for image recognition. In Proceedings of the IEEE Conference on Computer Vision and Pattern Recognition, Las Vegas, NV, USA, 27–30 June 2016; pp. 770–778.
26. Szegedy, C.; Ioffe, S.; Vanhoucke, V.; Alemi, A.A. Inception-v4, inception-resnet and the impact of residual connections on learning. In Proceedings of the Thirty-First AAAI Conference on Artificial Intelligence, San Francisco, CA, USA, 4–9 February 2017.
27. He, K.; Zhang, X.; Ren, S.; Sun, J. Spatial pyramid pooling in deep convolutional networks for visual recognition. *IEEE Trans. Pattern Anal. Mach. Intell.* **2015**, *37*, 1904–1916. [CrossRef] [PubMed]
28. Wang, C.-Y.; Liao, H.-Y.M.; Wu, Y.-H.; Chen, P.-Y.; Hsieh, J.-W.; Yeh, I.-H. CSPNet: A new backbone that can enhance learning capability of CNN. In Proceedings of the IEEE/CVF Conference on Computer Vision and Pattern Recognition Workshops, Seattle, WA, USA, 14–19 June 2020; pp. 390–391.
29. Hu, J.; Shen, L.; Sun, G. Squeeze-and-excitation networks. In Proceedings of the IEEE Conference on Computer Vision and Pattern Recognition, Salt Lake City, UT, USA, 18–22 June 2018; pp. 7132–7141.

30. Wang, C.Y.; Bochkovskiy, A.; Liao, H.Y.M. Scaled-yolov4: Scaling cross stage partial network. In Proceedings of the IEEE/CVF Conference on Computer Vision and Pattern Recognition, Nashville, TN, USA, 19–25 June 2021; pp. 13029–13038.
31. Available online: https://wenku.baidu.com/view/d031650ba9ea998fcc22bcd126fff705cd175c47.html (accessed on 10 June 2021).
32. Polat, E.; Mohammed, H.M.A.; Omeroglu, A.N.; Kumbasar, N.; Ozbek, I.Y.; Oral, E.A. Multiple barcode detection with mask r-cnn. In Proceedings of the 2020 28th Signal Processing and Communications Applications Conference (SIU), Gaziantep, Turkey, 5–7 October 2020; pp. 1–4.

Article

Multimodal Transformer Model Using Time-Series Data to Classify Winter Road Surface Conditions

Yuya Moroto [1], Keisuke Maeda [2], Ren Togo [3], Takahiro Ogawa [3] and Miki Haseyama [3,*]

1. Graduate School of Information Science and Technology, Hokkaido University, N-14, W-9, Kita-ku, Sapporo 060-0814, Japan; moroto@lmd.ist.hokudai.ac.jp
2. Data-Driven Interdisciplinary Research Emergence Department, Hokkaido University, N-13, W-10, Kita-ku, Sapporo 060-0813, Japan; maeda@lmd.ist.hokudai.ac.jp
3. Faculty of Information Science and Technology, Hokkaido University, N-14, W-9, Kita-ku, Sapporo 060-0814, Japan; togo@lmd.ist.hokudai.ac.jp (R.T.); ogawa@lmd.ist.hokudai.ac.jp (T.O.)
* Correspondence: mhaseyama@lmd.ist.hokudai.ac.jp

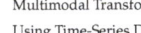

Citation: Moroto, Y.; Maeda, K.; Togo, R.; Ogawa, T.; Haseyama, M. Multimodal Transformer Model Using Time-Series Data to Classify Winter Road Surface Conditions. *Sensors* **2024**, *24*, 3440. https://doi.org/10.3390/s24113440

Academic Editor: Zhe-Ming Lu

Received: 27 March 2024
Revised: 18 May 2024
Accepted: 22 May 2024
Published: 27 May 2024

Copyright: © 2024 by the authors. Licensee MDPI, Basel, Switzerland. This article is an open access article distributed under the terms and conditions of the Creative Commons Attribution (CC BY) license (https://creativecommons.org/licenses/by/4.0/).

Abstract: This paper proposes a multimodal Transformer model that uses time-series data to detect and predict winter road surface conditions. For detecting or predicting road surface conditions, the previous approach focuses on the cooperative use of multiple modalities as inputs, e.g., images captured by fixed-point cameras (road surface images) and auxiliary data related to road surface conditions under simple modality integration. Although such an approach achieves performance improvement compared to the method using only images or auxiliary data, there is a demand for further consideration of the way to integrate heterogeneous modalities. The proposed method realizes a more effective modality integration using a cross-attention mechanism and time-series processing. Concretely, when integrating multiple modalities, feature compensation through mutual complementation between modalities is realized through a feature integration technique based on a cross-attention mechanism, and the representational ability of the integrated features is enhanced. In addition, by introducing time-series processing for the input data across several timesteps, it is possible to consider the temporal changes in the road surface conditions. Experiments are conducted for both detection and prediction tasks using data corresponding to the current winter condition and data corresponding to a few hours after the current winter condition, respectively. The experimental results verify the effectiveness of the proposed method for both tasks. In addition to the construction of the classification model for winter road surface conditions, we first attempt to visualize the classification results, especially the prediction results, through the image style transfer model as supplemental extended experiments on image generation at the end of the paper.

Keywords: deep learning; transformer; multimodal analysis; time-series processing; winter road surface condition

1. Introduction

In snow-covered and cold regions, which account for approximately 60% of the land area in Japan, numerous winter-related traffic accidents occur due to weather conditions, e.g., snowfall. Approximately 90% of these accidents are slip-related incidents associated with winter road surface conditions due to snow accumulation and ice formation [1]. In this context, road managers need to undertake snow and ice control operations, e.g., snow removal and the spreading of anti-freezing agents by detecting or predicting road surface conditions to prevent slip accidents [1,2].

Previous studies have investigated the detection or prediction of winter road surface conditions [3–8]. In the literature [3], the road surface condition was predicted based on the heat balance theory using digital geographical data, which represent the shape of the land, including roads on computers; however, this method requires the analysis of digital geographical data related to the road, and it is difficult to collect and accumulate such data for all roads. In another study [7], the automatic detection of winter road surface conditions was realized using deep learning models trained on images captured by vehicle-mounted cameras. Similarly, winter road surface conditions were classified using hierarchical deep learning models applied to images also captured by vehicle-mounted cameras [8]. Here, to use images captured by vehicle-mounted cameras, it is necessary to drive on the road to be analyzed with vehicles equipped with cameras. To reduce such efforts, in the literature [4], data obtained from sensors and fixed-point cameras installed along roads were adopted to detect or predict the winter road surface conditions using rule-based methods. In addition, a previous study [5] achieved detection by classifying road surface conditions using differential methods based on images captured by fixed-point cameras installed along the road (hereafter referred to as road surface images). However, due to the temporal variability of road surfaces and roadside features, methods based on differential approaches require manual updating of the reference images. Thus, there is a demand for models that can classify road surface conditions automatically and accurately to facilitate precise detection and prediction. Several studies have focused on the winter road surface condition classification using the images captured by vehicle-mounted cameras [9–11]. The purpose of these studies is to help with the construction of autonomous vehicles; however, our purpose is to assist road managers in reducing winter-related traffic accidents using fixed-point cameras.

The multimodal analysis, which uses several information sources, e.g., images and natural languages, has attracted significant attention for improving the representational ability of models [12–15]. For example, contrastive language image pre-training has been proposed as the pre-training framework for the multimodal analysis of vision and language [16]. Another example is to use the texts obtained from Twitter in addition to images for image sentiment analysis [17]. In this way, most works on multimodal analysis have used vision and language modalities; however, in the classification task of winter road surface conditions, the text information does not exist, and the other information is needed for multimodal analysis. Then, we previously proposed an automated classification method for road surface conditions using a multimodal multilayer perceptron (MLP) using images and auxiliary data [18]. Concretely, in that study, the features calculated from multiple modalities, including road surface images and auxiliary data related to the road surface conditions such as temperatures and traffic volume, were concatenated and input to the MLP to classify the road surface conditions. The cooperative use of multiple modalities allows for mutual complementation between modalities, and we improved classification accuracy compared to using a single modality. However, in the previous study, we focused on the construction of machine learning models using multiple modalities and performed multimodal analysis through a simple feature concatenation process. As a result, this approach may have inherent limitations in terms of classification accuracy. Thus, further improvements in classification accuracy can be expected by introducing the following processes.

1. Time-series Analysis

    In the field of glaciology, a previous study [19] reported that snow accumulation extremes exhibit time-series variability. In addition, Hirai et al. [20] suggested that changes in road surface conditions are related to the transitions of these conditions over the past several timesteps. Thus, rather than relying on data from a single timestep (as in our previous study), using time-series data to classify road surface conditions is expected to improve the detection and prediction accuracy.

2. Feature Integration using Attention Mechanisms

    In our previous study, feature integration was performed by concatenating the features derived separately from image and auxiliary data and then inputting them into an MLP. On the other hand, in the machine learning field, Transformers [21–24], which are the novel machine learning architecture focusing on the relationship of input data, have attracted significant attention for the remarkable performance based on the strong representational ability. With the advancement of such Transformers, recent research on feature integration has demonstrated that intermediate fusion, which combines features in the intermediate layers of neural networks using cross-attention, achieves higher accuracy than traditional feature integration methods [25–29]. Cross-attention is an attention mechanism [21] with several inputs, which facilitates the compensation of heterogeneous features calculated from multiple modalities. As a result, the cross-attention module enhances the representational ability after integration, and the use of feature integration based on cross-attention is expected to further improve classification accuracy.

In this paper, we propose a new method for classifying winter road surface conditions using a multimodal transformer (MMTransformer) capable of processing time-series data. In the proposed method, image and auxiliary features are extracted from data spanning multiple timesteps, and feature integration considering temporal changes is performed by applying cross-attention. With cross-attention, correlations are calculated feature-wise for input data across multiple timesteps, and attention is computed for each timestep. This procedure enables feature integration that accounts for temporal changes in road surface conditions. Finally, the classification of winter road surface conditions is realized using an MLP. By exploring methods for integrating multiple modalities and introducing time-series processing, we aim to achieve improvements in accuracy in the detection and prediction of road surface conditions.

In addition, the proposed method can learn the relationship between the input data and the corresponding teacher labels, which are the labels related to winter road surface conditions for training the model. By altering the teacher labels assigned to the input data during training, the proposed method can be adapted to both detection and prediction tasks. In experiments conducted on real-world data, we evaluated the effectiveness of the proposed method for both detection and prediction tasks with two sets of teacher labels. One experiment was conducted with the teacher labels being the road surface condition corresponding to the input data, and the subsequent experiment was conducted with the teacher labels being the road surface condition a few hours after the input data. This dual approach allows for a comprehensive assessment of the capabilities of the proposed method in detecting the current road surface conditions and predicting future road surface conditions.

In addition to the experiments on the classification of winter road surface conditions, we conducted supplemental extended experiments on image generation to visualize the classification results, particularly the prediction results in the Appendix A. To help road managers make decisions, it can be effective to incorporate classification results and road surface images that visualize the results. In this study, we generated such images using an image style transfer model conditioned by road surface conditions. Through these supplemental experiments and visualizing the transferred images, we confirmed the potential of the image transfer model for road surface images.

The primary contributions of this study are summarized as follows.

1. A multimodal transformer model based on time-series processing and attention mechanisms is constructed to classify road surface conditions.
2. Experiments conducted to evaluate the road surface condition detection and prediction tasks verify the effectiveness of the proposed classification model.
3. The results of the supplemental extended experiments in the Appendix A demonstrate the potential of the image transfer model for road surface images.

The remainder of this paper is organized as follows. Section 2 introduces the data used in this study. The proposed method for the classification of winter road surface conditions is explained in Section 3. Then, the experimental results are reported in Section 4, and the supplemental extended experiments are discussed in Appendix A. Finally, Section 5 concludes the paper.

2. Data

In the following, we describe the data used in this study. We utilized road surface images acquired using fixed-point cameras and auxiliary data related to the road surface conditions. Specifically, these data were provided by the East Nippon Expressway Company Limited and were acquired from 2017 to 2019. The road surface images were captured at 20-min intervals from 1 December at 00:00 to 31 March at 23:40 each year. In addition, each road surface image was labeled with one of the following seven categories related to road surface conditions.

- Dry
  The road surface is free of snow, ice, and wetness.
- Wet
  The road surface is wet due to moisture.
- Black sherbet
  Tire tread marks are present, the snow contains a high amount of moisture, and the color of the road surface is black.
- White sherbet
  Tire tread marks are present, the snow contains a high amount of moisture, and the color of the road surface is white.
- Snow
  Snow has accumulated on the road surface, and the snow does not contain a high amount of moisture.
- Compacted snow
  There is no black shine and no tire tread marks.
- Ice
  Snow and ice are present on the road surface, and it appears black and shiny.

These labels were assigned by three experienced road managers, and they divided the annotation task and assigned the labels through visual inspections. Example road surface images for each category are shown in Figure 1, and the locations where the road surface images were captured are shown in Figure 2. Here, the image size is 640 × 480 pixels. Please note that road surface images, including vehicles, were considered for analysis because the vehicles did not cover the entire road surface in the images.

Table 1 shows the contents of the auxiliary data and the corresponding data types. As shown in Table 1, the "location of road surface images" and "weather forecast" are discrete information, while other data contents are represented as continuous values. As shown in Figure 1 and Table 1, the images and auxiliary data differ significantly; thus, a feature integration mechanism is required to complement the deficiencies in each modality. Thus, we attempt to improve the classification accuracy of road surface conditions by integrating multiple modalities at several timesteps.

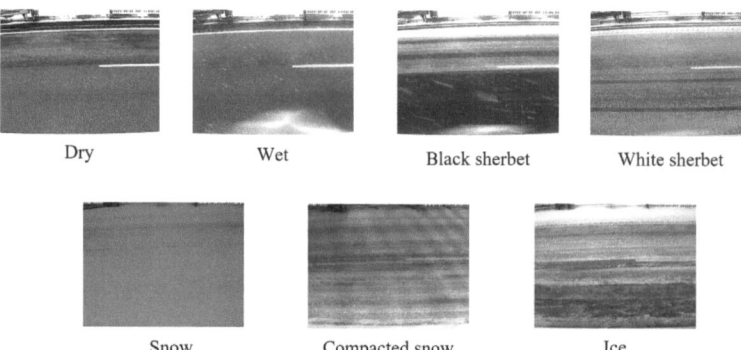

Figure 1. Road surface images for each winter road surface condition.

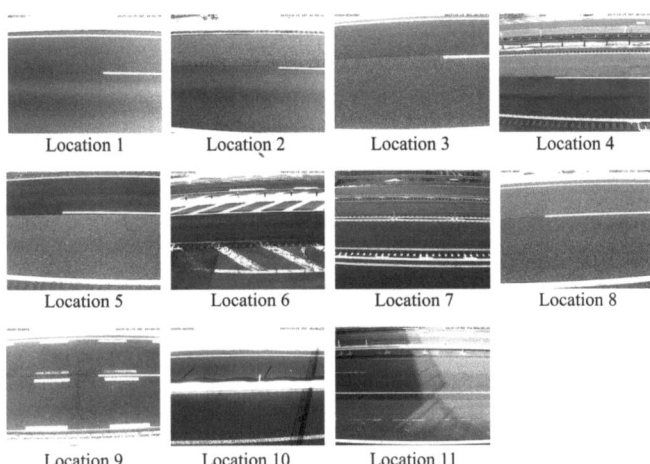

Figure 2. Locations where the road surface images were captured.

Table 1. Auxiliary data and corresponding data types.

Data Content	Data Type
Location of road surface images	Discrete
Temperature	Continuous
Road temperature	Continuous
Amount of snowfall	Continuous
Traffic volume	Continuous
Average of vehicle speed	Continuous
Weather forecast six hours ago	Discrete
Temperature forecast six hours ago	Continuous
Snowfall forecast six hours ago	Continuous
Weather forecast 12 h ago	Discrete
Temperature forecast 12 h ago	Continuous
Snowfall forecast 12 h ago	Continuous
Weather forecast 18 h ago	Discrete
Temperature forecast 18 h ago	Continuous
Snowfall forecast 18 h ago	Continuous
Weather forecast 24 h ago	Discrete
Temperature forecast 24 h ago	Continuous
Snowfall forecast 24 h ago	Continuous

3. Classification of Winter Road Surface Conditions Using MMTransformer

In this section, we describe the proposed method to classify winter road surface conditions based on the MMTransformer, which can process time-series data using images and auxiliary data at multiple timesteps as inputs. First, we construct encoders for both the image and auxiliary data at each timestep to extract relevant features. We then calculate the integrated features with the characteristics of both the image and auxiliary data by performing feature integration based on cross-attention. Finally, by inputting the integrated features into an MLP, we can classify the winter road surface conditions. An overview and flowchart of the proposed method are shown in Figures 3 and 4, respectively. Please note that the proposed model is trained in an end-to-end manner, which allows the image encoder to be fine-tuned and the parameters in the MLP to be optimized simultaneously. In the following, we explain the methods for feature extraction and feature integration based on cross-attention in Sections 3.1 and 3.2, respectively.

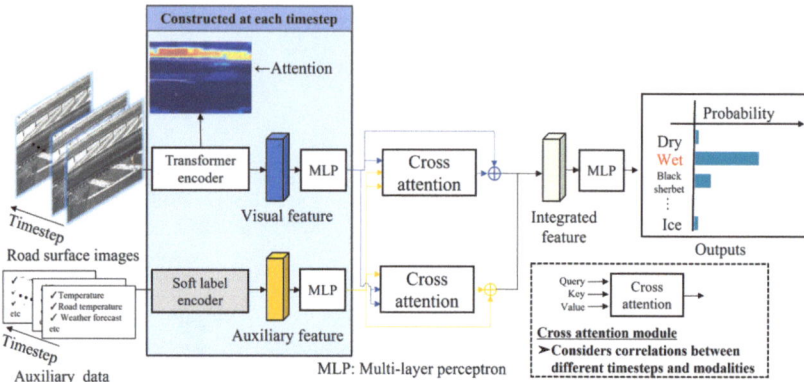

Figure 3. Overview of the proposed method.

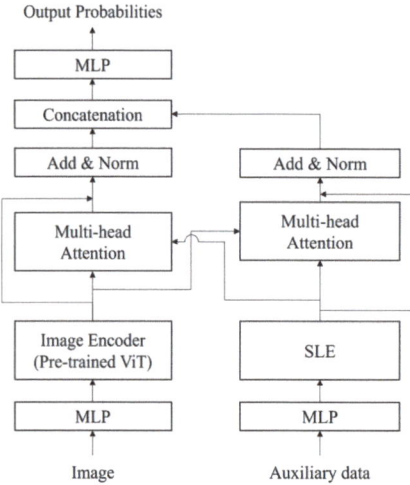

Figure 4. Flowchart of MMTransformer.

3.1. Feature Extraction

Here, we describe the method employed to construct the encoders used to extract the features from the image and auxiliary data.

3.1.1. Visual Features

The proposed method utilizes output values from the intermediate layers of a pretrained deep learning model as visual features. For the deep learning model, we employ the Vision Transformer (ViT) [24] or its derivative methods [22,23], which have achieved high classification accuracy in image classification tasks. Training a model based on the ViT requires a large amount of training data; thus, we fine-tune a model pretrained on ImageNet [30] to extract the visual features with high representational ability from the road surface images.

In the ViT, as shown in Figure 5, patches obtained by dividing the images and position embeddings are input sequentially to linear layers and the Transformer encoder. The output values are calculated by the MLP head after the Transformer encoder. During fine-tuning of the ViT, transfer learning is performed on the Transformer encoder by replacing the MLP head. Specifically, in the proposed method, the visual feature $x_t^{(\text{vis})} \in \mathbb{R}^{d_{\text{vis}}}$ for image V_t at timestep t ($t = 1, 2, \ldots, T$, where T is the number of timesteps) is calculated as follows:

$$x_t^{(\text{vis})} = f(E_{\text{vis}}(V_t)), \quad (1)$$

where $E_{\text{vis}}(\cdot)$ is the pretrained Transformer encoder in the ViT-based model, and $f(\cdot)$ is the MLP that calculates the visual features for input into the cross-attention mechanism. Thus, by employing an MLP head suitable for feature integration, it is possible to fine-tune the ViT-based model and train the cross-attention mechanism simultaneously.

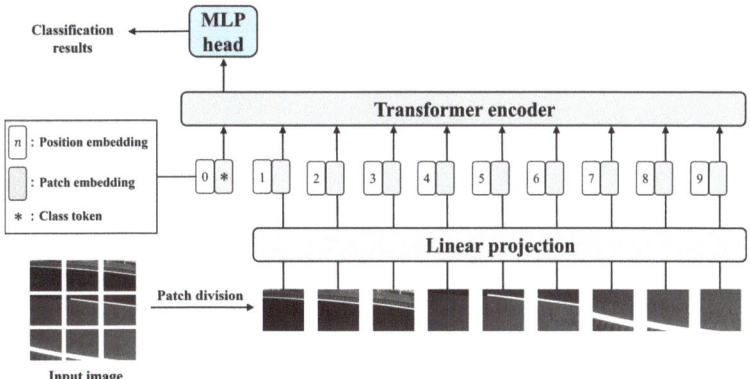

Figure 5. Architecture of the ViT [24].

3.1.2. Auxiliary Features

In the proposed method, the auxiliary data include both continuous quantitative variables, e.g., temperature and road temperature, and discrete qualitative variables, using nominal scales, e.g., location and weather conditions. Generally, in machine learning involving qualitative variables as inputs, one-hot encoding is used as a preprocessing method [31–33]. In one-hot encoding, elements equal to the number of items in the nominal scale are prepared, and the corresponding element is set to 1 (while others are set to 0). This procedure enables machine learning models to process qualitative variables. However, when one-hot encoded features $\{x_i\}_{i=0}^{n}$ are input to a neural network–based model, in the first layer of the forward propagation process, only the weights corresponding to the input elements with 1 are updated as follows:

$$a_{01} = \sum_{i=0}^{n} x_i W_{i0} + b_0, \quad (2)$$

where $\{W_{i0}\}_{i=0}^{n}$ represents the weights corresponding to x_i, and a_{01} is the output value at the 0th neuron in the first layer. As a result, the other weights corresponding to input elements with 0 are not updated, which makes it difficult to learn the correlations between the input elements. It has been reported that applying soft label encoding (SLE) to nominal scales in auxiliary data improves accuracy [33]. In SLE, the correlation between features can be learned by replacing the elements that are 0 in one-hot encoding with 0.1. Actually, in the literature [33], SLE (Figure 6) enabled the learning of correlations within auxiliary data and enhanced the representational ability. Thus, for the auxiliary data used in this study, applying SLE to the discrete qualitative variables is expected to improve the classification accuracy. Consequently, in the proposed method, SLE is applied to the discrete values, and a vector combined with continuous values is input to the MLP to calculate the auxiliary feature $x_t^{(\text{aux})} \in \mathbb{R}^{d_{\text{aux}}}$ at timestep t.

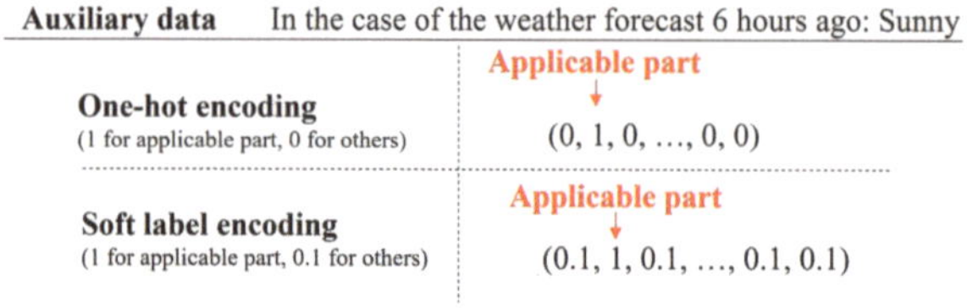

Auxiliary data	In the case of the weather forecast 6 hours ago: Sunny
	Applicable part ↓
One-hot encoding (1 for applicable part, 0 for others)	(0, 1, 0, ..., 0, 0)
	Applicable part ↓
Soft label encoding (1 for applicable part, 0.1 for others)	(0.1, 1, 0.1, ..., 0.1, 0.1)

Figure 6. Example of SLE.

3.2. Feature Integration Based on Cross-Attention Mechanism

This section explains the cross-attention-based feature integration method. In the cross-attention module, the importance of each element in the features is determined using the query $q \in \mathbb{R}^{T \times d'_m}$, key $k \in \mathbb{R}^{T \times d'_m}$ and the value $v \in \mathbb{R}^{T \times d'_m}$ ($m \in \{\text{vis}, \text{aux}\}$, $d'_m = d_m/h$). Here, h is a hyperparameter. The tuple (q, k, v) for each feature is calculated as follows:

$$q^m = X^m W^{(\text{q},m)\top}, \tag{3}$$

$$k^m = X^m W^{(\text{k},m)\top}, \tag{4}$$

$$v^m = X^m W^{(\text{v},m)\top}, \tag{5}$$

$$\text{s.t. } m \in \{\text{vis}, \text{aux}\}, \tag{6}$$

where $W^{(\text{q},m)} \in \mathbb{R}^{d'_m \times d_m}$, $W^{(\text{k},m)} \in \mathbb{R}^{d'_m \times d_m}$ and $W^{(\text{v},m)} \in \mathbb{R}^{d'_m \times d_m}$ are the trainable parameters. In addition, $X^m = [x_1^{m\top}, x_2^{m\top}, \ldots, x_T^{m\top}] \in \mathbb{R}^{T \times d_m}$. Next, using the tuple (q, k, v) among the heterogeneous features, the cross-attention $\text{CA}(\cdot, \cdot, \cdot)$ is calculated as follows:

$$\text{CA}\left(q^{m'}, k^m, v^m\right) = \left[\text{head}_1^{(m',m)}, \text{head}_2^{(m',m)}, \ldots, \text{head}_h^{(m',m)}\right] W^{(\text{o},m)}, \tag{7}$$

$$\text{head}_i^{(m',m)} = \text{Softmax}\left(\frac{q^{m'} k^{m\top}}{\sqrt{d_y^m}}\right), \tag{8}$$

$$\text{s.t. } m' \neq m, \ i = 1, 2, \ldots, h, \tag{9}$$

where $W^{(\text{o},m)} \in \mathbb{R}^{hd'_m \times d_{m'}}$ is the trainable parameter. Finally, feature integration is performed by applying residual connections to each feature and the output values of the cross-attention mechanism as follows:

$$\hat{X}^m = X^m + \text{CA}(q^m, k^{m'}, v^{m'}), \tag{10}$$

$$\hat{X}^{\text{int}} = [\hat{X}^{\text{vis}}, \hat{X}^{\text{aux}}]. \tag{11}$$

In the proposed method, vectorization is performed by applying mean pooling to the integrated feature $\hat{X}^{\text{int}}$, which is then input to the MLP to output the final classification results. Thus, using cross-attention-based feature integration, the proposed method corrects features using heterogeneous data and processes time-series data across multiple timesteps. As a result, the proposed method improves the detection and prediction accuracy of winter road surface conditions.

4. Experiments

Experiments were conducted to verify the effectiveness of the proposed classification method based on MMTransformer. In the following, Section 4.1 describes the experimental dataset, Section 4.2 explains the experimental settings, and Section 4.3 presents the experimental results and a corresponding discussion.

4.1. Experimental Dataset

Here, we describe the dataset used in the experiments. The experiments utilized the winter road surface images and auxiliary data discussed in Section 4.1 to verify the effectiveness of the proposed method on real-world data. In addition, the seven categories (dry, wet, black sherbet, white sherbet, snow, compacted snow, and ice) were reorganized into three new categories, i.e., dry/wet, sherbet, and snow/compacted snow/ice, to detect and predict the winter road surface conditions from a practical perspective. The experiments were designed to confirm the effectiveness of using data across multiple timesteps to detect and predict winter road surface conditions. The classifications of road surface conditions were made for $\{0, 1, 3\}$ hours later when inputting data at T $(= \{1, 3, 5\})$ timesteps. Here, the data at one timestep were acquired at 20-min intervals. Please note that the input data were used on a per-timestep basis, and the teacher labels were used on an hourly basis. The number of samples for each road surface condition and the experimental settings are shown in Tables 2–4. In the multi-timestep experimental settings, missing data were imputed using the average values from the data at other timesteps. In addition, data from 2017 and 2018 were used as the training data without distinction of the location, and data from 2019 were used as the test data. Also, note that the number of samples in each category varied significantly in the training data; thus, to suppress the reduction in classification accuracy due to the imbalanced number of samples, random extraction was performed such that the number of samples belonging to each category was approximately equal. As a result, the number of samples in the training data was smaller than that of test data through such an undersampling operation.

Table 2. Breakdown of experimental data used to immediately predict (detect) the road surface condition (0 h later).

	Number of Timesteps Used as Input					
	1		3		5	
Road Surface Condition	Training	Test	Training	Test	Training	Test
Dry/Wet	6000	35,771	6000	36,388	6000	36,460
Sherbet	5829	2474	5921	2505	5939	2506
Snow/Compacted snow/Ice	4614	418	4740	420	4747	420
Sum	16,443	38,663	16,661	39,313	16,686	39,386

Table 3. Breakdown of experimental data used to predict the road surface condition one hour later.

	Number of Timesteps Used as Input					
	1		3		5	
Road Surface Condition	Training	Test	Training	Test	Training	Test
Dry/Wet	6000	35,738	6000	36,370	6000	36,444
Sherbet	5838	2477	5935	2504	5948	2506
Snow/Compacted snow/Ice	4604	416	4730	420	4739	420
Sum	16,442	38,631	16,665	39,294	16,687	39,370

Table 4. Breakdown of experimental data used to predict the road surface condition three hours later.

	Number of Timesteps Used as Input					
	1		3		5	
Road Surface Condition	Training	Test	Training	Test	Training	Test
Dry/Wet	6000	35,703	6000	36,332	6000	36,407
Sherbet	5836	2475	5942	2504	5954	2506
Snow/Compacted snow/Ice	4604	416	4730	420	4741	420
Sum	16,428	38,593	16,667	39,256	16,695	39,333

4.2. Experimental Settings

Here, we describe the experimental settings. The MLP used in the proposed method comprised three layers, and the feature dimensions of the images and auxiliary data were set to $d_{\text{vis}} = 16$ and $d_{\text{aux}} = 16$, respectively. For the Transformer encoder in the proposed method, we employed the ViT-B/16 model [24], which was pretrained on ImageNet [30]. For the loss function, cross-entropy loss was used, and for the optimization method, the Adam optimizer [34] with a learning rate of 0.001 was employed. During the training, the batch size was set to 8, and the number of epochs was set to 10. Moreover, we set $h = 4$ as the hyperparameter.

To verify the effectiveness of the cross-attention–based feature integration implemented in the proposed method, we compared a method (Concatenation) that does not employ cross-attention by replacing Equation (11) with the following expression:

$$\hat{X}^{\text{int}} = [X^{\text{vis}}, X^{\text{aux}}]. \tag{12}$$

To evaluate the performance of the detection and prediction results, accuracy, macro precision, macro recall, and macro F1 metrics were considered, which are frequently used in the machine learning field for multiclass classification tasks. Each evaluation metric is calculated as follows:

- Accuracy

$$\text{Accuracy} = \frac{\sum_{l=1}^{L} \text{TP}_l}{\sum_{l=1}^{L} (\text{TP}_l + \text{FP}_l)}. \tag{13}$$

- Macro Precision

$$\text{Macro Precision} = \frac{1}{L} \sum_{l=1}^{L} \text{Precision}_l, \tag{14}$$

$$\text{Precision}_l = \frac{\text{TP}_l}{\text{TP}_l + \text{FP}_l}. \tag{15}$$

- Macro Recall

$$\text{Macro Recall} = \frac{1}{L} \sum_{l=1}^{L} \text{Recall}_l, \tag{16}$$

$$\text{Recall}_l = \frac{\text{TP}_l}{\text{TP}_l + \text{FN}_l}. \tag{17}$$

- Macro F1

$$\text{Macro F1} = \frac{1}{L} \sum_{l=1}^{L} \text{F1}_l, \tag{18}$$

$$\text{F1}_l = \frac{2 \times \text{Recall}_l \times \text{Precision}_l}{\text{Recall}_l + \text{Precision}_l}. \tag{19}$$

Here, TP_l and FN_l represent the number of true positive samples and false negative samples for the lth category, respectively, and FP_l denotes the number of false positive samples for the lth category.

4.3. Results and Discussion

4.3.1. Effectiveness of Time-Series Analysis

The experimental results obtained with different numbers of timesteps in the input data are shown in Tables 5–7. Under all experimental conditions, the increase in the number of timesteps resulted in a higher macro F1 score, and we confirmed the effectiveness of using multiple timesteps when detecting and predicting the winter road surface conditions. On the other hand, when comparing MMTransformer w/5 with MMTransformer w/3 in Table 5, the macro Precision score decreased. Similarly, when comparing MMTransformer w/5 with MMTransformer w/3 in Table 7, the macro Recall score decreased. These score decreases were caused by differences in FP_l for macro Precision and in FN_l for macro Recall; however, both FP_l and FN_l should be evaluated for the classification model. Thus, we mainly focused on the harmonic mean of macro Precision and macro Recall, i.e., macro F1, and discussed the difference in the performance based on the macro F1. Thus, the effectiveness of time-series analysis with input data at multiple timesteps in the proposed method has been verified.

Table 5. Experimental results obtained when varying the number of timesteps in the experiment to immediately predict (detect) the road surface condition.

Method	Accuracy	Macro Precision	Macro Recall	Macro F1
MMTransformer w/1 timestep	0.954	0.689	0.768	0.702
MMTransformer w/3 timesteps	0.958	0.710	0.791	0.735
MMTransformer w/5 timesteps	0.956	0.698	0.808	0.740

Table 6. Experimental results obtained when varying the number of timesteps in the experiment to predict the road surface condition one hour later.

Method	Accuracy	Macro Precision	Macro Recall	Macro F1
MMTransformer w/ 1 timestep	0.941	0.633	0.774	0.678
MMTransformer w/ 3 timesteps	0.944	0.636	0.791	0.683
MMTransformer w/ 5 timesteps	0.948	0.667	0.799	0.717

Table 7. Experimental results obtained when varying the number of timesteps in the experiment to predict the road surface condition three hours later.

Method	Accuracy	Macro Precision	Macro Recall	Macro F1
MMTransformer w/ 1 timestep	0.919	0.560	0.746	0.612
MMTransformer w/ 3 timesteps	0.926	0.579	0.747	0.627
MMTransformer w/ 5 timesteps	0.936	0.625	0.737	0.666

4.3.2. Effectiveness of Cross-Attention Mechanism

The experimental results comparing the proposed method with other methods are shown in Tables 8–10. As can be seen, the macro F1 score of the proposed method surpasses that of the compared methods, which confirms the effectiveness of the MMTransformer. Specifically, by comparing MMTransformer and Concatenation, we verified that the cross-attention-based feature integration is effective for the classification of winter road surface conditions. On the other hand, when comparing MMTransformer w/5 with Concatenation w/5 in Table 10, the macro Recall score decreased. As well as macro Precision in Table 5 and macro Recall in Table 7, we mainly focused on the harmonic mean of macro Precision and macro Recall, i.e., macro F1, and discussed the difference in the performance based on the macro F1.

Table 8. Comparison of results in experiments to immediately predict (detect) road surface conditions.

Method	Accuracy	Macro Precision	Macro Recall	Macro F1
Concatenation w/5 timesteps	0.957	0.697	0.796	0.722
MMTransformer w/5 timesteps	0.956	0.698	0.808	0.740

Table 9. Comparison of results in experiments to predict road surface conditions one hour later.

Method	Accuracy	Macro Precision	Macro Recall	Macro F1
Concatenation w/5 timesteps	0.944	0.647	0.799	0.701
MMTransformer w/5 timesteps	0.948	0.667	0.799	0.717

Table 10. Comparison of results in experiments to predict road surface conditions three hours later.

Method	Accuracy	Macro Precision	Macro Recall	Macro F1
Concatenation w/5 timesteps	0.927	0.590	0.756	0.644
MMTransformer w/5 timesteps	0.936	0.625	0.737	0.666

Thus, the effectiveness of using feature integration based on the cross-attention mechanism as the feature integration method has been verified. In addition, confusion matrices for the classification results of Concatenation w/5 timesteps and MMTransformer w/5 timesteps are shown in Figures 7–9. In Figures 7 and 8, the number of samples classified correctly for the dry/wet and snow/compacted snow/ice categories is approximately the same for both the MMTransformer and Concatenation. For the sherbet category, the MMTransformer outperformed the Concatenation considerably in terms of the number of correctly classified samples. In Figure 9, the number of correctly classified samples for the sherbet and snow/compacted snow/ice categories is similar; however, the MMTransformer outperformed the Concatenation considerably in the dry/wet category. These results confirm that the MMTransformer can predict winter road surface conditions more accurately than the Concatenation. However, when predicting the winter road surface conditions three hours later, as shown in Figure 9, there was no significant improvement in terms of classification accuracy for the important sherbet and snow/compacted snow/ice categories, which are critical for the effective detection and prediction of winter road surface conditions. Thus, improving the accuracy of predictions for winter road surface conditions at later times remains a challenge for future work.

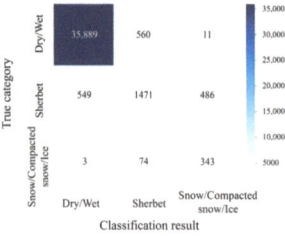   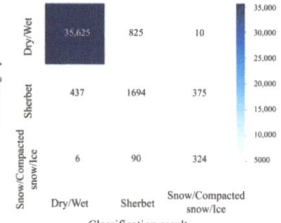

(a) Concatenation w/5 timesteps   (b) MMTransformer w/5 timesteps

Figure 7. Confusionmatrix for the experiment to immediately predict (detect) the road surface condition (corresponding to Table 8).

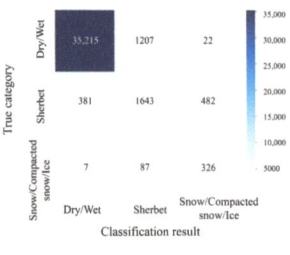   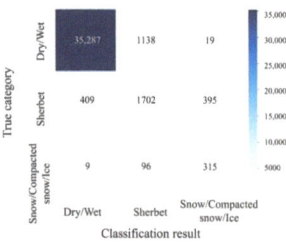

(a) Concatenation w/5 timesteps   (b) MMTransformer w/5 timesteps

Figure 8. Confusion matrix for the experiment to predict the road surface condition one hour later (corresponding to Table 9).

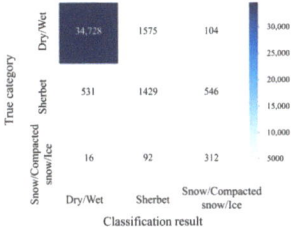
(a) Concatenation w/5 timesteps

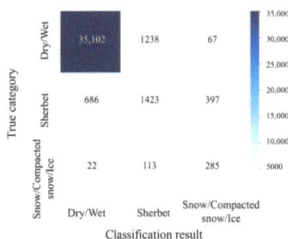
(b) MMTransformer w/5 timesteps

Figure 9. Confusionmatrix for the experiment to predict the road surface condition three hours later (corresponding to Table 10).

4.3.3. Qualitative Evaluation through Visualization

In the MMTransformer, the output values obtained from the ViT model's intermediate layers are used as image features. The ViT model employs an attention mechanism that recognizes important regions in images automatically and applies weighting to these regions. To achieve this, attention rollout [35], which presents the regions focused on by ViT through visualizing the weights in the attention mechanism, has been proposed. The regions presented by attention rollout are expected to serve as a basis for the rationale behind the classification results obtained by the ViT. In the proposed method, by observing the regions for winter road surface images, it is possible to gain insights into the relationship between the winter road surface images and winter road surface conditions and to use this information to enhance the performance of the classification model.

Figure 10 shows a visualization example obtained by applying attention rollout to the ViT encoder in MMTransformer, where redder regions are of higher interest in MMTransformer, and bluer regions are of lower interest. Here, the visualization was performed for MMTransformer w/5 timesteps in the experimental setting to detect the winter road surface conditions. As can be seen, there is more attention on the snow at the roadside at 20:00 and 20:40, and there is consistent attention to certain parts of the road surface over all timesteps. These observations imply that MMTransformer w/5 timesteps recognizes the presence of snow on the roadside but correctly identifies the road surface condition as sherbet due to the lesser amount of snow compared to the snow/compacted snow/ice conditions. From this result, it can be inferred that MMTransformer w/5 timesteps performs detection and prediction by focusing on the snow accumulation on the surface of the road in the images. Thus, by outputting the visualization results for the input images, we can gain insights into the relationship between the winter road surface images and the road surface conditions, and these insights can be used to enhance the performance of detection and prediction models.

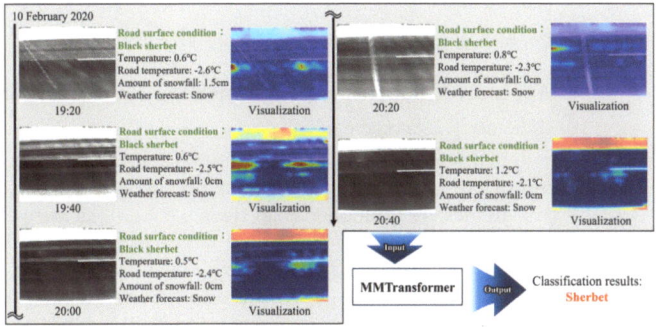

Figure 10. Example visualizationobtained by applying attention rollout to the ViT model, i.e., the image encoder in MMTransformer.

5. Conclusions

This paper has proposed the MMTransformer method, which uses time-series data to detect and predict winter road surface conditions. The proposed method enhances the representational ability of the integrated features by performing feature correction through mutual complementation between modalities based on a cross-attention-based feature integration method for multiple modalities, e.g., road surface images and auxiliary data. In addition, by introducing time-series processing for the input data at multiple timesteps, the proposed method can integrate features in consideration of the temporal changes in winter road surface conditions. As a result, the proposed method improves the classification accuracy of winter road surface conditions by introducing a new integration for multiple modalities and time-series processing.

Experiments confirmed that the proposed MMTransformer method achieves high accuracy in classifying winter road surface conditions and is effective for both the detection and prediction tasks by varying the teacher labels. In addition, using attention rollout for visualization, we expected to provide additional insights into the relationship between road surface images and road surface conditions. In this way, as the experimental findings, it was implied that attention rollout works well for the multimodal classification model of winter road surface conditions. The visualization in the image encoder can be utilized to enhance the classification model when detecting and predicting road surface conditions, and the experimental findings discussed in this paper have demonstrated the potential of this technique.

On the other hand, confusion matrices indicate that performance improvement was slight for the data belonging to sherbet or snow/compacted snow/ice categories since the road surface images belonging to sherbet or snow/compacted snow/ice categories were visually similar to those of each other category. Such limitations caused by visual similarity can be solved by effectively leveraging non-visual information, including auxiliary data, which remains in future works.

Author Contributions: Conceptualization, Y.M., K.M., R.T., T.O. and M.H.; methodology, Y.M., K.M. and T.O.; software, Y.M.; validation, Y.M.; data curation, Y.M.; writing—original draft preparation, Y.M.; writing—review and editing, K.M., R.T., T.O. and M.H.; visualization, Y.M.; funding acquisition, K.M., R.T., T.O. and M.H. All authors have read and agreed to the published version of the manuscript.

Funding: This work was partly supported by the JSPS KAKENHI Grant Numbers JP21H03456, JP23K11211, and JP22KJ0006.

Institutional Review Board Statement: Not applicable.

Informed Consent Statement: Not applicable.

Data Availability Statement: Experimental data cannot be disclosed.

Acknowledgments: In this research, we used the data provided by East Nippon Expressway Company Limited.

Conflicts of Interest: The authors declare no conflict of interest.

Appendix A. Supplemental Extended Experiments on Image Generation

Appendix A.1. Background

In this section, to visualize the classification results, especially in terms of the prediction results, we conducted supplemental extended experiments focusing on image generation.

In the classification of road surface conditions, we assume that the workflow of the classification model presents only the classification results of road surface conditions to road managers. Such a workflow makes it difficult to visualize the detailed state of the road surface. Thus, by visually presenting the road surface conditions a few hours later in addition to the classification results, it is expected that more informed decisions will be made with the knowledge and experience of road managers. In this way, the visualization of classification results facilitates effective decision support for snow and ice removal operations.

In the computer vision field, tasks involving the style transformation of images have traditionally been addressed [36,37]. Such style transfer tasks involve learning the relationships between domains to transform a target image into a desired image style. For example, by learning the relationship between a domain of images capturing a horse and a domain of images capturing a zebra, the image style transfer model can output an image where the patterns on the body of the horse are transformed into that of the zebra. Similarly, for road surface images, it is possible to transfer an input image to an image with the style of the predicted road surface condition using the image style transfer model. As a result, the image generation reflecting the style of the predicted road surface conditions can be realized using image style transfer with input road surface images. The generated images hold promise in terms of providing visual decision support for road managers making snow and ice removal decisions.

In the supplemental extended experiments, we first attempted to generate images using the style of specific road surface conditions using the image style transfer model. Specifically, since there are multiple categories of road surface conditions, we performed multidomain style transfer for each category as a domain. Here, we used StarGAN v2 [38] as the style transfer model. The StarGAN v2 model is a well-known multidomain style transfer model that achieves efficient multidomain style transfer by training a single generator to handle multiple domains to acquire domain-specific features.

Appendix A.2. Image Style Transfer

In this subsection, we summarize the method used to transform the road surface conditions in the road surface images using the StarGAN v2 model. An overview of the image style transfer process using the StarGAN v2 model is shown in Figure A1. When the input image and domain are denoted $x \in \mathcal{X}$ and $y \in \mathcal{Y}$, respectively, StarGAN v2 attempts to transform the input image x into the style of each domain y using a single generator G. Here, $\mathcal{X}$ and $\mathcal{Y}$ represent the set of images and the set of domains after transformation, respectively. To generate images that reflect the style of each domain from a single generator, domain-specific style features are input along with the input image, and the StarGAN v2 model controls the style of the image output by the generator G. In the following, we explain the modules used in the StarGAN v2 model, i.e., the generator, mapping function, style encoder, discriminator, and the objective function for optimization.

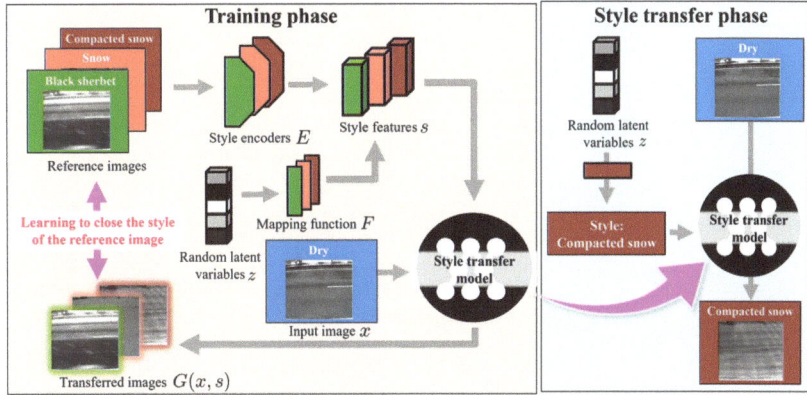

Figure A1. Overview of image generation using the style transfer model. It should be noted that the discriminator D is used to close the styles of the reference images and those of the transferred images.

In the StarGAN v2 model, the generator G transforms the input image x into image $G(x,s)$ using style features s obtained from either the mapping function F or the style encoder E. By incorporating adaptive instance normalization [39,40] into the generator, StarGAN v2 enables style transfer using the style features s. As a result, by calculating the style code s to represent domain-specific features, it is possible to generate images that reflect the style of multiple domains using only a single generator (without the need to construct separate generators for each domain).

The mapping function F calculates the style features s from the random latent variables z. Specifically, by utilizing an MLP with multiple output branches corresponding to each road surface condition, the style features are calculated as $s = F(z)$. This multitask architecture enables efficient calculation of the style features.

The style encoder E extracts the style features s from the image x as $s = E(x)$. Using the style features calculated by inputting a reference image into the style encoder, it is possible to transform the input image into an image that reflects the style of the reference image.

The discriminator D distinguishes between images that belong to the target domain and images that are transformed by the generator when an image is input. Here, efficient learning is achieved by adopting a multitasking architecture similar to the mapping function F and style encoder E.

In the StarGAN v2 model, to enable a single generator to output images corresponding to the styles of multiple domains, the entire model is trained by optimizing the following objective function:

$$\min_{G,F,E} \max_{D} \mathcal{L}_{adv} + \lambda_{sty}\mathcal{L}_{sty} - \lambda_{ds}\mathcal{L}_{ds} + \lambda_{cyc}\mathcal{L}_{cyc}, \tag{A1}$$

where λ_{sty}, λ_{ds}, and λ_{cyc} are hyperparameters, and $\mathcal{L}_{adv}$ is the adversarial loss used to acquire domain-specific style features and enhance the quality of the generated images. In addition, $\mathcal{L}_{sty}$ is the style reconstruction loss, which is used to enable the extraction of style features that correspond to each domain from images. This reconstruction loss is inspired by the literature [41,42]; however, the main difference lies in the ability to extract style features for multiple domains using a single style encoder. $\mathcal{L}_{ds}$ is the diversity regularization loss [43,44] used to ensure the diversity of the generated images, and $\mathcal{L}_{cyc}$ is the cycle consistency loss [45–47], which is used to preserve domain-invariant features in the input image in the transformed image. Using these different losses, it is possible to generate images that correspond to the styles of multiple domains using only a single generator.

Appendix A.3. Experimental Results

We conducted the supplemental extended experiment using the StarGAN v2 model to transform the surface conditions in the road surface images using a style transfer model. The road surface conditions targeted in this experiment and the number of training images labeled for each condition are shown in Table A1. Please note that the number of road surface conditions differed from that described in Section 4.1 to confirm that the image style transfer models can represent diverse road surface conditions. Here, the purpose of this section is to confirm the potential of applying image style transfer models to road surface images, and we experimentally used as many road surface conditions as possible. In addition, the labels were assigned not by the classification models, e.g., MMTransformer, but by experienced road managers (Section 2) to evaluate the image style transfer model without misclassification effects.

Table A1. Number of training images labeled with each road surface condition.

Road Surface Condition	Dry	Wet	Black Sherbet	White Sherbet	Snow	Ice	Compacted Snow	Sum
	39,807	45,778	13,568	2910	2313	320	2045	106,741

In this experiment, the hyperparameters λ_{sty}, λ_{ds}, and λ_{cyc} were all set to 1, and the dimensionality of the random latent variables was set to 16. In addition, the dimensionality of the style features was set to 64. The model was optimized using the Adam optimizer [34] with 100,000 epochs and a batch size of 8. The learning rates for D, E, and G were set to 0.0001, and the learning rate for F was set to 0.000001.

Figure A2 shows examples of road surface condition transfer in road surface images and the corresponding compared images. Here, the compared images are road surface images labeled with the same conditions as the transferred images. The experimental results confirm that the transferred images visually resemble the compared images. In addition, the ability to acquire visually distinct images accurately supports the potential to generate road surface images with transferred road surface conditions by training a style transfer model on road surface images.

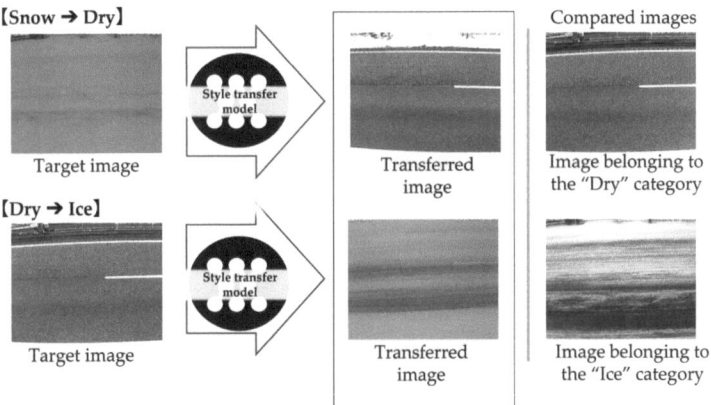

Figure A2. Examples of road surface condition transformation in road surface images using StarGAN v2 model. For reference and comparison, a road surface image with the same label as the transferred image is also shown.

Appendix A.4. Conclusions

In addition to the construction of the classification model for winter road surface conditions, we conducted supplemental extended experiments on image generation to visualize the classification results, especially the prediction results. The experimental results demonstrate that the generated images reflect the specified styles. Thus, the classification results can be represented as images using image style transfer models to help road managers make decisions. However, comparative experiments and quantitative evaluations were not conducted in this study, although we have supported the potential of using an image style transfer model for road surface images. Thus, the construction of an image style transfer model specific to road surface images and its evaluation remains an issue for future work.

References

1. Nakai, S.; Kosugi, K.; Yamaguchi, S.; Yamashita, K.; Sato, K.; Adachi, S.; Ito, Y.; Nemoto, M.; Nakamura, K.; Motoyoshi, H.; et al. Study on advanced snow information and its application to disaster mitigation: An overview. *Bull. Glaciol. Res.* **2019**, *37*, 3–19. [CrossRef]
2. Kogawa, K.; Tsuchihashi, H.; Sato, J.; Tanji, K.; Yoshida, N. Development of winter road surface condition prediction system to support snow and ice work decisions. In Proceedings of the JSSI and JSSE Joint Conference Snow and Ice Research (Japanese), Sapporo, Japan, 1–5 October 2022; p. 125.
3. Saida, A.; Fujimoto, A.; Tokunaga, R.; Hirasawa, M.; Takahashi, N.; Ishida, T.; Fukuhara, T. Verification of HFN forecasting accuracy in Hokkaido using route-based forecasting model of road snow/ice conditions. In Proceedings of the JSSI and JSSE Joint Conference Snow and Ice Research (Japanese), Nagoya, Japan, 28 September–2 October 2016; p. 4.

4. Uchida, M.; Gotou, K.; Okamoto, J. Web systems for sensing and predicting road surface conditions in winter season. *Yokogawagiho* **2000**, *44*, 21–24.
5. Yamada, M.; Tanizaki, T.; Ueda, K.; Horiba, I.; Sugie, N. A System of Discrimination of the Road Condition by means of Image Processing. *IEEJ Trans. Ind. Appl.* **2000**, *120*, 1053–1060. [CrossRef]
6. Ohiro, T.; Takakura, K.; Sakuraba, T.; Hanatsuka, Y.; Hagiwara, T. Development of Advanced Anti-icing Spray System using Automated Road Surface Condition Judgement System. *JSTE J. Traffic Eng.* **2019**, *5*, B_7–B_15.
7. Li, J.; Masato, A.; Sugisaki, K.; Nakamura, K.; Kamiishi, I. Efficiency improvement of winter road surface interpretation by using artificial intelligence model. *Artif. Intell. Data Sci.* **2020**, *1*, 210–216.
8. Takase, T.; Takahashi, S.; Hagiwara, T. A Study on identification of a winter road surface state in highway based on machine learning using in-vehicle camera images. *IEICE Tech. Rep.* **2020**, *44*, 31–34.
9. Cordes, K.; Broszio, H. Camera-Based Road Snow Coverage Estimation. In Proceedings of the IEEE/CVF International Conference on Computer Vision (ICCV), Paris, France, 2–6 October 2023; pp. 4011–4019.
10. Ojala, R.; Seppänen, A. Lightweight Regression Model with Prediction Interval Estimation for Computer Vision-based Winter Road Surface Condition Monitoring. *IEEE Trans. Intell. Veh.* **2024**, 1–13. [CrossRef]
11. Xie, Q.; Kwon, T.J. Development of a highly transferable urban winter road surface classification model: A deep learning approach. *Transp. Res. Rec.* **2022**, *2676*, 445–459. [CrossRef]
12. Xu, P.; Zhu, X.; Clifton, D.A. Multimodal learning with transformers: A survey. *IEEE Trans. Pattern Anal. Mach. Intell.* **2023**, *45*, 12113–12132. [CrossRef]
13. Yin, S.; Fu, C.; Zhao, S.; Li, K.; Sun, X.; Xu, T.; Chen, E. A survey on multimodal large language models. *arXiv* **2023**, arXiv:2306.13549.
14. Jabeen, S.; Li, X.; Amin, M.S.; Bourahla, O.; Li, S.; Jabbar, A. A review on methods and applications in multimodal deep learning. *ACM Trans. Multimed. Comput. Commun. Appl.* **2023**, *19*, 1–41. [CrossRef]
15. Das, R.; Singh, T.D. Multimodal sentiment analysis: A survey of methods, trends, and challenges. *ACM Comput. Surv.* **2023**, *55*, 1–38. [CrossRef]
16. Radford, A.; Kim, J.W.; Hallacy, C.; Ramesh, A.; Goh, G.; Agarwal, S.; Sastry, G.; Askell, A.; Mishkin, P.; Clark, J.; et al. Learning transferable visual models from natural language supervision. In Proceedings of the International Conference on Machine Learning (ICML), Online, 18–24 July 2021; pp. 8748–8763.
17. Vadicamo, L.; Carrara, F.; Cimino, A.; Cresci, S.; Dell'Orletta, F.; Falchi, F.; Tesconi, M. Cross-media learning for image sentiment analysis in the wild. In Proceedings of the IEEE International Conference on Computer Vision (ICCV), Venice, Italy, 22–29 October 2017; pp. 308–317.
18. Moroto, Y.; Meada, K.; Togo, R.; Ogawa, T.; Haseyama, M. Winter road surface condition classification using deep learning with focal loss based on text and image information. *Artif. Intell. Data Sci.* **2022**, *3*, 293–306.
19. Ito, T. Time series analyses on the maximum depth of snow cover in Akita city. *J. Jpn. Soc. Snow Ice* **1979**, *41*, 267–275. [CrossRef]
20. Hirai, S.; Makino, H.; Yamazaki, I.; Ookubo, Y. Adaptation of image road surface sensors to winter road management work. In Proceedings of the ITS Symposium, Tokyo, Japan, 1–2 December 2005; pp. 1–6.
21. Vaswani, A.; Shazeer, N.; Parmar, N.; Uszkoreit, J.; Jones, L.; Gomez, A.N.; Kaiser, L.; Polosukhin, I. Attention is all you need. In Proceedings of the 31st Conference on Neural Information Processing Systems (NeurIPS 2017), Long Beach, CA, USA, 4–9 December 2017; pp. 6000–6010.
22. Zhang, Z.; Zhang, H.; Zhao, L.; Chen, T.; Arik, S.Ö.; Pfister, T. Nested hierarchical transformer: Towards accurate, data-efficient and interpretable visual understanding. In Proceedings of the AAAI Conference Artificial Intelligence (AAAI), Online, 22 February–1 March 2022; Volume 36, pp. 3417–3425.
23. Liu, Z.; Lin, Y.; Cao, Y.; Hu, H.; Wei, Y.; Zhang, Z.; Lin, S.; Guo, B. Swin transformer: Hierarchical vision transformer using shifted windows. In Proceedings of the IEEE/CVF International Conference on Computer Vision (ICCV), Virtual, 11–17 October 2021; pp. 10012–10022.
24. Dosovitskiy, A.; Beyer, L.; Kolesnikov, A.; Weissenborn, D.; Zhai, X.; Unterthiner, T.; Dehghani, M.; Minderer, M.; Heigold, G.; Gelly, S.; et al. An Image is Worth 16x16 Words: Transformers for Image Recognition at Scale. In Proceedings of the International Conference on Learning Representations (ICLR), Virtual Event, Austria, 3–7 May 2021.
25. Kim, J.H.; Jun, J.; Zhang, B.T. Bilinear attention networks. In Proceedings of the Advances in Neural Information Processing Systems (NeurIPS), Montreal, QC, Canada, 3–8 December 2018; Volume 31.
26. Ishihara, K.; Nakano, G.; Inoshita, T. MCFM: Mutual cross fusion module for intermediate fusion-based action segmentation. In Proceedings of the IEEE International Conference on Image Processing (ICIP), Bordeaux, France, 16–19 October 2022; pp. 1701–1705.
27. Joze, H.R.V.; Shaban, A.; Iuzzolino, M.L.; Koishida, K. MMTM: Multimodal transfer module for CNN fusion. In Proceedings of the IEEE/CVF Computer Vision and Pattern Recognition Conference (CVPR), Seattle, WA, USA, 14–19 June 2020; pp. 13289–13299.
28. Bose, R.; Pande, S.; Banerjee, B. Two headed dragons: Multimodal fusion and cross modal transactions. In Proceedings of the IEEE International Conference on Image Processing (ICIP), Anchorage, AK, USA, 19–22 September 2021; pp. 2893–2897.
29. Kim, J.H.; On, K.W.; Lim, W.; Kim, J.; Ha, J.W.; Zhang, B.T. Hadamard Product for Low-rank Bilinear Pooling. In Proceedings of the International Conference on Learning Representations (ICLR), Toulon, France, 24–26 April 2017.
30. Deng, J.; Dong, W.; Socher, R.; Li, L.J.; Li, K.; Fei-Fei, L. Imagenet: A large-scale hierarchical image database. In Proceedings of the IEEE/CVF Conference on Computer Vision and Pattern Recognition (CVPR), Miami, FL, USA, 20–25 June 2009; pp. 248–255.

31. Chen, J.; Liang, D.; Zhu, Z.; Zhou, X.; Ye, Z.; Mo, X. Social media popularity prediction based on visual-textual features with xgboost. In Proceedings of the ACM International Conference on Multimedia (ACMMM), Nice, France, 21–25 October 2019; pp. 2692–2696.
32. Zheng, H.T.; Chen, J.Y.; Liang, N.; Sangaiah, A.K.; Jiang, Y.; Zhao, C.Z. A deep temporal neural music recommendation model utilizing music and user metadata. *Appl. Sci.* **2019**, *9*, 703. [CrossRef]
33. Cai, G.; Zhu, Y.; Wu, Y.; Jiang, X.; Ye, J.; Yang, D. A multimodal transformer to fuse images and metadata for skin disease classification. *Vis. Comput.* **2023**, *39*, 2781–2793. [CrossRef]
34. Kingma, D.P.; Ba, J. Adam: A method for stochastic optimization. *arXiv* **2014**, arXiv:1412.6980.
35. Abnar, S.; Zuidema, W. Quantifying Attention Flow in Transformers. *arXiv* **2020**, arXiv:2005.00928.
36. Liu, L.; Xi, Z.; Ji, R.; Ma, W. Advanced deep learning techniques for image style transfer: A survey. *Signal Process. Image Commun.* **2019**, *78*, 465–470. [CrossRef]
37. Zhao, C. A survey on image style transfer approaches using deep learning. *J. Phys. Conf. Ser.* **2020**, *1453*, 012129. [CrossRef]
38. Choi, Y.; Uh, Y.; Yoo, J.; Ha, J.W. Stargan v2: Diverse image synthesis for multiple domains. In Proceedings of the IEEE/CVF Conference on Computer Vision and Pattern Recognition (CVPR), Seattle, WA, USA, 14–19 June 2020; pp. 8188–8197.
39. Huang, X.; Belongie, S. Arbitrary style transfer in real-time with adaptive instance normalization. In Proceedings of the IEEE/CVF International Conference on Computer Vision (ICCV), Venice, Italy, 22–29 October 2017; pp. 1501–1510.
40. Karras, T.; Laine, S.; Aila, T. A style-based generator architecture for generative adversarial networks. In Proceedings of the IEEE/CVF Conference on Computer Vision and Pattern Recognition (CVPR), Long Beach, CA, USA, 15–20 June 2019; pp. 4401–4410.
41. Huang, X.; Liu, M.Y.; Belongie, S.; Kautz, J. Multimodal unsupervised image-to-image translation. In Proceedings of the European Conference Computer Vision (ECCV), Munich, Germany, 8–14 September 2018; pp. 172–189.
42. Zhu, J.Y.; Zhang, R.; Pathak, D.; Darrell, T.; Efros, A.A.; Wang, O.; Shechtman, E. Toward multimodal image-to-image translation. In Proceedings of the Advances in Neural Information Processing Systems (NeurIPS), Long Beach, CA, USA, 4–9 December 2017; Volume 30.
43. Yang, D.; Hong, S.; Jang, Y.; Zhao, T.; Lee, H. Diversity-sensitive conditional generative adversarial networks. *arXiv* **2019**, arXiv:1901.09024.
44. Mao, Q.; Lee, H.Y.; Tseng, H.Y.; Ma, S.; Yang, M.H. Mode seeking generative adversarial networks for diverse image synthesis. In Proceedings of the IEEE/CVF Conference on Computer Vision and Pattern Recognition (CVPR), Long Beach, CA, USA, 15–20 June 2019; pp. 1429–1437.
45. Choi, Y.; Choi, M.; Kim, M.; Ha, J.W.; Kim, S.; Choo, J. Stargan: Unified generative adversarial networks for multi-domain image-to-image translation. In Proceedings of the IEEE/CVF Conference on Computer Vision and Pattern Recognition (CVPR), Salt Lake City, UT, USA, 18–22 June 2018; pp. 8789–8797.
46. Kim, T.; Cha, M.; Kim, H.; Lee, J.K.; Kim, J. Learning to discover cross-domain relations with generative adversarial networks. In Proceedings of the International Conference on Machine Learning (ICML), Sydney, Australia, 6–11 August 2017; pp. 1857–1865.
47. Zhu, J.Y.; Park, T.; Isola, P.; Efros, A.A. Unpaired image-to-image translation using cycle-consistent adversarial networks. In Proceedings of the IEEE/CVF Computer Vision and Pattern Recognition Conference (CVPR), Honolulu, HI, USA, 21–26 July 2017; pp. 2223–2232.

Disclaimer/Publisher's Note: The statements, opinions and data contained in all publications are solely those of the individual author(s) and contributor(s) and not of MDPI and/or the editor(s). MDPI and/or the editor(s) disclaim responsibility for any injury to people or property resulting from any ideas, methods, instructions or products referred to in the content.

Article

Research on Waste Plastics Classification Method Based on Multi-Scale Feature Fusion

Zhenxing Cai, Jianhong Yang *, Huaiying Fang, Tianchen Ji, Yangyang Hu and Xin Wang

Key Laboratory of Process Monitoring and System Optimization for Mechanical and Electrical Equipment (Huaqiao University), Fujian Province University, Xiamen 361021, China
* Correspondence: yjhong@hqu.edu.cn

Abstract: Microplastic particles produced by non-degradable waste plastic bottles have a critical impact on the environment. Reasonable recycling is a premise that protects the environment and improves economic benefits. In this paper, a multi-scale feature fusion method for RGB and hyperspectral images based on Segmenting Objects by Locations (RHFF-SOLOv1) is proposed, which uses multi-sensor fusion technology to improve the accuracy of identifying transparent polyethylene terephthalate (PET) bottles, blue PET bottles, and transparent polypropylene (PP) bottles on a black conveyor belt. A line-scan camera and near-infrared (NIR) hyperspectral camera covering the spectral range from 935.9 nm to 1722.5 nm are used to obtain RGB and hyperspectral images synchronously. Moreover, we propose a hyperspectral feature band selection method that effectively reduces the dimensionality and selects the bands from 1087.6 nm to 1285.1 nm as the features of the hyperspectral image. The results show that the proposed fusion method improves the accuracy of plastic bottle classification compared with the SOLOv1 method, and the overall accuracy is 95.55%. Finally, compared with other space-spectral fusion methods, RHFF-SOLOv1 is superior to most of them and achieves the best (97.5%) accuracy in blue bottle classification.

Keywords: plastic bottles recycling; hyperspectral image; multi-scale feature fusion

Citation: Cai, Z.; Yang, J.; Fang, H.; Ji, T.; Hu, Y.; Wang, X. Research on Waste Plastics Classification Method Based on Multi-Scale Feature Fusion. *Sensors* **2022**, *22*, 7974. https://doi.org/10.3390/s22207974

Academic Editors: Erik Blasch and Yufeng Zheng

Received: 20 September 2022
Accepted: 17 October 2022
Published: 19 October 2022

Publisher's Note: MDPI stays neutral with regard to jurisdictional claims in published maps and institutional affiliations.

Copyright: © 2022 by the authors. Licensee MDPI, Basel, Switzerland. This article is an open access article distributed under the terms and conditions of the Creative Commons Attribution (CC BY) license (https://creativecommons.org/licenses/by/4.0/).

1. Introduction

The common raw materials of disposable plastic bottles are polyethylene terephthalate (PET) and polypropylene (PP). PET and PP plastic bottles are widely used in the beverage-packaging industry due to their advantages of non-friability and safety. More than 480 billion plastic bottles were sold worldwide in 2016 [1]. Not only do waste plastics pollute the environment, but the microplastics produced by them are also more harmful to bio-safety. If waste plastic bottles are not recycled in time, they will take up considerable amounts of land and cause water and soil pollution [2]. Therefore, the recycling of waste plastic bottles is becoming an important issue from the perspective of protecting the environment and improving economic efficiency.

Since plastic bottles of different colors and materials have different recycling values, they are generally classified according to colors [3,4] and materials [5–7]. The advanced classification technologies of waste plastic bottles include deep-learning-based computer vision techniques and spectral detection techniques.

According to research conducted in recent years, computer vision techniques based on deep learning are widely used in various engineering fields, such as fruit picking [8] and the fault diagnosis of structures [9]. In the waste-sorting field, classifying plastic bottles based on computer vision is also effective. Jaikumar P et al. used the mask region proposal convolutional neural network (Mask R-CNN) to perform object detection and instance segmentation of waste plastic bottles in a customized dataset containing 192 images. Data augmentation was used to improve the model's performance in order to address the limitations of small datasets, and the final model achieved a 59.4 mean average precision (mAP) [10]. Some researchers used convolutional neural networks (CNN) and support vector machines (SVM) to classify plastic bottle images in a dataset containing a total of 1100 images and demonstrated that the CNN outperformed the SVM classifier in terms of testing accuracy [11].

There are several types of plastic resins in plastic bottles; if they are mixed during recycling, they will cause serious problems and the product quality will suffer [12]. Therefore, it is critical to classify plastic bottles according to materials. Optical and spectroscopic methods are used to identify the materials of plastics. Masoumi et al. [13] proposed a plastic identification and separation system based on near-infrared light (NIR). This system achieved the separation of polyethylene terephthalate (PET), high-density polyethylene (HDPE), polyvinyl chloride (PVC), polypropylene (PP), and polystyrene (PS) by calculating the reflectance ratio between 1656 nm and 1724 nm. With the widespread use of CNNs, some researchers have started to focus on combining NIR with a CNN to achieve a highly robust and accurate method of plastic classification, while preprocessing methods such as principal component analysis (PCA) are used for the dimensionality reduction of feature vectors [7].

However, the problem faced in waste plastic bottles' classification is that computer vision technology based on deep learning mainly relies on RGB images. Distinguishing between blue and transparent plastic bottles on a black conveyor belt is difficult due to the influence of the bottles' quality and the nearby light source. For spectral detection techniques, a disadvantage of NIR is that it is strongly affected by dark colors [14], which not only leads to the shift of reflection but also eliminates the peak value, and hues do not affect spectral reflectance [13,15]. Therefore, although spectral detection techniques can effectively distinguish plastic bottles according to materials, they can not classify them more precisely according to colors.

Light blue PET bottles, transparent PET bottles, and transparent PP bottles on black conveyor belts are difficult to classify via computer vision techniques based on deep learning, and it is difficult to distinguish light blue PET bottles and transparent PET bottles by spectral detection techniques. In this paper, a multi-scale feature fusion method for RGB and hyperspectral images (HSI) based on Segmenting Objects by Locations [16] (RHFF-SOLOv1) is proposed, and the classification accuracy of waste plastic bottles is improved by the feature fusion of RGB and HSI. This paper first introduces the dual camera platform for image acquisition and preprocessing methods. Then, the network structure of RHFF-SOLOv1 is described. In addition, a spectral feature interval selection method is proposed. Finally, the proposed RHFF-SOLOv1 is compared with other spatial-spectral fusion methods. The results show that our method is superior to most of them and achieves the best accuracy of 97.5% in the classification of blue PET bottles.

2. Materials and Methods

2.1. Samples Preparation and Data Collection

In this study, waste plastic bottles were collected from the recycled products of a company, including blue PET bottles, transparent PET bottles, and transparent PP bottles. Twenty-five samples were collected for each category as training and test datasets. The caps and labels were removed, as these materials are different from the bottle.

For data acquisition, RGB images were collected by a line-scan camera produced by Dalsa, and hyperspectral images were collected by NIR spectral camera with a wavelength range of 935.9–1722.5 nm with 224 bands. In this study, using a variable-speed conveyor as the bottle conveying device, the speed of the belt in the experiment was set to 0.36 m/s. The dual camera platform used for image acquisition is shown in Figure 1.

The dual camera acquisition platform mode of image acquisition was used to collect RGB and hyperspectral images of waste plastic bottles simultaneously. These samples were randomly placed on the conveyor belt and repeatedly collected. In order to address the limitations of custom datasets, we also tried to improve the generalization performance of our model by using data augmentation and Copy-Paste [17]. The final datasets included

768 images in the training set and 300 images in the test set. The size of RGB images is 640 × 640, while the size of hyperspectral images is 640 × 640 pixels with 224 bands in the wavelength range of 935.9–1722.5 nm. According to the color and material of plastic bottles, they were divided into three categories: transparent PET bottles (Trans_PET), blue PET bottles (Blue_PET), and transparent PP bottles (Trans_PP). All experiments were conducted on a workstation equipped with a Giga Texel Shader eXtreme (GTX) 3090 graphics-processing unit (GPU) and 24 G of memory.

Figure 1. Dual camera platform mode of image acquisition.

2.2. Data Processing

The spectrum data of samples is easily affected by various factors, such as the light source and the background. In order to eliminate the effects of dark current and the noise caused by the uneven intensity of the light source, it is necessary to use the black-and-white correction method to convert the collected data into reflectivity, as shown in Formula (1):

$$r = \frac{DN - DN_b}{DN_w - DN_b} \tag{1}$$

where DN denotes the collected raw data, while DN_w denotes the standard white frame data, which is obtained by collecting standard whiteboards. DN_b represents the standard black frame data obtained by covering the camera lens, and r denotes the obtained reflectivity data.

2.3. Network Structure of RHFF-SOLOv1

The proposed multi-scale feature fusion method for RGB and hyperspectral images based on SOLOv1 (RHFF-SOLOv1) consists of feature extraction and fusion. In terms of feature extraction, Resnet50 and feature pyramid networks (FPN) are used as the backbone network for the RGB imagery feature extraction branch, and in the branch of hyperspectral image feature extraction, "Multi-scale Filter Bank" [18] is used to obtain the spatial-spectral feature map, which is input into the hyperspectral imagery convolutional neural network (HSI-CNN) to extract multi-scale feature maps. In terms of feature fusion, the feature maps output by FPN and HSI-CNN is concatenated according to its size. Finally, the fused feature maps are input into the detection head of SOLOv1 to predict the position and category of plastic bottles. The proposed RHFF-SOLOv1 integrates multi-scale features of RGB and HSI, and the classification accuracy of plastic bottles is improved. The network structure of RHFF-SOLOv1 is shown in Figure 2.

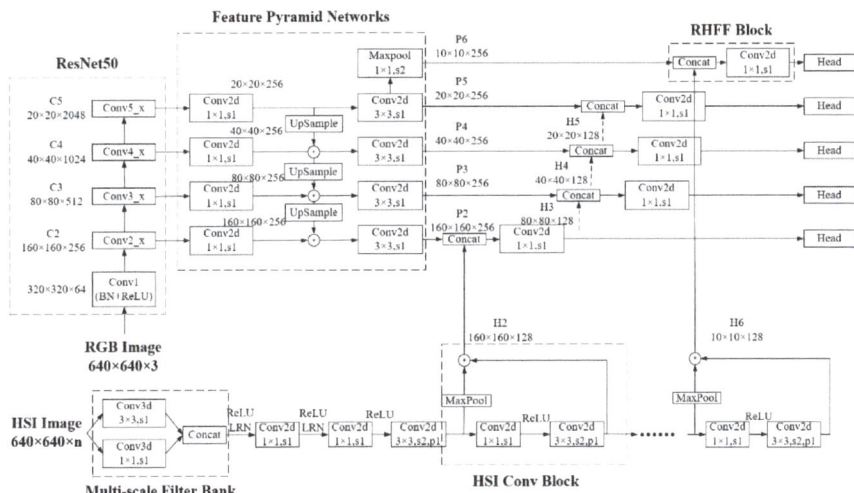

Figure 2. Network structure of RHFF-SOLOv1.

2.3.1. RGB Imagery Feature Extraction Branch

Resnet50 and FPN are used as the backbone network of the RGB image feature extraction branch, while the output from the second feature extraction stage to the fifth feature extraction stage of Resnet50 is connected with FPN, and FPN outputs multi-scale feature maps. This has the advantage of combining low-detail level information with high-level semantic information.

2.3.2. Hyperspectral Imagery Feature Extraction Branch

In the hyperspectral imagery feature extraction branch, a 3D CNN block with a kernel of $n \times 3 \times 3$ and $n \times 1 \times 1$ is used to form a "Multi-scale Filter Bank", where the former is used to extract spatial information, and the latter is used to extract spectral information. Then, the spatial-spectral feature map is used as the input of HSI-CNN, which contains five hyperspectral image convolution blocks (HSI Conv Block). The network structure of HSI CNN is shown in Figure 3.

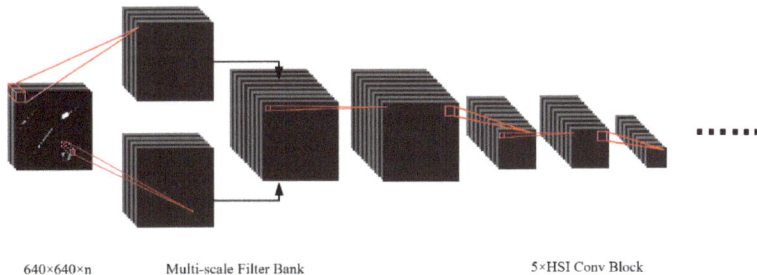

Figure 3. Network structure of HSI-CNN.

2.3.3. Multi-Scale Feature Fusion and Detection Head

Before feature fusion, maximum pooling is used to compress the shallow feature map that is output by HSI Conv Block, and it is added to the deep feature map. The feature fusion method of RGB and hyperspectral images is used to concatenate the feature maps output from FPN and HSI-CNN by size. The multi-scale feature fusion maps are used as the input of the detection head of SOLOv1 and a 2D CNN layer with the kernel of 1×1

is used to reduce the channel from 384 to 256. The HSI Conv Block and RHFF Block are shown in Figure 4.

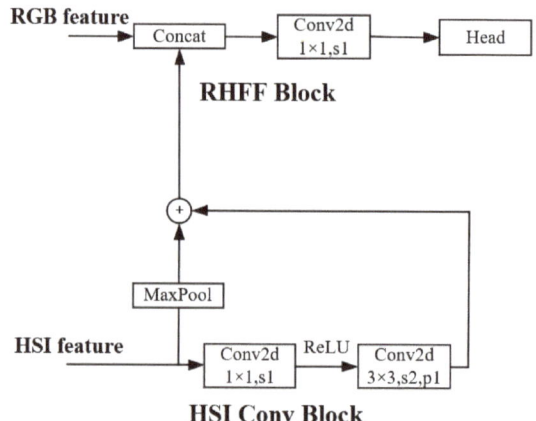

Figure 4. HSI Conv Block and RHFF Block.

The detection head of SOLOv1 divides the image into S × S grids; the grid where the center of the instance is located is responsible for predicting the instance category and mask, so the output of the detection head has a category branch and mask branch. The category branch predicts the category of the instance, and the final output is S × S × C, where C represents the number of classes. In the mask branch, the mask of the instance is predicted by the decoupled head, which predicts X and Y to obtain the mask through an "Element-Wise" operation. The detection head of SOLOv1 is shown in Figure 5.

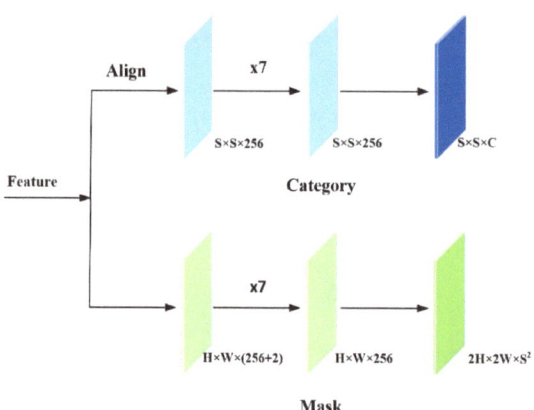

Figure 5. Detection head of SOLOv1.

3. Results

3.1. Hyperspectral Band Selection

In a hyperspectral image, each pixel includes 224 bands and contains many spectral features that will increase the complexity of the model. Therefore, reducing the number of spectral bands in hyperspectral images is necessary.

Figure 6 shows the average spectral curves of Tans_PET, Blue_PET, and Trans_PP. It can be seen that the spectral curves of Trans_PET and Blue_PET are similar, and only the peaks are different. Moreover, the spectral curve of Trans_PP exhibits a large fluctuation amplitude, and it has two obvious absorption valleys at 1200.5 nm and 1394.5 nm. In this paper, a method of hyperspectral feature band selection is proposed, 224 bands are divided into several spectral intervals, and the interval with the most apparent feature is selected.

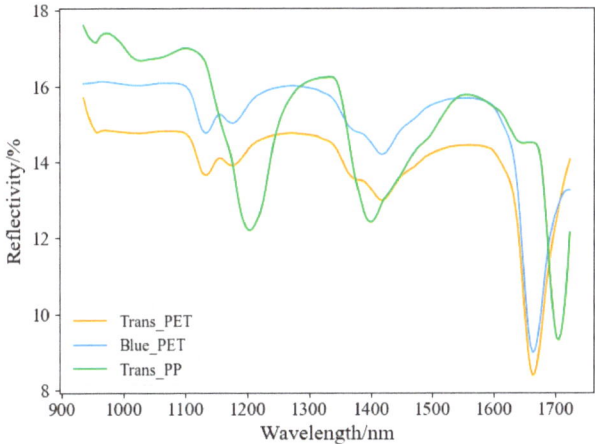

Figure 6. Mean spectral curve of waste plastic bottles.

Specifically, the mean spectral curves of the waste plastic bottles were smoothed by the Savitzky-Golay filter, and the extreme points were calculated. In order to find the spectral band with the best classification result among the three types, several split points were chosen from among the extreme points, including 1087.6 nm, 1285.1 nm, 1419.2 nm, 1542.6 nm, and 1666.1 nm. The spectral feature intervals are shown in Figure 7.

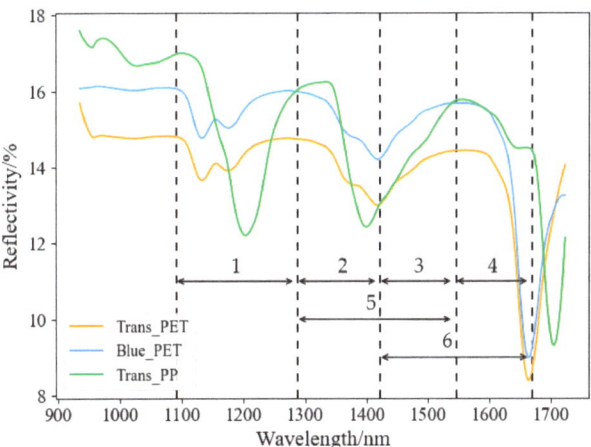

Figure 7. Spectral feature intervals. Interval 1: 1087.6 nm–1285.1 nm; Interval 2: 1285.1 nm–1419.2 nm; Interval 3: 1419.2 nm–1542.6 nm; Interval 4: 1542.6 nm–1666.1 nm; Interval 5: 1285.1 nm–1542.6 nm; Interval 6: 1419.2 nm–1666.1 nm.

The bands before 1087.6 nm and after 1666.1 nm were not retained due to noise and inconspicuous features. Finally, different classifiers such as SVM, 1D convolutional neural networks (1D-CNN), and Random Forest (RF) were used to classify the plastic bottles based on the spectral feature intervals. Table 1 shows the overall accuracy of the classification results (%).

Table 1. The overall accuracy of classification results based on feature intervals.

Classifiers	Interval 1	Interval 2	Interval 3	Interval 4	Interval 5	Interval 6
SVM	71.9	70.0	59.5	70.8	70.6	75.7
1D-CNN	88.4	81.8	74.7	79.2	84.8	86.4
RF	76.3	75.6	68.5	73.6	75.4	77.3

The results show that the classification performance is better when interval 1 (1087.6 nm–1285.1 nm) and interval 6 (1419.2 nm–1666.1 nm) are selected, which means that these feature intervals contain important features for the classification of Trans_PET, Blue_PET, and Trans_PP.

3.2. Evaluating Indicators

The indicators used to evaluate the classification results include precision (P), recall (R), overall accuracy (OA), average accuracy (AA), and Kappa (k). The calculation methods are shown in Formulas (2)–(7). TP is the True Positive, FN is the False Negative, TN is the True Negative, and FP is the False Positive.

$$P = \frac{TP}{TP + FP} \times 100\% \tag{2}$$

$$R = \frac{TP}{TP + FN} \times 100\% \tag{3}$$

$$OA = \frac{TP}{n} \times 100\% \tag{4}$$

$$AA = \frac{\text{sum}(R)}{C} \times 100\% \tag{5}$$

$$k = \frac{p_o - p_e}{1 - p_e} \times 100\% \tag{6}$$

$$P_e = \frac{\sum_{i=1}^{C} a_i \times b_i}{n^2} \tag{7}$$

in which C is the number of classes, n is the total number of samples, the number of samples of each category is a_i, and the number of prediction samples of each category is b_i. In Formula (6), p_o is OA, and k is used to evaluate the consistency between the predicted results and the actual classification results.

3.3. Comparison between Different Intervals

In order to select the best spectral feature interval, this study used the hyperspectral image of feature intervals 1 and 6 as the input of RHFF-SOLOv1 for comparative experiments. The HSI size of interval 1 is 640 × 640 × 57, and the HSI size of interval 6 is 640 × 640 × 71. The comparison of the experimental results is shown in Table 2.

Table 2. RHFF-SOLOv1 classification results (%) based on different spectral feature intervals.

	Interval 1	Interval 6
Blue_PET	97.60	97.60
Trans_PET	94.80	94.12
Trans_PP	96.00	95.43
OA	95.55	95.01
AA	95.69	95.06
Kappa × 100	93.28	92.47

The results show that the characteristic of spectral interval 1 is better than that of spectral interval 6; so, interval 1 was selected as the hyperspectral image for this study.

3.4. Comparison between RHFF-SOLOv1 and Other Methods

In order to verify the feasibility and effectiveness of RHFF-SOLOv1, this paper first compares the classification results of SOLOv1 and RHFF-SOLOv1. Table 3 shows the comparative experimental results (%).

Table 3. The classification results of SOLOv1 and RHFF-SOLOv1.

	SOLOv1	RHFF-SOLOv1
Blue_PET	87.80	97.60
Trans_PET	87.88	94.80
Trans_PP	92.16	96.00
OA	86.66	95.55
AA	87.51	95.69
Kappa × 100	80.10	93.28

Compared with SOLOv1, the results of OA and AA of the proposed RHFF-SOLOv1 were improved by 8.89% and 8.18%, respectively.

Recently, a large number of studies have applied CNNs to hyperspectral image classification, and these methods can be applied to different applications. Our proposed method is compared with the spatial-spectral fusion networks such as ContextualNet [18], 3D-CNN [19], the Spectral-Spatial Residual Network (SSRN) [20], and Hybrid SpectralNet (HybridSN) [21] to demonstrate that RHFF-SOLOv1 is also superior to spectral detection technology. Since these networks are based on the classification of pixels in HSI, in order to ensure the same number of prediction instances in all experiments, 300 hyperspectral images based on interval 1 in the test set are selected as their datasets, with a size of 640 × 640 × 57. Each HSI is divided into pixels with a ratio of $Dataset_{train}:Dataset_{test}$ = 3:7, and the class results with the most pixels in the instance constitute the category of the instance. The comparison of the experimental results (%) is shown in Table 4.

Table 4. Classification results between RHFF-SOLOv1 and other fusion methods.

	ContextualNet	3D-CNN	SSRN	HybridSN	RHFF-SOLOv1
Blue_PET	79.02	89.89	93.26	92.11	97.60
Trans_PET	86.43	94.49	95.67	98.11	94.80
Trans_PP	97.47	99.48	98.99	96.48	96.00
OA	86.52	94.07	95.69	95.42	95.55
AA	87.68	94.44	96.06	95.69	95.69
Kappa × 100	75.59	91.01	93.47	93.06	93.28

From the results, the proposed RHFF-SOLOv1 improves the OA, AA, and Kappa of waste plastic bottles' classification results by fusing the hyperspectral image feature map and RGB feature map. Compared with other spatial-spectral fusion networks, RHFF-SOLOv1 is superior, except with respect to SSRN in the evaluation indicators. Among them, RHFF-SOLOv1 achieves the highest accuracy in terms of accuracy for determining blue PET bottles.

Figure 8 shows the classification maps for the bottles using SOLOv1, ContextualNet, 3D-CNN, SSRN, HybridSN, and RHFF-SOLOv1. It can be seen that the prediction charts obtained by HybridSN and RHFF-SOLOv1 are of good quality and can correctly identify blue PET bottles more efficiently than other methods.

The proposed RHFF-SOLOv1 can also achieve good classification results when waste plastic bottles are cluttered and stacked. However, compared to other methods, our method cannot segment the contours of some plastic bottles completely during the segmentation of the instances. The classification maps of the cluttered bottles are shown in Figure 9.

Alongside improving the classification accuracy by multi-scale feature fusion, the determination of a method for more accurately segmenting the contours of the instances is also a future research direction.

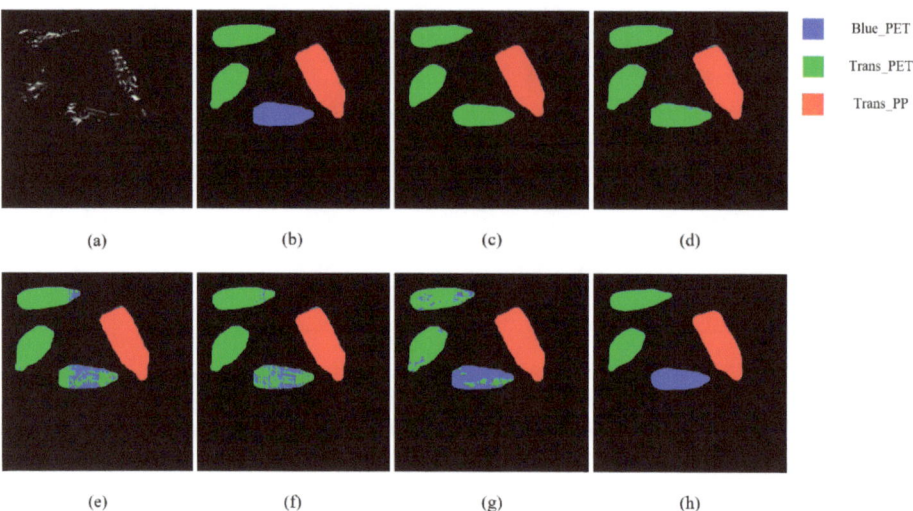

Figure 8. Classification maps for bottles. (**a**) RGB image. (**b**) Ground truth. (**c–h**) Predicted classification maps for SOLOv1, ContextualNet, 3D-CNN, SSRN, HybridSN, and proposed RHFF-SOLOv1, respectively.

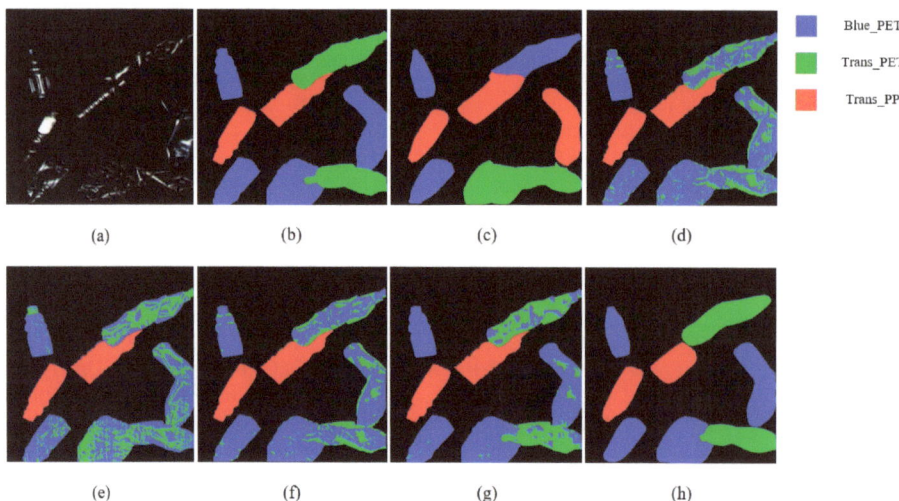

Figure 9. Classification maps for cluttered bottles. (**a**) RGB image. (**b**) Ground truth. (**c–h**) Predicted classification maps for SOLOv1, ContextualNet, 3D-CNN, SSRN, HybridSN, and proposed RHFF-SOLOv1, respectively.

4. Conclusions

This study proposes a multi-scale feature fusion method of RGB and hyperspectral images based on SOLOv1 (RHFF-SOLOv1). The spectral interval 1 (1087.6 nm–1285.1 nm) was selected as the spectral feature of waste plastic bottles classification through the hyperspectral band selection method. To effectively combine spatial and spectral information, multi-scale feature maps of RGB and HSI were extracted and integrated using RHFF-SOLOv1. In terms of bottle classification accuracy, our proposed method achieves the best (97.5%) accuracy with respect to blue PET bottle classification only. However, the OA, AA, and Kappa results of the proposed method are better than SOLOv1 and the state-of-the-art spatial-spectral fusion network, except with respect to SSRN, which confirms the feasibility and effectiveness of RHFF-SOLOv1. With the development of multi-sensor fusion technology, multiple sources of information can be integrated by effective fusion methods to address the difficulty of classification.

Author Contributions: Conceptualization, Z.C., J.Y. and H.F.; methodology, Z.C.; Z.C., T.J., Y.H. and X.W. designed the platform and software; Z.C. and Y.H. built the platform; T.J. and Y.H. implemented the software; Z.C., Y.H. and X.W. collected data; Z.C. analyzed data and drafted the manuscript. All authors have read and agreed to the published version of the manuscript.

Funding: This research was financially supported by the Major Special Program of Science and Technology of Fujian Province (2020YZ017022), the Science and Technology Project of Xiamen (2021FCX012501190024), and the Key Technologies Research and Development Program of Shenzhen (JSGG20201103100601004).

Institutional Review Board Statement: Not applicate.

Informed Consent Statement: Not applicate.

Data Availability Statement: The data presented in this study are available on request from the corresponding author.

Conflicts of Interest: The authors declare no conflict of interest.

References

1. Laville, S.; Taylor, M. *A Million Bottles a Minute: World's Plastic Binge 'as Dangerous as Climate Change'*; The Guardian: Manchester, UK, 2017.
2. Younos, T. Bottled water: Global impacts and potential. *Potable Water* **2014**, *30*, 213–227.
3. He, X.; He, Z.; Zhang, S.; Zhao, X. A novel vision-based PET bottle recycling facility. *Meas. Sci. Technol.* **2016**, *28*, 025601. [CrossRef]
4. Wang, Z.; Peng, B.; Huang, Y.; Sun, G. Classification for plastic bottles recycling based on image recognition. *Waste Manag.* **2019**, *88*, 170–181. [CrossRef] [PubMed]
5. Scavino, E.; Wahab, D.A.; Hussain, A.; Basri, H.; Mustafa, M.M. Application of automated image analysis to the identification and extraction of recyclable plastic bottles. *J. Zhejiang Univ.-SCIENCE A* **2009**, *10*, 794–799. [CrossRef]
6. Özkan, K.; Ergin, S.; Işık, Ş.; Işıklı, I. A new classification scheme of plastic wastes based upon recycling labels. *Waste Manag.* **2015**, *35*, 29–35. [CrossRef] [PubMed]
7. Maliks, R.; Kadikis, R. Multispectral data classification with deep CNN for plastic bottle sorting. In Proceedings of the 2021 6th International Conference on Mechanical Engineering and Robotics Research (ICMERR), Krakow, Poland, 11–13 December 2021; IEEE: Piscataway, NJ, USA, 2021; pp. 58–65.
8. Tang, Y.; Zhou, H.; Wang, H.; Zhang, Y. Fruit detection and positioning technology for a Camellia oleifera C. Abel orchard based on improved YOLOv4-tiny model and binocular stereo vision. *Expert Syst. Appl.* **2022**, *211*, 118573. [CrossRef]
9. Tang, Y.; Zhu, M.; Chen, Z.; Wu, C.; Chen, B.; Li, C.; Li, L. Seismic performance evaluation of recycled aggregate concrete-filled steel tubular columns with field strain detected via a novel mark-free vision method. *Structures* **2022**, *37*, 426–441. [CrossRef]
10. Jaikumar, P.; Vandaele, R.; Ojha, V. Transfer learning for instance segmentation of waste bottles using Mask R-CNN algorithm. In *Intelligent Systems Design and Applications, Proceedings of the International Conference on Intelligent Systems Design and Applications, Online, 13–15 December 2021*; Springer: Cham, Germany, 2020; pp. 140–149.
11. Fadlil, A.; Umar, R.; Nugroho, A.S. Comparison of machine learning approach for waste bottle classification. *Emerg. Sci. J.* **2022**, *6*, 1075–1085. [CrossRef]
12. Carvalho, M.T.; Ferreira, C.; Portela, A.; Santos, J.T. Application of fluidization to separate packaging waste plastics. *Waste Manag.* **2009**, *29*, 1138–1143. [PubMed]
13. Masoumi, H.; Safavi, S.M.; Khani, Z. Identification and classification of plastic resins using near infrared reflectance spectroscopy. *Int. J. Mech. Ind. Eng.* **2012**, *6*, 213–220.

14. Stiebel, T.; Bosling, M.; Steffens, A.; Pretz, T.; Merhof, D. An inspection system for multi-label polymer classification. In Proceedings of the 2018 IEEE 23rd International Conference on Emerging Technologies and Factory Automation (ETFA), Turino, Italy, 4–7 September 2018; IEEE: Piscataway, NJ, USA, 2018; Volume 1, pp. 623–630.
15. Scheirs, J. *Polymer Recycling*; Wiley: Hoboken, NJ, USA, 1998.
16. Wang, X.; Kong, T.; Shen, C.; Li, L. *SOLO: Segmenting Objects by Locations*; Springer: Cham, Germany, 2019.
17. Ghiasi, G.; Cui, Y.; Srinivas, A.; Qian, R.; Lin, T.-Y.; Cubuk, E.D.; Le, Q.V.; Zoph, B. Simple copy-paste is a strong data augmentation method for instance segmentation. In *Computer Vision and Pattern Recognition*; IEEE: Piscataway, NJ, USA, 2021.
18. Lee, H.; Kwon, H. Contextual deep CNN based hyperspectral classification. In *Geoscience & Remote Sensing Symposium*; IEEE: Piscataway, NJ, USA, 2016.
19. Ying, L.; Haokui, Z.; Qiang, S. Spectral–spatial classification of hyperspectral imagery with 3D convolutional neural network. *Remote Sens.* **2017**, *9*, 67. [CrossRef]
20. Zhong, Z.; Li, J.; Luo, Z.; Chapman, M. Spectral-spatial residual network for hyperspectral image classification: A 3-D deep learning framework. *IEEE Trans. Geosci. Remote Sens.* **2017**, *56*, 847–858. [CrossRef]
21. Roy, S.K.; Krishna, G.; Dubey, S.R.; Chaudhuri, B.B. HybridSN: Exploring 3-D–2-D CNN feature hierarchy for hyperspectral image classification. *IEEE Geosci. Remote Sens. Lett.* **2020**, *17*, 277–281. [CrossRef]

MDPI AG
Grosspeteranlage 5
4052 Basel
Switzerland
Tel.: +41 61 683 77 34

Sensors Editorial Office
E-mail: sensors@mdpi.com
www.mdpi.com/journal/sensors

Disclaimer/Publisher's Note: The title and front matter of this reprint are at the discretion of the Guest Editors. The publisher is not responsible for their content or any associated concerns. The statements, opinions and data contained in all individual articles are solely those of the individual Editors and contributors and not of MDPI. MDPI disclaims responsibility for any injury to people or property resulting from any ideas, methods, instructions or products referred to in the content.

www.ingramcontent.com/pod-product-compliance
Lightning Source LLC
LaVergne TN
LVHW072328090526
838202LV00019B/2369